Analysis of Natural Compounds Exercising Health Importance from Food

Analysis of Natural Compounds Exercising Health Importance from Food

Editors

Jiuliang Zhang
Rui Zhang
Jingren He

MDPI

Basel • Beijing • Wuhan • Barcelona • Belgrade • Novi Sad • Cluj • Manchester

Editors
Jiuliang Zhang
Huazhong Agricultural University
Wuhan
China

Rui Zhang
Wuhan Polytechnic University
Wuhan
China

Jingren He
Wuhan Polytechnic University
Wuhan
China

Editorial Office
MDPI AG
Grosspeteranlage 5
4052 Basel, Switzerland

This is a reprint of articles from the Special Issue published online in the open access journal *Foods* (ISSN 2304-8158) (available at: https://www.mdpi.com/journal/foods/special_issues/B34VC71864).

For citation purposes, cite each article independently as indicated on the article page online and as indicated below:

Lastname, A.A.; Lastname, B.B. Article Title. *Journal Name* **Year**, *Volume Number*, Page Range.

ISBN 978-3-7258-2303-1 (Hbk)
ISBN 978-3-7258-2304-8 (PDF)
doi.org/10.3390/books978-3-7258-2304-8

Contents

foods

MDPI

Review

Himalayan Sources of Anthocyanins and Its Multifunctional Applications: A Review

Mustafa Ahmed [1], Ipsheta Bose [1], Gulden Goksen [2] and Swarup Roy [1,3,*]

[1] School of Bioengineering and Food Sciences, Shoolini University, Solan 173229, India; mustafaahmed9560@gmail.com (M.A.); ipsheta18@gmail.com (I.B.)
[2] Department of Food Technology, Vocational School of Technical Sciences at Mersin Tarsus Organized Industrial Zone, Tarsus University, 33100 Mersin, Turkey; guldengoksen@tarsus.edu.tr
[3] Department of Food Technology and Nutrition, School of Agriculture, Lovely Professional University, Phagwara 144411, India
* Correspondence: swaruproy2013@gmail.com

Abstract: Anthocyanins, the colored water-soluble pigments, have increasingly drawn the attention of researchers for their novel applications. The sources of anthocyanin are highly diverse, and it can be easily extracted. The unique biodiversity of the Himalayan Mountain range is an excellent source of anthocyanin, but it is not completely explored. Numerous attempts have been made to study the phytochemical aspects of different Himalayan plants. The distinct flora of the Himalayas can serve as a potential source of anthocyanins for the food industry. In this context, this review is an overview of the phytochemical studies conducted on Himalayan plants for the estimation of anthocyanins. For that, many articles have been studied to conclude that plants (such as *Berberis asiatica*, *Morus alba*, *Ficus palmata*, *Begonia xanthina*, *Begonia palmata*, *Fragaria nubicola*, etc.) contain significant amounts of anthocyanin. The application of Himalayan anthocyanin in nutraceuticals, food colorants, and intelligent packaging films have also been briefly debated. This review creates a path for further research on Himalayan plants as a potential source of anthocyanins and their sustainable utilization in the food systems.

Keywords: anthocyanin; himalayan plant; natural colorant; food; nutraceuticals; smart packaging

Citation: Ahmed, M.; Bose, I.; Goksen, G.; Roy, S. Himalayan Sources of Anthocyanins and Its Multifunctional Applications: A Review. *Foods* **2023**, *12*, 2203. https://doi.org/10.3390/foods12112203

Academic Editors: Jiuliang Zhang, Rui Zhang and Jingren He

Received: 2 May 2023
Revised: 28 May 2023
Accepted: 29 May 2023
Published: 30 May 2023

1. Introduction

Nature has lavishly gifted mankind with a plethora of medicinal herbs that have long served as a provider of conventional medicine to treat many human illnesses. Plant natural products are a diverse class of chemical entities with extensive structural and chemical diversity, as well as biochemical specialization and a wide range of biological activity. These have been utilized as medications, additives, insecticides, agrochemicals, fragrance and flavor components, food additives, and pesticides with great care. For thousands of years, plants have served as the basis for many traditional medicines used around the world, and they continue to provide people with novel treatments. They are one of the most abundant sources of chemicals. The complex relationship between plants and people has a long evolutionary history. For centuries people have used plants for food, fiber, medicines, and energy [1]. In the twenty-first century, plants are being used to derive pharmaceuticals, multicomponent herbal products, nutraceuticals, and functional foods [2].

The Himalayas, one of the youngest and largest hill systems on this planet, is well known for its floral and faunal diversity [3]. The Himalayan biodiversity hotspot is shelter to a high percentage of flora and animals, along with several environmental assets, yet the people who live here struggle to meet basic requirements such as food and nourishment. Wild plants and their fruits, on the other hand, contribute greatly to the existence of the indigenous communities in the Himalayan region. Plants contain phytochemicals which are secondary metabolites produced for disease protection and contribute to their color, aroma, and flavor [4]. Phytochemicals have multiple uses in pharmaceuticals and agricultural products and as coloring agents and additives in the food industry [5]. One of the most numerous and widespread phytochemicals are flavonoids. These are a category of health-promoting phytochemicals with variable phenolic structures. Based on their chemical structure, flavonoids are further assorted into several subgroups which include flavones, chalcones, flavonols, isoflavones, and anthocyanins [6], of these subgroups, this review is majorly focused on anthocyanins.

Anthocyanins are a class of water-soluble pigments (colored substances) present in plant sources, particularly fruits and vegetables. They are responsible for giving fruits, vegetables, and tubers their various colors, such as purple, red, and blue. These pigments from the phenol group have been glycosylated. The term 'Anthocyanins' originated from ancient Greek language words 'Anthos' and 'Kyanous'. The first implies flowers, while the second implies a deep blue color [7]. Many millennia ago, anthocyanins were consumed by humans through the means of dried leaves, fruits, roots, or seeds and were employed in traditional herbal treatments utilized by North American Indians, Europeans, and Chinese. For decades, anthocyanins have been a part of traditional herbal medicine due to their highly valued medicinal properties. Historically, anthocyanin-rich mixes and extracts were used for treating hypertension, pyrexia, liver diseases, diarrhea, dysentery, and ailments of the urinary system such as renal calculi, urinary tract infections, and rhinitis (common cold). Anthocyanins have wide applications, particularly in the food industry due to their characteristic property of imparting vibrant colors. Food color is an important sensory property that impacts consumer preference. Keeping in mind the safety concerns pertaining to the practice of using synthetic food colorants such as lemon yellow, the use of nature-derived compounds such as anthocyanins and their acylated counterparts as food colorants have become more popular owing to their edible nature and possible health benefits. Several studies on the exploration and utilization of anthocyanins and their related acylated variants as food colorants were conducted. Because the color of anthocyanins and their related forms depends mainly on the surrounding pH conditions, they were investigated as pH-sensitive dyes for their utilization in food packaging. Since food decomposition is generally followed by a change in pH, variations in the tints of sensitive anthocyanin dyes can be used for the determination of the quality and stability of packaged foods [8]. Choi et al. created an intelligent pH-sensitive color indicator by adding anthocyanin pigment extract from purple sweet potato on a matrix of agar-starch and utilized it to detect pH variations in meat products by evaluating the colors at various pHs as extent of deterioration. The color of the indicator, which ranges from red to green, may represent the degeneration of the meat due to pH fluctuations. Anthocyanins have also been added to food products and demonstrating their physiological effect is a field of ongoing research [9].

Because of their enhanced stability and fat solubility, anthocyanins have proven to be capable of being used as a food dye, functionalization, and active and smart packaging in food processing. However, there are still challenges and limitations regarding the stability and widespread use of these pigments in food processing industries. Taking the published studies into account, this review aims to compile available information concerning plant sources of anthocyanins distributed across the Indian Himalayan region with a special focus on the multifunctional applications of anthocyanins obtained from these sources. Additionally, some emphasis has been given to their chemical, physical and functional properties, and extraction techniques. The contemporary review critically examines the outcomes of phytochemical studies published on Himalayan plants. There are many works published on anthocyanin and its application but based on the literature survey we concluded there is no review report published on Himalayan anthocyanin and insight into this work is expected to offer more interest in exploring the potential of unexplored anthocyanin from nature and its use in multi-dimension application.

2. Chemistry of Anthocyanins

The structure of anthocyanin is given in Figure 1. The C6-C3-C6 unit (xanthine cation) carbon skeleton of anthocyanins is made up of anthocyanidin (aglycone units) coupled to sugar, which is often present at the 3-position on the C-ring, as well as hydroxyl and methoxyl groups [10]. However, the substitution pattern in the B-ring has a significant impact on anthocyanin stability, which can either become better with more methoxyl groups or develop worse with more hydroxyl groups.

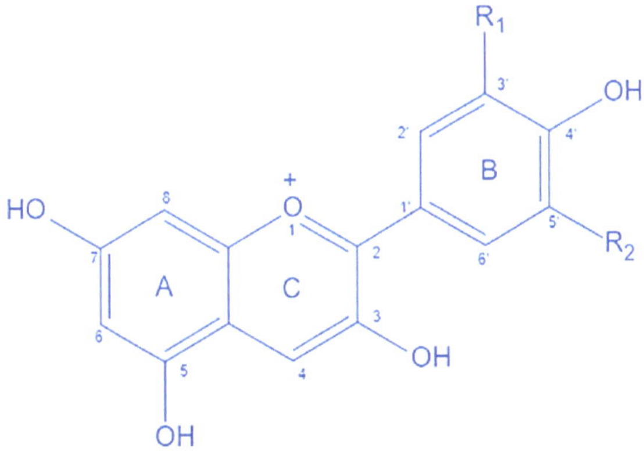

Figure 1. Structure of anthocyanin.

Anthocyanins are hydrophilic natural food colorants with molecular weights of around 400 to 1200 Da (g/mol). In polar solvents such as methanol, ethanol, and water, anthocyanins are soluble. Anthocyanin's chemical makeup is based on the salts of the flavylium (2-phenyl benzopyrylium) glycoside. Anthocyanin's structure can be diverse, reliant on the nature, position, and the total aliphatic or aromatic acids involved in the sugars. Commonly, anthocyanins have a sole glucoside unit, nevertheless, there are several types of anthocyanins with two or multiple sugars bound at several positions or involved as oligosaccharide side chains. The stability of anthocyanins can be increased through acylation and glycosylation. Additionally, anthocyanins can be affected by pH, enzymes, light, chemical structure, solvents, oxygen, heat, concentrations, and a few other factors which may influence their stability [11]. Acylated anthocyanins are prospective substitutes for anthocyanins due to their distinctive properties and added benefits over non-acylated counterparts. Acylation brings a structural change in anthocyanins that alters how they function [12–14]. Acylated anthocyanins have complicated acylation and glycosylation patterns. They occur naturally in many plants [15]. However, anthocyanins can also be acylated in vitro by chemical and enzymatic methods. These in-vitro acylated anthocyanins are more stable and highly soluble in organic solvents as compared with non-acylated anthocyanins [14]. It has been found that deterioration by heat (thermal degradation) has a damaging result on anthocyanins, which mainly depends on their structure [16]. Thermal degradation of anthocyanins during food processing is a major problem for the industry [17]. Anthocyanins can initially release glycosyl moieties on heating. The resulting anthocyanin aglycone is subsequently broken down into phenolic acids and phloroglucinaldehyde [18]. Heat-induced degradation kinetics depend on the specific chemical makeup of anthocyanins and the properties of the food matrix. For instance, degradation kinetics can be altered by the presence of phenols, ascorbic acid, micellar systems, and soluble solids [19].

Due to the instability and ease of degradation, the use of anthocyanins in a wide spectrum of applications is restricted. As a result, it is difficult to use anthocyanins in food products on an industrial scale [20]. Anthocyanin-containing foods can only go into the blood circulation for better absorption and metabolism after passing through the lumen of the intestine. As a result, improving anthocyanin stability is required to optimize metabolism and absorption and consequently upgrade their health merits. Anthocyanins have been reported to be protected against unfavorable environmental conditions by encapsulated delivery systems [21]. One of the most significant challenges in manipulating anthocyanins is that they are extremely susceptible to deterioration, especially when separated. Because several physicochemical factors have been shown to interact with anthocyanin stability, basic anthocyanin chemistry is needed to comprehend the limitations of a certain extraction procedure.

3. Extraction of Anthocyanins

Various traditional and developing procedures have been used to extract anthocyanins from plant sources. Each extraction method is unique and has a profound result on the stability, purity, yield, and concentration of the extracted anthocyanins [22]. Natural anthocyanins are particularly unstable and prone to degradation, resulting in loss of bioactivity and color. Factors affecting the rate of decomposition include heat, light, oxygen, pH, enzymes, co-pigments, and water activity [23]. In the literature on anthocyanins, two main extraction processes were discovered: methods that optimize the extraction to define and detect anthocyanins, and methods that are expandable for application in the food sector [24]. Several factors, particularly the solvents, must be considered in the anthocyanin analysis. Due to their polarity, anthocyanin molecules can be extracted using a variety of polar solvents such as methanol, alcohol, and acetone water. Since anthocyanins are highly reactive compounds, choosing an appropriate solvent for extraction is critical. Traditional techniques for extracting anthocyanin include soaking, stirring, ultrasound-assisted extraction, and enzyme-driven extraction. The growing interest in anthocyanin's antioxidant activity drives a greater desire for more efficient extraction methods, such as lower solvent usage, lower environmental repercussions, higher extraction yield, and shorter extraction timeframes [25].

4. Himalayan Sources of Anthocyanins

Himalayan plants are excellent sources of anthocyanin. Fruits, flowers, leaves, and roots of plants are natural sources of anthocyanins. Cherries, berries, and cereals are among the most popular anthocyanin found in natural goods. Numerous anthocyanins and their compounds extracted from natural sources have been mentioned in the scientific literature. A brief discussion of anthocyanins from Himalayan sources is provided below. Figure 2 shows some of the indigenous Himalayan sources of anthocyanin. On the other hand, the classification of various Himalayan anthocyanins is presented in Table 1.

Table 1. Classification of different Himalayan plant sources of anthocyanins (H = Hindi, E = English, NA = not available).

Name of the Species	Family	Common Name	References
Berberis asiatica	Berberidaceae	H: Chitra/Chotra E: Indian Barberry, Tree Turmeric	[26]
Morus alba	Moraceae	H: Shahtoot E: Mul berry/Silkworm Mulberry	[27]
Berberis lycium	Berberidaceae	E: Indian Barberry, H: Kashmal	[28]
Myrica esculenta	Myricaceae	E: Box berry H: Kaphal	[29]
Duchesnea indica	Rosaceae	H: Kiphaliya E: Indian strawberry/Mock strawberry	[26]
Lycium ruthencium	Solanaceae	E: Black wolfberry	[30]
Fragaria nubicola	Rosaceae	E: Himalayan strawberry	[31]
Ficus palmata	Moraceae	H: Abjiri/Bedu E: Wild Fig	[26]
Gaultheria trichophylla	Ericaeae	E: Himalayan snowberry	[32]
Genus begonia	Begoniacece	NA	[33]

Figure 2. Various sources of plant-based Himalayan anthocyanin.

4.1. *Berberis asiatica*

With more than 500 species found worldwide, the genus Berberis is widely recognized for its therapeutic potential [34]. *Berberis asiatica* is the most widely distributed species in the Western Himalayas and Northeast India. It is found in Jammu and Kashmir, Himachal Pradesh, Uttarakhand, Arunachal Pradesh, and Assam.

Morphological aspects: It is an evergreen shrub with a height of 1.2–1.8 m and the stem is up to 10 cm in diameter. It has a rough, furrowed, and corky bark. Twigs are yellowish with a smooth surface. Leaves are dark green, oblong, elliptic, and usually have large distant spinous teeth. The size of the leaves is 2.5–6.3 by 1.3–3.8 cm. Prominent primary and secondary pale reticulate venation above and glaucous beneath can be observed. Petiole is usually absent but sometimes visible up to 10 mm. Simple racemose type of inflorescence which is 3 cm long. Pedicels are 4–19 mm long, slender, often glaucous. This plant yields fruits that are 7–10 mm long, ovoid, and dark black in color [35].

Functional and therapeutic roles: The root is widely utilized in many indigenous therapeutic practices to treat a wide range of health problems such as rheumatism, diabetes, jaundice, gastric disorders, skin conditions, malarial fever, and as a tonic, among others. The presence of phenols in this plant is thought to be responsible for its wide range of health-promoting properties. These phenols provide antioxidant action through a variety of mechanisms, including the removal of free radicals, metal chelation, hydrogen donors, and gene expression alterations [36]. The fruits are eaten raw and taste juicy with an acidic flavor. Traditionally, these fruits are provided to children as mild laxatives and the juice may be helpful in dental ailments. The species is employed in pharmacological, nutraceutical, and cosmetic products [37].

Phytochemical aspects: The *Berberis asiatica* plant is a reservoir of various bioactive chemical compounds such as phenolics, alkaloids, flavonoids, anthocyanins, tannins, vitamins, and minerals [38]. As per physicochemical investigations, it was revealed that berberine, an alkaloid found in the plant was 2.4%, tannins 1.73%, total ash 2.650%, starch 16.444%, acid insoluble ash 0.266%, and alcohol soluble extractive 11.833%. [35]. Berberine is the plant's main alkaloid, it is a quaternary ammonium salt that comes under the protoberberine class of isoquinoline alkaloids. It is known for its biological activity against many chronic diseases [39].

Anthocyanins: The anthocyanin content expressed as mg of cyanidin-3-glucoside equivalents (CGE) per g of fresh weight (mg CGE/g fw) in these fruits were found to be 24.59 mg/100 g fw [40]. To date, there are very limited studies available on the anthocyanin composition of the *Berberis asiatica* plant.

4.2. *Morus alba*

The Genus, *Morus* (Mulberry) is a genus of flowering plants with approximately 150 species of which *Morus alba* L. is the most important [41]. It is a plant with significant economic value that is extensively cultivated in Centra, East, and South Asia for sericulture. *M. alba* species are thought to have evolved on the low Himalayan slopes that border China and India [42]. With around 626,000 hectares of land under Mulberry cultivation, China comes first in terms of Mulberry production, followed by India at 280,000 hectares of land under cultivation [43].

Morphological aspects: It is a deciduous tree of medium-sized found in China and hilly areas of the Himalayas at an elevation of up to 3300 m. The tree grows up to a height of 20–30 m and a girth of 1.8 m. It can be reduced to a low-growing bush to make it easier to gather leaves and fruits. Dark grey-brown lenticels run horizontally down the bark. The petioles are long and thin, the leaves are alternating, glossy green collects at the apex, and the margins are notched or toothed at the end (serrated). The leaves' lengths range from 5.0 to 7.5 cm, and they have a wide range of shapes. It has been observed that the tress are dioecious in temperate and sub-tropical climatic zones but in some cases, they change from one sex to another. Male and female catkins are made up of discrete, pendulous, greenish blooms. The characteristic feature of the mulberry fruit is that it has multiple drupes formed by each flower to make a sorosis. The weight of the fruit is approximately 3.49 g with a moisture content of nearly 71.5%. The color of the fruit varies with its maturity. In the early stages, the color is green which turns orange to red in the intermediate stage. When it is fully matured, the color is purplish black [44–46].

Functional and therapeutic roles: Traditionally, it is used in kidney disorders and purification, curing weakness, anemia, fatigue, and early graying of hair. Due to its calming properties, the plant is beneficial in monitoring illness according to Chinese traditional medicine. It functions by reducing the symptoms of sore throat, cough, and fever, protecting the liver, enhancing vision, helping in urination, and stabilizing blood pressure [47]. The *Morus alba* plant has been also known to show excellent therapeutic activity against diabetes mellitus due to the presence of a special alkaloid 1-deoxynojirimycin (DNJ) [48].

Phytochemical aspects: It has been revealed that *Morus alba* majorly contains alkaloids (Calystegins B2, C1, and 1-Deoxynojirimycin), terpenoids (betulinic acid, ursolic acid, and uvaol), flavonoids (astragalin, atalantoflavone, anthocyanins and chalcones such as Morachalcones B, C), phenolic acids, stilbenoids, and coumarins [44]. These bioactive components make it highly beneficial for multifunctional applications.

Anthocyanins: The content of anthocyanin in the fruit was found to be 21.97 mg CGE/100 g fw [40].

4.3. *Ficus palmata*

The common wild fig (*Ficus palmata*) grows on hot, dry slopes of Northwest hills, on clay-loam soils in the states of Uttarakhand, Punjab, and Kashmir in India, Nepal, the Arabian Peninsula, Ethiopia, Somalia, Sudan, Pakistan, Afghanistan, Iran, and Southern parts of Egypt [49].

Morphological aspects: It is a deciduous tree with an average height of 6 to 10 m. Leaves are broad, ovate, alternate, and membranous with a size of around 12.93 cm long and 14.16 cm broad. It has small flowers which are unisexual, monoecious, and greenish-white in color. It yields deep violet to black colored fruits having a diameter of approximately 2.5 cm and measuring 6 g in weight. Both fruit and seeds are edible. Due to the presence of white latex below the fruit's outer covering, the fruit may have a little acidic or bitter taste and smell (astringency) [50].

Functional and therapeutic roles: In the Uttarakhand region of the Indian Himalayas, fruits of *Ficus palmata* are well-known for their traditional use in treating ailments such as stomach ulcers, digestive problems, bronchitis, eczemas, hemorrhoids, and as a diuretic. In traditional medicine, *Ficus palmata* is commonly used for its digestive, hypoglycemic, insulinase, anti-tumor, anti-ulcer, antidiabetic, lipid-lowering, anti-carcinogenic, antifungal, and inflammation-reducing properties [51]. It can also be used as a dietary component in the treatment of constipation and ailments related to the gallbladder and respiratory system. The plant's sap is often administered as the medication for warts [50].

Phytochemical aspects: Numerous phytoconstituents, including alkaloids, steroids, lipids and fixed oils, flavonoids, tannins, proteins, and sugars, are found in the bark, root, and fruits. It is a promising source of flavonoids, polyphenolic compounds with powerful antioxidant qualities that aid in the prevention and treatment of a variety of oxidative stress-related disorders, including neurological and hepatic conditions [52].

Anthocyanins: The anthocyanin content in the fruits was estimated to be 19.27 mg CGE/100 g fw [40].

4.4. *Berberis lycium*

Berberis lycium is a native plant to the Himalayan mountains and is extensively dispersed in temperate and semi-temperate climatic zones of India, Nepal, Bangladesh, and Pakistan.

Morphological aspects: It has strong woody branches covered by thin, brittle bark. The plant is a small-sized, stiff deciduous shrub that can reach heights of 1.0 to 2.5 m. Leaves are 2.5 to 7.5 by 8.18 mm in size and pale green in color. The plant's stem has a few noticeable spinous teeth that are alternately placed. The inflorescence is composed of corymbose racemes, which have 11–16 flowers per cluster. Insects pollinate flowers, which are often hermaphrodite, borne in axillary clusters, and pale yellow in color. The plant's blooming and fruiting season lasts from March through July. The first two weeks of March mark the start of the flowering season, which lasts until April. In the second week of May, the fruit starts to ripen, and it does not fully mature until June. Fruits are incredibly nutrient-rich in the form of 8 mm long globose ovoid berries with bluish-purple color when they are fully mature [53,54].

Functional and therapeutic role: Since the beginning of time, ancient societies of the Himalayan region of Jammu and Kashmir, and Himachal Pradesh have utilized these berries as a food source [55]. The herb is employed in traditional medical practices to cure several ailments and disorders. People use the various parts of the plant, such as the leaves, stem, roots, fruits, and flowers, as food and medicines. The plant is well known for preventing diseases of the skin, eyes, and abdomen. Recent research work has revealed that it has good antimicrobial, antidiabetic properties [56].

Phytochemical aspects: Berberine, palmatine, and jatrorrhizine are the major alkaloid found in this species with berberine being the most prominent one. It was discovered that the root extract of *B. lycium* contained 80% dry weight of berberine and very small levels of other alkaloids [57,58]. There are several important flavonoids and phenolic compounds present in this plant, such as chlorogenic acid, hydroxybenzoic acid, quercetin, rutin, and mandelic acid. Ascorbic acid, β-carotene, and anthocyanins are also found in the fruit in significant amounts [59].

Anthocyanins: The *B. lycium* fruit's anthocyanin content was reported to be 20.58 mg CGE/100 g fw [60]. Characterization of *Berberis lycium* anthocyanins by LC-MS and UV spectral analysis revealed that there are a total of twelve anthocyanins in the purified extract, but among all the anthocyanins, delphinidin-3-glucoside (35.3%) and cyanidin-3-glucoside (47.2%) were the most prominent. The glycosides of cyanidin, pelargonidin, malvidin, and peonidin were the other ten anthocyanins that were also characterized. Overall, it was discovered that the species has a potential future for use in food systems [61].

4.5. *Myrica esculenta*

In India, the genus Myrica is represented by *Myrica esculenta* species [62]. It is an evergreen dioecious small tree of Indian origin and is widely disseminated along the mid-Himachal Pradesh foothills track, which extends from Ravi eastward to Assam and includes the states of Sikkim, Manipur, Arunachal Pradesh, Uttaranchal, and the Khasi, Naga, Jaintia, and Lushai Hills of Meghalaya, all of which are located between 900 and 2100 m above sea level. It is also found in Singapore, China, Pakistan, Japan, Nepal, and the Malayan Islands [63].

Morphological aspects: It is a medium to large woody tree with a trunk diameter of 92.5 cm and a height between 12 and 15 m. Most leaves are clustered at the ends of branches and are lanceolate with an entire or serrated edge, pale green on the underside, and dark green on the top side. While the inflorescence of staminate flowers is a compound raceme, the inflorescence of pistillate flowers is small, sessile, solitary, and bracteate; the sepals and petals are either absent or not visible; the catkin's axillary length is 4.2 cm; it bears about 25 flowers in a thread-like fashion. The flowering season begins in February and lasts through the second week of April, however, the first week of March marked the peak flowering period. Fruits are drupe, red to dark brown in color, oval, and about 2–7 mm in diameter. Fruits taste sweet and sour [29].

Functional and therapeutic roles: Local population use this traditional ayurveda plant in a variety of ways due to the diverse therapeutic benefits of the plant parts. Natural antioxidants included in *M. esculenta* fruit are a significant source of protection against oxidative stress and certain degenerative illnesses [64]. Research has demonstrated that flavonoids and phenolic acids found in the plant are responsible for its antioxidant potential. Analgesic, antiasthmatic, anticancer, antidepressant, antidiabetic, anti-inflammatory, chemopreventive, hepatoprotective, and wound healing effects are some of the pharmacological activities that have been reported recently [65]. Moreover, they are non-toxic and have no harmful materials which makes them useful for versatile applications.

Phytochemical aspects: *Myrica esculenta* plant has been discovered as a rich source of flavonoids, flavonols, and phenolic chemicals. Other bioactive components of the plant include alkaloids, glycosides, diarylheptanoids, ionones, steroids, saponins, triterpenoids, and volatile compounds. [29,63]. Myricetin and quercetin are the major flavonoids whereas flavonoid glycosides such as Myricitrin (myricetin 3-O-rhamnoside) were also detected. Arjunolic Acid was the major saponin. Corchoionoside C; (6S,9R)-roseoside was the ionone [66,67].

Anthocyanins: The fruit had an anthocyanin content of 7.17 mg CGE/100 g [60]. The major anthocyanins detected were malvidin 3-(6″-acetylglucoside), cyanidin 3-O-(6″-acetylglucoside), delphinidin-3-O-arabinoside, and cyanidin-3.5-di-O-beta-D-glucoside [68].

4.6. *Duchesnea indica*

Duchesnea indica, generally known as a mock strawberry/Indian strawberry, is a straggling herb belonging to the family Rosaceae [69]. It is a perennial plant having widespread distribution in Asia, Europe, and America.

Morphological aspects: Stoloniferous species *Duchesnea indica* (Andrews) Focke have bostryx-like cymose inflorescences with incredibly long petioles, yellow petals, terminal styles, anthers with two thecae, and achenes. It can be identified by its perennial habit, long, creeping flower stem, roots in nodes, basal leaves with three to seven lobes, and style that is shorter than or equal to the carpel [70]. Ripened fruits are red, shining, spongy, and 1–2.5 cm in diameter. Achenes are small, ovoid, glabrous, and pitted [71]. Its fruits are also used as a cooling agent, tonic, and in eye infections. This plant's ash has long been used to cure burns and skin problems. Additionally, leaf water extract has been utilized as an anthelmintic [72].

Functional and therapeutic role: For thousands of years, it is employed as a traditional herbal medicine in Asia, mostly to cure leprosy, congenital fever, and tissue inflammation. These days, it is clinically employed for cancer treatment or as a key component of formulae for Chinese herbal medicines used to treat cancers, particularly gynecological cancers. A recent report has shown that extracts of the *Duchesnea indica* plant lowered the growth of SKOV-3 ovarian cancer cells by inducing apoptosis via the mitochondrial pathway and stopping cell cycle progression in the S phase [73].

Phytochemical aspects: This plant has been found to contain a number of phenolic chemicals, such as flavonoids, ellagic acids, and phenolic acids. A total of 27 phenolic compounds have been reported in the species, and they fall into four categories: flavonols, ellagitannins, ellagic acid and its derivatives, hydroxybenzoic acid, and hydroxycinnamic acid. In addition to that, brevifolin carboxylate, caffeic acid, brevifolin, methyl brevifolin carboxylate, and kaempferol O-robinobioside were also reported [74].

Anthocyanins: The anthocyanin content in its fruit was found to be 7.06 mg CGE/100 g [60]. Another study found the total amount of anthocyanin on a cyanidin-3-rutinoside basis to be 205 mg/g of fresh stoned fruit, with cyanidin 3-O-rutinoside accounting for major composition [75]. The detailed anthocyanin composition has been described in Table 2.

4.7. *Lycium ruthencium*

Genus Lycium has long been recognized as a source of nutrients and medications throughout Southeast Asia, specifically in China. This genus contains 97 species and 6 variations of perennial flowering plants belonging to the family Solanaceae, which are primarily found in South America, South Africa, China, and a few species in temperate Asia and Europe. *Lycium ruthencium*, locally known as "Khizer", is a plant that is extensively distributed in the Trans-Himalayan Ladakh region at a height of 3063–3196 m above mean sea level. It grows primarily on the sides of highways in the Nubra valley's Hunder and Udmaru sections [76].

Morphological aspects: It is a thorny, perennial shrub with a long lifespan that is a member of the Solanaceae family. It has small sessile leaves, zigzag-shaped stems, internodes with little thorns, and deep roots. It can reach a height of 2 m. It has pale purple funnel-shaped flowers which are hermaphrodite in nature. The shoots are short with a length of approximately 5–10 mm bearing one or two flowers. The fruits are 6–9 mm long, color ranges from black to purple berries commonly called goji berries. Seeds germinate in the late spring or early summer, while the bushes bloom in the months of June and July and bear fruit in the months of August and September [77].

Functional and therapeutic roles: These berries are utilized in various herbal remedies as a medication or a coloring component in India [78]. It has been reported that historically, the fruits have been used as treatments for a variety of illnesses, including menstruation-related problems, hypertension, urethral stones, tinea, furuncles, and blindness in camels in mountain communities, particularly in the Chinese and Tibetan medical systems [79–81]. While using leaves as a diuretic to clear urine blockage [82].

Phytochemical aspects: The primary chemical components that have been identified are alkaloids, phenolic acids, anthocyanins, pro-anthocyanidins, fatty acids, coumarins, polysaccharides, essential oils, carotenoids, and cinnamate derivates. The major bioactive chemical components of *L. ruthenicum* Murr are anthocyanins, which belong to the flavonoids class of phytochemicals and are believed to be mostly responsible for the alleged therapeutic benefits [83].

Anthocyanins: The anthocyanin content in the fruits was estimated to be 9.28 ± 1.19 to 82.58 ± 0.95 mg CGE per g of dry weight (C3GE/g DW) [77].

4.8. *Gaultheria trichophylla*

Gaultheria trichophylla (family-Ericaceae), is a highly valuable wild Himalayan plant. It is a decumbent, aromatic shrub that forms a mat at higher elevations, between 3200 and 5300 m above sea level. Its distribution is restricted to the Trans-Himalayan regions of Pakistan, China, India, and Nepal [84]. The Himalayas are home to this species; hence it is often called 'Himalayan snowberry'. It produces blueberries that the local community uses as a source of refreshing food. Dried branches are used by people in the Trans Himalaya region of Uttarakhand to make incense fire during religious ceremonies [85].

Morphological aspects: It is a small herb with a wiry, slender, and dark brown stem of length 4–7 cm. Leaves are oval, with log setae with a dark green upper surface and whitish green lower surface. Leaves are 4–8 mm long with a diameter of 2.3–4 mm. The plant yields oval blue fruits with a length of about 1.4 cm and a diameter of 1 cm [32].

Functional and therapeutic roles: *Gaultheria trichophylla* fruits have a long history of usage in traditional medical practices, particularly for the treatment of pain and inflammation [32,86].

Phytochemical aspects: *Gaultheria trichophylla*'s wild edible fruits have been shown to be extremely nutrient-dense and abundant in polyphenols and antioxidants. The results of the phytochemical analysis show that fruits from the Milam bugyal region have higher levels of total phenolic compounds at 3.71 mg gallic acid equivalents per g of fresh weight (GAE/g FW), flavonoids 1.75 mg quercetin equivalent per g of fresh weight (QE/g FW), tannins 2.62 mg tannic acid equivalent per g fresh weight (TAE/g FW), and flavonols 1.03 mg catechin equivalents per g fresh weight (CE/g FW) [87]. For this plant, so far, no potential studies of its anthocyanin profiling and composition were reported. So, further research needs to be conducted in this area.

4.9. *Species of Genus Begonia*

With 1870 species, the pantropical genus Begonia is the sixth-largest genus of flowering plants. [88]. Begonia is also a significant source of phytochemicals due to its diverse range of taxa and a high degree of morphological variety. Begonia plants have a high content of phenolics such as total phenolic and flavonoids. Begonia is also one of the largest genera of vascular plants, encompassing approximately 1800 species. Additionally, numerous cultivars are grown specifically for their ornamental value as flowers. The northeastern region of India, along with Myanmar, is a significant area for the genus Begonia. Numerous species have been recently described from this region, and many more are currently being studied and evaluated (Taram et al., 2023). Earlier authors have reported the presence of flavonoids in Begonia species. Five flavonols and two flavones were isolated from the leaves of *B. erythrophylla*. These phytochemicals were identified as 3-methyl ethers of kaempferol, 3-methyl ethers of quercetin, quercetin, and its 3-O-rutinoside and 3-O-rhamnoside, and 7-O-glycoside of luteolin. [33]. Additionally, numerous cultivars are grown specifically for their ornamental value as flowers. The red-colored leaves of *B. xanthina*, *B. palmata*, and *B. megaptera* were found to contain a high amount of anthocyanin, which can be used to create biobased color. The amount of anthocyanin in *B. xanthina*, *B. palmata*, and *B. megaptera* was found to be 88.08 mg, 68.26 mg, and 20.08 mg CGE/g fw [89,90]. It is important to conduct comprehensive phytochemical surveys of more wild Begonia species to further explore their anthocyanin content and composition.

4.10. *Fragaria nubicola*

Fragaria nubicola, sometimes known as Himalayan strawberry, is a perennial herb that is typically found in shady areas near the edges of forests between the altitudes of 2100 and

4000 m above sea level [31]. Native to the Himalayas, this species is found in Afghanistan, India, Nepal, Tibet, Myanmar, Pakistan, South-Western China, and Bhutan.

Morphological aspects: This plant is a small herb, stoloniferous, with a height of 2–4 cm. Leaves are trifoliate, lateral leaflets are in the form of distinctly petiolate, elliptic, or obovate. Petioles and stems are resisted from spreading. One to numerous flowers can be seen in the inflorescence. The flowering season is from May to August. Flowers are large, sometimes more than 2.5 cm in diameter [91]. Its edible fruits have an anthocyanin-tinged monopodial stolon and are broadly ovoid or compressed ovoid in shape, measuring 5.5–16.5 mm long by 7.0–17.5 mm in diameter [92].

Therapeutic roles: Fresh fruits of the plant in combination with dried leaves of *Potentilla peduncularis* and dried roots of *Geumelatum elatum* are made into a fine paste and usually taken orally to treat fever and rhinitis [93]. Tibetan doctors utilize it to treat neuropsychiatric conditions and nerve inflammation [94]. The unripe fruit is chewed to treat acne, while plant juice is applied to ease heavy menstrual bleeding [95].

Phytochemical aspects: The plant is abundant in phenolic compounds and ellagic acid, both of which are recognized as powerful antioxidants [96]. Phenolics (1.18–3.08), proanthocyanidins (0.53–1.01), flavonoids (0.99–2.63), flavonols (0.95–1.80), and tannins (0.73–1.42) mg per g gallic acid equivalent (GAE) are the various phytochemicals identified in this plant [97].

Anthocyanins: The total monomeric anthocyanins in the berries were found to be 1.46 mg CGE/g fw [31].

Table 2. Anthocyanin composition of some Himalayan plant sources.

Species	Anthocyanin Composition	Concentration	References
Berberis lycium	1. Delphinidin-3-glucoside	35.32%	[61]
	2. Cyanidin-3-lathyroside	0.08%	
	3. Cyanidin-3-glucoside	47.24%	
	4. Cyanidin-3-rutinoside	0.53%	
	5. Cyanidin-3-galactoside	1.62%	
	6. Pelargonidin-3-pentoxilhexoside	2.26%	
	7. Malvidin-3,5-dihexoside	4.21%	
	8. Pelargonidin-hexoside	0.58%	
	9. Pelargonidin-3,5-diglucoside	1.05%	
	10. Cyanidin-3,5-dihexoside	6.12%	
	11. Peonidin-3-rutinoside	0.77%	
	12. Pelargonidin-3-rutinoside	0.22%	
Morus alba	1. Cyanidin-3-glucoside	1. 301.75 mg/g MAE	[98]
	2. Cyanidin-3-rutinoside	2. 108.79 mg/g MAE	
	3. Pelargonidin-3-glucoside	3. NA	
	4. Pelargonidin-3-rutinoside	4. NA	
	5. Cyanidin 3-O-(6″-O-α-rhamnopyranosyl-β-D-glucopyranoside)	5. 270 mg/g CMA	
	6. Cyanidin 3-O-(6″-O-a-rhamnopyranosyl-β-D-galactopyranoside)	6. 57 mg/g CMA	
	7. Cyanidin 3-O-β-D-galactopyranoside	7. 233 mg/g CMA	
	8. Cyanidin 7-O-β-D-glucopyranoside	8. 33 mg/g CMA	
	9. Petunidin 3-O-β-glucopyranoside	9. 33 mg/g CMA	

Table 2. *Cont.*

Species	Anthocyanin Composition	Concentration	References
Myrica esculen.	1. Malvidin3-(6″-acetylglucoside) 2. Cyanidin3-O-(6″- acetylglucoside) 3. Delphinidin-3-O-arabinoside 4. Cyanidin-3.5-di-O-beta-D-glucoside	0.205 ± 0.4 mg CGE/100 g FW′ 0.342 ± 0.4 mg CGE/100 g FW′ 0.421 ± 0.4 mg CGE/100 g FW′ 0.421 ± 0.4 mg CGE/100 g FW′	[68]
Duchesnea indica	1. Cyanidin 3-O-rutinoside 2. Peonidin 3-O-rutinoside 3. Petunidin 3-O-rutinoside	125 mg/g 70 mg/g 10 mg/g	[75]
Lycium ruthencium	1. Petunidin-3-Orutinoside (trans-p-coumaroyl)-5-O-glucoside 2. Petunidin 3. Malvidin 4. Delphinidin	10 mg CGE/g DW 5.71 mg CGE/g DW 0.47 mg CGE/g DW 0.29 mg CGE/g DW	[77]

MAE = mulberry anthocyanin extract. NA = not available. CMA = crude mulberry anthocyanin. CEE = crude ethanol extract. FW = frozen weight. FW′ = fruit weight. DW = dry weight. CGE = cyanidin-3-glycoside equivalent.

5. Physical and Functional Properties

Anthocyanin possesses various important physical and functional properties, and this section briefly discusses it. Reactive oxygen species (ROS), Free radicals, and/or reactive nitrogen species (RNS) are essential for the human body to function normally. Redox homeostasis is responsible for maintaining a balance of these radicals in our body. In certain circumstances, the body may experience oxidative stress brought on by an imbalance of these radicals. This stress results in the progression of chronic degenerative diseases such as cancer, aging, and coronary heart disease [99]. Anthocyanins have been defined as substances that scavenge free radicals and modulate oxidative stress to curb or inhibit oxidation. Anthocyanins function as H-atom donors or as single electron transfer agents. The antioxidant activity of these compounds can be determined using several analytical methods based on both these working mechanisms. The antioxidant nature of anthocyanins is subject to the overall concentration, structure, and environmental conditions [100].

Due to their high reactivity to pH changes, anthocyanins display distinct chemical structures and colors in solutions with varying pH values. Furthermore, as the pH value rises, the absorption peak of anthocyanin solutions in the UV-VIS spectrum exhibits a noticeable bathochromic shift [101]. Anthocyanin-rich solutions turn from red to pink, purple/blue, and yellow as the pH rises. This color change is the result of anthocyanin's structural transformation from red flavylium cation to purple/blue quinoidal base, colorless carbinol pseudo base, and yellow chalcone [102].

Apart from these physical properties, anthocyanins show numerous functional properties in plants, food, and human nutrition. Anthocyanins provide different hues to plants for attracting animals to perform pollination and seed dispersal [103]. Anthocyanins have been known to protect plants from reddening caused by UV-B light. However, hydroxycinnamoylated structures are necessary for absorption to take place in the 280–320 nm region, and many simply glycosylated anthocyanins would not perform this function. More recent research has shown that anthocyanins shield the organelles that contain chlorophyll and thereby avert photoinhibition by protecting chloroplasts against high-intensity lights [104]. Three primary roles for anthocyanins in plants have been proposed: (i) absorbing harmful radiations, (ii) working as modes of transportation for monosaccharides and (iii) as osmotic regulators during drought and low-temperature periods [105–107]. Moreover, anthocyanins are regarded as essential phytochemicals for human nutrition [108]. However, the actual health benefits of anthocyanins are determined by their final bioavailability and bio accessibility in the human body. In some clinical studies, individuals were given anthocyanin-rich diets or extracts for consumption, and it was investigated that less than

1% of ingested anthocyanin was available in a standard blood test of those individuals. This study denotes the low bioavailability of anthocyanins [109]. This is attributed to several factors such as (i) degradation in the oral cavity due to interaction with human enzymes, microbiota, and spontaneous binding to salivary proteins, (ii) change in form and stability due to varying pH, water content, and gas composition along the gastrointestinal tract, and (iii) modification by mammalian enzymes during circulation [110–112]. Researchers have developed novel techniques such as nano-encapsulation and exosome, polysaccharide-based, lipid-based, and protein-based complexes, for improved bioavailability and controlled release of anthocyanins [113].

6. Applications of Anthocyanins

Anthocyanins have wide applications in the industry, especially as food colorants, as they can produce distinct colors (Figure 3). They also possess exceptional health-promoting properties that have been integrated into human diets for many years. In addition to serving as a source of nutrients, anthocyanins are also found in traditional herbal remedies for conditions such as high blood pressure, pyrexia, liver abnormalities, diarrhea, dysentery, urinary issues, and the common cold [8].

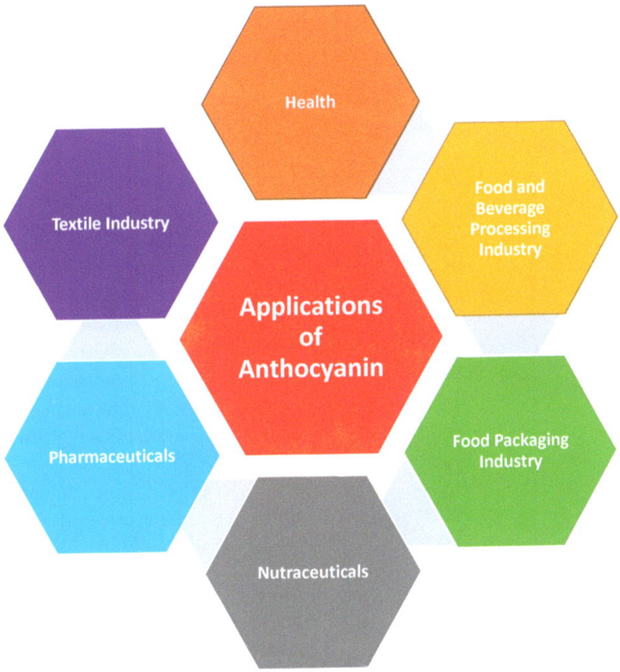

Figure 3. Various applications of anthocyanins.

6.1. Food and Beverage Processing Industries

The demand for foods and extracts high in anthocyanins have gained attention owing to the health benefits they offer. These extracts can now be used in a wider range of food applications owing to their improved stability and potential health effects [114]. Anthocyanins are extensively being used as natural colorants since synthetic colors used in the food industry were found to have a detrimental impact on human health. Consumers and food manufacturers are searching for natural food colorants over synthetic dyes [115]. According to a 2007 trial conducted on children aged 3 and between 8–9 years, mixtures of artificial colorants such as tartrazine (E102), sunset yellow (E110), carmoisine (E122), ponceau 4R (E124), allura red AC (E129), and quinoline yellow (E104), when combined with sodium benzoate (E211) statistically caused considerable rise of hyperactivity in children [116]. Because synthetic dyes are more stable with regard to light, oxygen, temperature, and pH, the replacement of synthetic dyes with natural food colorants presents certain obstacles [117]. Natural dyes do not particularly have stability and can be characterized by their own physiological activities [118]. Anthocyanins with acylated substituents have been discovered to be more stable during the production and storage of food goods, according to published research. Researchers around the world are evaluating promising sources of acylated anthocyanins. Red cabbage, radish, and black carrot were found to contain high amounts of acylated anthocyanins [119]. The cyanidin and delphinidin glycoside anthocyanins found in fruits of *Morus alba* are used as natural food coloring agents. These reddish-purple to purple pigments have strong antioxidant properties such as FRAP and DPPH [120,121]. *Morus alba* and its anthocyanins are highly suggested as a natural, functional food colorant due to their high antioxidant and color-enhancing characteristics since they can produce better color quality with value-added attributes for the product and are safer to take as compared to synthetics [122]. Himalayan flora offers a diverse range of anthocyanins, these anthocyanins can be extracted and stabilized for their uses as a natural coloring agent in food industries. Anthocyanins can be stabilized for application in food systems. Additions of co-pigment chemicals, such as polymers, phenolic compounds, and metals, are among the stabilizing methods. Various encapsulation processes, hard-panned candy-coating procedures for the colors blue, green, and brown, and the exclusion of O2 during production and storage can also be taken into consideration [123].

Anthocyanins have potential applications in the nutriment sector for the development of various fortified foods and beverages, which may have positive health effects [124]. Berries, grapes, red and purple vegetables, as well as their processed products, including drinks, are food sources rich in anthocyanins. (Wines, juices). Anthocyanin-rich feedstocks are used to obtain red wine which is an important source of antioxidants when consumed in moderate amounts. Anthocyanins hold great importance in wine research as they play a major role in affecting wine properties such as wine color, mouthfeel, aging, stability, and overall quality [125]. Himalayan sources of anthocyanins and their closely related species are already being used to prepare food and beverages with high nutritional value and functional characteristics. *Morus alba* is highly regarded as a delightful dessert fruit and is commonly utilized in making jelly, refreshing cold drinks, and alcoholic beverages. It is frequently incorporated into ice cream, gastriques, sorbet, and various baked goods, particularly pies. Mulberry leaves are commonly combined with other herbs such as ligustrum, chrysanthemum, and apricot seeds. They are used in various food and beverage applications such as wine, fruit juice, canned food, and jams. Mulberry leaves are especially useful in creating desserts, dairy products, shakes, and tea. They are also employed in the preparation of different types of beverages [126]. Shrubs belonging to the *Berberis genus* are part of the Berberidaceae family. They naturally grow in central and southern Europe, western Asia, and northwest Africa. The fruit called Zereshk is highly popular in Iran. It is commonly used in cooking and making jam, leading to a yearly production of about 22,000 tons of fresh, seedless barberries. These fruits are also used to make beverages, syrups, candy, and other sweet treats that are enjoyed in Iran. Additionally, the leaves and fruits are used to create food flavorings and herbal teas [127].

Numerous attempts have been made to fortify anthocyanins in real food systems. Bread fortified with 1%, 2%, and 4% of anthocyanin-rich black rice extract powder (ABREP) as a nutraceutical ingredient had digestion rates that were lowered by 12.8%, 14.1%, and 20.5%, respectively. These findings suggested that adding anthocyanins to bread could open new production possibilities for functional bread by reducing the rate of digestion and so improving consumer health [128]. It has been observed that using anthocyanins to fortify and/or color some foods has produced functional foods with health advantages. Adding extracts rich in anthocyanins to baked goods such as cookies, biscuits, and macarons can curb damage to the food during baking and increase its antioxidant capacity beyond that of synthetic additives, all without impacting the food's acceptability [129–132]. Moreover, anthocyanins have proved to have great stability during storage in products such as kefir, yogurt, and various beverages, and might be thought of as ideal foods for anthocyanin fortification [133–135].

6.2. Health

Anthocyanins have significant effects on human health. Research has shown that the intake of these pigments reduces the rate of ailments of the heart and blood vessels, cancer, hyperlipidemia, and chronic diseases [136]. Anthocyanins have the capacity to terminate or neutralize free radicals and reactive chemical species, controlling signaling pathways, reducing the extent to pro-inflammatory factors, and lowering the occurrence of neurodegenerative diseases, cancer, and cardiovascular ailments [137]. Anthocyanins are known to provide a variety of health benefits, but researchers have paid more attention to the antioxidant properties of diets sufficient in anthocyanins and their role in improving cardiovascular health. *M. alba* root, also called San Baipi in traditional Chinese medicine, is commonly used to cure conditions such as cough, asthma, and similar ailments. In a study, fourteen compounds were extracted from *M. alba* and evaluated for their potential effects against HIV [138] Antioxidants may confer a beneficial impact on human health at several levels since free radical damage has a major function in the genesis of many chronic diseases. Traditionally, the leaves, fruits, and latex of *Ficus palmata* have been used to treat several disorders such as hypoglycemia, hyperlipidemia, gastrointestinal disorders, tumors, ulcers, diabetes, and fungal infections [139]. This plant has been documented in several scientific reports for its medicinal applications in treating a range of diseases and disorders. These include gastrointestinal issues, diabetes, tumors, hypoglycemia, ulcers, lipid level reduction, and antimicrobial properties [50]. In many in vitro models, anthocyanins have proven to be strong antioxidants and free radical scavengers. The anthocyanins that possess the highest antioxidant capacities among those tested are the 3 glucosides of delphinidin (gallocatechol structure) [140]. Existing clinical trials on this group of plants primarily focus on their effects on conditions related to cardiovascular diseases, associated risk factors, neurodegenerative diseases, and inflammation. One notable group of clinical trials conducted investigated a specific combination of *B. aristata* and Silybum marianum. This combination was chosen to address the low bioavailability of *B. aristata*, while *S. marianum* was included to enhance its absorption in the intestines. In a 52-week double-blind placebo-controlled study involving 136 obese patients with type-2 diabetes mellitus (T2DM) and metabolic syndrome, various parameters were analyzed. These parameters included fasting blood glucose, insulin levels, total cholesterol, HDL and LDL cholesterol, triglycerides, and body mass index [141–143]. The wild edible fruits of Fragaria nubicola, belonging to the Rosaceae family, are known for their significant content of antioxidants and polyphenolics. These fruits hold great potential as a valuable source for the nutraceutical and food industries.

Protection from oxidative stress is closely related to anthocyanin's role in preventing cardiovascular diseases [144]. Anthocyanins have the capability to act upon different cells responsible for the occurrence of atherosclerosis which is one of the major causes of cardiovascular dysfunction [145]. It has been found that anthocyanins have anti-inflammatory as well as chemo preventive effects. Different cell cycles and growth-related pathways have been recognized as potential targets for anthocyanins used in in vitro and on lab animals. Red wine's anthocyanin-rich fraction has been demonstrated to have a potency for inhibiting the growth of gastric adenocarcinoma (AGS) and HCT-15 human colon cancer cells [146,147]. Anthocyanins have already been found to assist in improving vision and weight management [148]. In a study conducted in 2000, oral doses of anthocyanins obtained from blackcurrant, depending on the dose, were competent in lowering the threshold for dark adaptation [149]. Bilberry anthocyanins when interacting with rhodopsin or phosphodiesterase improved night vision [150]. Anthocyanins also have a potency for regulating adipocyte function, thus, preventing obesity [151]. Anthocyanins also have antidiabetic and anti-obesity effects and act they act as neuroprotective agents. Numerous studies have demonstrated that anthocyanin consumption has a beneficial impact on human health [152].

6.3. Active and Intelligent Packaging Films

Although conventional had great importance in the historical development of food distribution networks, it is no more adequate due to the complexity of modern society. Innovative packaging with enhanced functionality is highly desired due to consumer demands for foods that have fewer chemical preservatives and are minimally processed, better regulation, food safety concerns, and globalized markets [153]. Conventional food packaging films are made of synthetic polymers, which are not biodegradable and can only shield food from certain external environmental factors such as moisture, light, air, microbes, and mechanical harm [154,155]. Therefore, researchers have given more attention to the development of active packaging technology that works by releasing functional components into the packaged food or the external environment to maintain its quality and extend shelf life. On the other side, intelligent packaging helps to monitor the freshness of the food during packaging. Functional components such as organic acids, enzymes, microbial toxins, natural extracts, and ionic compounds are deliberately included in the packaging material to develop these active packaging systems [156,157]. According to the legal definition of the EU (EC, 2009), intelligent packaging includes a component that permits the monitoring of the state of packed food or the environment around the food during transport and storage. Hence, intelligent packaging is a system that gives the user accurate and reliable details on the state of the food, the environment, and/or the integrity of the package. Intelligent packaging conveys information to the consumer based on its capacity to sense, detect, or record changes in the food product or its environment [158].

Researchers have recently become interested in the development of active and intelligent packaging by incorporating anthocyanins with biopolymers. These packaging films are developed by casting or extrusion methods. The matrix of anthocyanin-rich films can be several polysaccharides such as starch, cellulose, pectin, chitosan, and agar. In addition to that, proteins such as zein, soy protein, and soy protein isolate can be used [159]. The anthocyanin-added packaging film/label is found to be highly effective to determine the freshness of meat and fish products, dairy products, and fruits and vegetables.

The excellent antioxidant and antimicrobial properties of anthocyanins enable these films to keep food product quality high and prolong their shelf lives [160–162]. Free radicals and other reactive species play an important role in causing food spoilage and loss of nutrients [163]. Food packaging films having antioxidant activity can help extend the shelf life of food [164]. Thus, the addition of anthocyanin-rich extracts can significantly increase the films' antioxidant activity [165]. Berberis plants have significant potential in the food processing industry, although there is limited research on their direct application in food products. One example is the use of seed oil and fruit extracts from *B. crataegina* to enhance the physical and functional properties of chitosan-based edible film. The resulting films were studied for their physical, chemical, and biological properties. The findings revealed that the chitosan-fruit extract film exhibited superior thermal stability, antimicrobial, antioxidant, and anti-quorum sensing activities compared to other films. These results suggest that incorporating *B. crataegina* fruit extract into a chitosan-based edible film can improve its overall characteristics, making it a valuable ingredient in food production [166]. Foodborne diseases have serious consequences and may result in mortality and economic loss due to spoilage. Transnational food supply can become a pathway for the spread of foodborne diseases. Therefore, the globalization of food supply is under threat of an outbreak of foodborne diseases [167]. To prevent pathogen growth and food spoiling, antimicrobial activity is a crucial characteristic of food packaging films prolonging the food's shelf life [168]. Anthocyanins were found to have considerate antimicrobial potential [167]. The biopolymer-based film having no anthocyanin exhibited minimal to no antimicrobial activity. However, with the addition of anthocyanins-rich extracts, the antimicrobial nature of the films can be greatly increased [169]. A ternary blend film has been made using chitosan, gelatin, poly (vinyl alcohol), and *Duchesnea indica* extract (DIE) using a casting method. In compliance with the antimicrobial properties of anthocyanins and other phytochemicals, with the release of DIE from the film matrix, the film demonstrated a clear inhibitory action on *S. aureus*, up to 4.9 logs CFU/mL. Such films can be used as a coating material in fresh fruits and vegetables to extend their microbial safety [170]. Additionally, because anthocyanins can change their chemical structures and exhibit different colors depending on pH, anthocyanin-rich films can be used to evaluate the quality of packaged foods [171]. Further research is required on the preparation of anthocyanin-based active and intelligent packaging systems by utilizing Himalayan plants.

6.4. Nutraceuticals and Functional Foods

Nutraceuticals are foods or components of foods that shows health benefits, such as the ability to treat or prevent disease. Because they avoid side effects and contain naturally occurring dietary supplements, nutraceuticals offer an advantage over pharmaceuticals [172]. Anthocyanins are gaining popularity as one of the most favorable ingredients in nutraceutical preparations. Many studies have been conducted on the nutraceutical properties of anthocyanins and anthocyanidin [173]. Anthocyanidins are aglycones from which anthocyanins are formed. These substances have a flavylium cation backbone that can be hydroxylated in a variety of ways to form various anthocyanidins (usually on carbons C3, C5, C6, and C7 as well as C3′, C4′, and C5′). The charged oxygen atom is found on the C ring of flavonoids, which preserves their ring nomenclature while having an oxonium group in their structure. Their brilliant colors such as red, blue, and purple, are caused by the buildup of these substances and their capacity to absorb light that is opposite to chlorophyll. Because it lessens the extremely appealing green coloring of some plants, this color variation may act as a defense strategy for those plants against possible herbivorous predators [174]. Due to the flavylium skeleton, which permits the delocalization of radical electrons on the sp2 orbitals of the oxonium group, anthocyanins have special properties. The oxidation of anthocyanins' phenolic hydroxyl groups is intimately related to their antioxidant action. The stabilization of oxidation products resulting from one-electron oxidation and the production of semiquinones depend heavily on the para- and ortho-phenolic groups. The antioxidant qualities of anthocyanin extracts produced from various culinary and agricultural sources have been the subject of numerous investigations. To assess the radical scavenging properties of anthocyanins, these investigations frequently use techniques based on single electron transfer (SET) processes, hydrogen atom transfer (HAT), or a combination of the two. The DPPH and ABTS+ antioxidant tests are examples of SET and HAT assays, which involve direct electron transfer [175–177]. Anthocyanins have significant antioxidant, anti-inflammatory, anti-fibrotic, and anti-apoptotic properties and have been suggested for the prevention and treatment of several disorders [178]. Anthocyanin's antidiabetic properties have been well investigated. Petunidin-3-O-p-coumaryl-rutinoside-5-O-glucoside is the major anthocyanin found in the extract of the Blue Congo variant of purple potato, in a study, it was found to reduce the fasting sugar levels in diabetic rats induced with streptozotocin [179]. In Zucker Diabetic Fatty rats, the mulberry fruit's anthocyanin extract significantly lowered glucose levels, demonstrating its anti-diabetic effects [180]. Anthocyanin's anticancer properties have been well-researched and examined [181]. The anti-cancer activities of Vitelotte potato (*Solanum tuberosum* L.) anthocyanins have been well investigated in the research. The implications for food consumption and as functional food ingredients for potential cancer prevention and treatment make these findings interesting [182]. Studies have also shown that anthocyanins in purple tea had anti-inflammatory, anticancer, and antioxidant properties [183]. The Himalayan region's wild edible fruit plants are a rich source of diverse antioxidant bioactive compounds including anthocyanins. The food and flavor industries are looking for novel food ingredients to create dietary supplements. Therefore, these plants have the potential to serve as sources of anthocyanins for nutraceutical development [40]. *Berberis lycium* berries are a rich source of anthocyanins. They are consumed in raw form, particularly by the rural population [184].

In a study, the phytochemical, antioxidant, and antimicrobial profiling of *Berberis asiatica* fruit was conducted in four different processed forms (dried fruit, pulp, juice, and pomace). It was concluded that fruit can be utilized for nutraceutical development, and pomace works best for fruit waste utilization in food and nutraceutical applications [185]. In another study, the fruits of *Ficus palmata* were investigated using different extraction methods for their comparative phytochemical and biological activities. The results indicated that the fruit is a reservoir of polyphenols such as anthocyanins and may be suggested for the development of new nutraceuticals and related products [186]. These studies indicate that the underutilized anthocyanin-rich Himalayan plants are potential candidates for emerging nutraceutical industries. The sustainable utilization of these plants can help in the nutritional security of consumers and would also benefit the local people.

7. Conclusions and Future Perspectives

Anthocyanins are natural colorants that are gaining popularity due to their diverse color palette as well as safe and favorable health effects. Anthocyanins have a high potential for application in food, pharmaceutical, cosmetic, and related industries. Anthocyanin can be isolated and purified from an endless number of natural resources, but the Himalayan sources of anthocyanin are relatively undiscovered. Most anthocyanin research is currently focused on identifying different sources of anthocyanin, as well as purification and extraction. Natural food colorants are recently being preferred by consumers since they have fewer adverse effects than synthetic/artificial substances. Himalayan plants are rich sources of anthocyanins. Anthocyanins have wide applications in the health, nutraceuticals, food, and other sectors. Anthocyanin has been emerging as an excellent choice for making intelligent packaging film as a freshness indicator. Anthocyanin-based food colorants are universally authorized, even though there are significant regulatory variances between countries. Human-nature relationships have long been woven with the importance of anthocyanins in healing a variety of ailments. However, a better knowledge of the processes and mode of action that led to these effects has yet to be developed. A collaborative effort is currently required by academia, industry, and consumer welfare organizations to examine the benefits of these pigments and develop sustainable food colorants for human health and well-being. Moreover, the stability of anthocyanin and its degradation is a challenge that needs to be improved. Ultimately, the potential and identification of these Himalayan sources of anthocyanins, in addition to accurate study and utilization may help achieve sustainable development.

Author Contributions: Conceptualization, M.A. and S.R.; validation, M.A., I.B. and S.R.; formal analysis, M.A. and I.B.; investigation, M.A. and S.R.; writing—original draft preparation, M.A., I.B. and S.R.; writing—review and editing, S.R. and G.G.; visualization, S.R. and G.G.; supervision, S.R. All authors have read and agreed to the published version of the manuscript.

Funding: This research received no external funding.

Data Availability Statement: Data is contained within the article.

Conflicts of Interest: The authors declare no conflict of interest.

References

1. Schaal, B. Plants and People: Our Shared History and Future. *Plants People Planet* **2019**, *1*, 14–19. [CrossRef]
2. Raskin, I.; Ribnicky, D.M.; Komarnytsky, S.; Ilic, N.; Poulev, A.; Borisjuk, N.; Brinker, A.; Moreno, D.A.; Ripoll, C.; Yakoby, N.; et al. Plants and Human Health in the Twenty-First Century. *Trends Biotechnol.* **2002**, *20*, 522–531. [CrossRef]
3. Kumar, V.; Chopra, A.K. Impact of Climate Change on Biodiversity of India with Special Reference to Himalayan Region-An Overview. *J. Appl. Nat. Sci.* **2009**, *1*, 117–122. [CrossRef]
4. Delgoda, R.; Murray, J.E. Evolutionary Perspectives on the Role of Plant Secondary Metabolites. In *Pharmacognosy*; Elsevier: Amsterdam, The Netherlands, 2017; pp. 93–100. [CrossRef]
5. Oz, A.T.; Kafkas, E. Phytochemicals in Fruits and Vegetables. In *Superfood and Functional Food—An Overview of Their Processing and Utilization*; InTech: Sydney, Australia, 2017; pp. 175–184. [CrossRef]
6. Panche, A.N.; Diwan, A.D.; Chandra, S.R. Flavonoids: An Overview. *J. Nutr. Sci.* **2016**, *5*, e47. [CrossRef]

7. Hurst, R.D.; Lyall, K.A.; Wells, R.W.; Sawyer, G.M.; Lomiwes, D.; Ngametua, N.; Hurst, S.M. Daily Consumption of an Anthocyanin-Rich Extract Made from New Zealand Blackcurrants for 5 Weeks Supports Exercise Recovery Through the Management of Oxidative Stress and Inflammation: A Randomized Placebo Controlled Pilot Study. *Front. Nutr.* **2020**, *7*, 6. [CrossRef]
8. Roy, S.; Rhim, J.-W. Preparation of gelatin/carrageenan-based color-indicator film integrated with shikonin and propolis for smart food packaging applications. *ACS Appl. Bio Mater.* **2021**, *4*, 770–779. [CrossRef]
9. Roy, S.; Rhim, J.-W. Fabrication of cellulose nanofiber-based functional color indicator film incorporated with shikonin extracted from root. *Food Hydrocoll.* **2021**, *114*, 106566. [CrossRef]
10. Ghosh, S.; Sarkar, T.; Das, A.; Chakraborty, R. Micro and Nanoencapsulation of Natural Colors: A Holistic View. *Appl. Biochem. Biotechnol.* **2021**, *193*, 3787–3811. [CrossRef] [PubMed]
11. Ren, S.; Jiménez-Flores, R.; Giusti, M.M. The Interactions between Anthocyanin and Whey Protein: A Review. *Compr. Rev. Food Sci. Food Saf.* **2021**, *20*, 5992–6011. [CrossRef] [PubMed]
12. Cruz, L.; Fernandes, I.; Guimarães, M.; de Freitas, V.; Mateus, N. Enzymatic Synthesis, Structural Characterization and Antioxidant Capacity Assessment of a New Lipophilic Malvidin-3-Glucoside–Oleic Acid Conjugate. *Food Funct. J.* **2016**, *7*, 2754–2762. [CrossRef] [PubMed]
13. Luo, C.-L.; Zhou, Q.; Yang, Z.-W.; Wang, R.-D.; Zhang, J.-L. Evaluation of Structure and Bioprotective Activity of Key High Molecular Weight Acylated Anthocyanin Compounds Isolated from the Purple Sweet Potato (*Ipomoea Batatas* L. Cultivar Eshu No. 8). *Food Chem.* **2018**, *241*, 23–31. [CrossRef]
14. Yang, W.; Kortesniemi, M.; Ma, X.; Zheng, J.; Yang, B. Enzymatic Acylation of Blackcurrant (*Ribes nigrum*) Anthocyanins and Evaluation of Lipophilic Properties and Antioxidant Capacity of Derivatives. *Food Chem.* **2019**, *281*, 189–196. [CrossRef] [PubMed]
15. Matsufuji, H.; Kido, H.; Misawa, H.; Yaguchi, J.; Otsuki, T.; Chino, M.; Takeda, M.; Yamagata, K. Stability to Light, Heat, and Hydrogen Peroxide at Different PH Values and DPPH Radical Scavenging Activity of Acylated Anthocyanins from Red Radish Extract. *J. Agric. Food Chem.* **2007**, *55*, 3692–3701. [CrossRef]
16. Cavalcanti, R.N.; Santos, D.T.; Meireles, M.A.A. Non-Thermal Stabilization Mechanisms of Anthocyanins in Model and Food Systems—An Overview. *Food Res. Int.* **2011**, *44*, 499–509. [CrossRef]
17. Ursu, M.S.; Aprodu, I.; Milea, Ş.A.; Enachi, E.; Râpeanu, G.; Bahrim, G.E.; Stănciuc, N. Thermal Degradation Kinetics of Anthocyanins Extracted from Purple Maize Flour Extract and the Effect of Heating on Selected Biological Functionality. *Foods* **2020**, *9*, 1593. [CrossRef] [PubMed]
18. Sadilova, E.; Carle, R.; Stintzing, F.C. Thermal Degradation of Anthocyanins and Its Impact on Color Andin Vitro Antioxidant Capacity. *Mol. Nutr. Food Res.* **2007**, *51*, 1461–1471. [CrossRef]
19. Cisse, M.; Vaillant, F.; Acosta, O.; Dhuique-Mayer, C.; Dornier, M. Thermal Degradation Kinetics of Anthocyanins from Blood Orange, Blackberry, and Roselle Using the Arrhenius, Eyring, and Ball Models. *J. Agric. Food Chem.* **2009**, *57*, 6285–6291. [CrossRef] [PubMed]
20. Rashwan, A.K.; Karim, N.; Xu, Y.; Xie, J.; Cui, H.; Mozafari, M.R.; Chen, W. Potential Micro-/Nano-Encapsulation Systems for Improving Stability and Bioavailability of Anthocyanins: An Updated Review. *Crit. Rev. Food Sci. Nutr.* **2021**, 1–24. [CrossRef]
21. Tarone, A.G.; Cazarin, C.B.B.; Marostica Junior, M.R. Anthocyanins: New Techniques and Challenges in Microencapsulation. *Food Res. Int.* **2020**, *133*, 109092. [CrossRef]
22. Farooq, S.; Shah, M.A.; Siddiqui, M.W.; Dar, B.N.; Mir, S.A.; Ali, A. Recent Trends in Extraction Techniques of Anthocyanins from Plant Materials. *J. Food Meas. Charact.* **2020**, *14*, 3508–3519. [CrossRef]
23. Weber, F.; Boch, K.; Schieber, A. Influence of Copigmentation on the Stability of Spray Dried Anthocyanins from Blackberry. *LWT-Food Sci. Technol.* **2017**, *75*, 72–77. [CrossRef]
24. Ongkowijoyo, P.; Luna-Vital, D.A.; Gonzalez de Mejia, E. Extraction Techniques and Analysis of Anthocyanins from Food Sources by Mass Spectrometry: An Update. *Food Chem.* **2018**, *250*, 113–126. [CrossRef] [PubMed]
25. del Pilar Garcia-Mendoza, M.; Espinosa-Pardo, F.A.; Baseggio, A.M.; Barbero, G.F.; Maróstica Junior, M.R.; Rostagno, M.A.; Martínez, J. Extraction of Phenolic Compounds and Anthocyanins from Juçara (Euterpe Edulis Mart.) Residues Using Pressurized Liquids and Supercritical Fluids. *J. Supercrit. Fluids* **2017**, *119*, 9–16. [CrossRef]
26. Chand, R.; Singh, A.N.; Nirmala, C. Ethnoecological Survey of Underutilized Plant Diversity of Hamirpur District, Himachal Pradesh, India: An Edibility Assessment. *Environ. Ecol. Res.* **2017**, *5*, 13–29. [CrossRef]
27. Muhammad, F.; Javed, I.; Akhtar, M.; Khaliq, T.; Aslam, B.; Waheed, A.; Yasmin, R.; Zafar, H.; Zafar, M. White Mulberry (*Morus alba*): A Brief Phytochemical and Pharmacological Evaluations Account. *Int. J. Agric. Biol.* **2013**, *15*, 612–620.
28. Parra, S.; Gaur, K.; Ranawat, S.; Rather, M. An Overview on Various Aspects of Plant *Berberis Lycium* Royale. *Am. J. Pharmacol. Sci.* **2018**, *6*, 19–24. [CrossRef]
29. Shri, R.; Sood, P. A Review on Ethnomedicinal, Phytochemical and Pharmacological Aspects of *Myrica esculenta*. *Indian J. Pharm. Sci.* **2018**, *80*, 2–13. [CrossRef]
30. Liu, Z.; Liu, B.; Wen, H.; Tao, Y.; Shao, Y. Phytochemical Profiles, Nutritional Constituents and Antioxidant Activity of Black Wolfberry (*Lycium Ruthenicum* Murr.). *Ind. Crops Ad Prod.* **2020**, *154*, 112692. [CrossRef]
31. Bhutia, P.O.; Kewlani, P.; Pandey, A.; Rawat, S.; Bhatt, I.D. Physico-Chemical Properties and Nutritional Composition of Fruits of the Wild Himalayan Strawberry (*Fragaria nubicola* Lindle.) in Different Ripening Stages. *J. Berry Res.* **2021**, *11*, 481–496. [CrossRef]
32. Alam, F.; Najum us Saqib, Q. Pharmacognostic Standardization and Preliminary Phytochemical Studies of *Gaultheria trichophylla*. *Pharm. Biol.* **2015**, *53*, 1711–1718. [CrossRef]

33. Iwashina, T.; Saito, Y.; Kokubugata, G.; Peng, C.-I. Flavonoids in the Leaves of Hillebrandia and Begonia Species (Begoniaceae). *Biochem. Syst. Ecol.* **2020**, *90*, 104040. [CrossRef]
34. Ahrendt, L.W.A. Berberis and Mahonia. *Bot. J. Linn. Soc.* **1961**, *57*, 1–410. [CrossRef]
35. Srivastava, S.K.; Singh Rawat, A.K.; Mehrotra, S. Pharmacognostic Evaluation of the Root of *Berberis asiatica*. *Pharm. Biol.* **2004**, *42*, 467–473. [CrossRef]
36. Leopoldini, M.; Russo, N.; Toscano, M. The Molecular Basis of Working Mechanism of Natural Polyphenolic Antioxidants. *Food Chem.* **2011**, *125*, 288–306. [CrossRef]
37. Potdar, D.; Hirwani, R.R.; Dhulap, S. Phyto-Chemical and Pharmacological Applications of *Berberis aristata*. *Fitoterapia* **2012**, *83*, 817–830. [CrossRef]
38. Belwal, T.; Pandey, A.; Bhatt, I.D.; Rawal, R.S.; Luo, Z. Trends of Polyphenolics and Anthocyanins Accumulation along Ripening Stages of Wild Edible Fruits of Indian Himalayan Region. *Sci. Rep.* **2019**, *9*, 5894. [CrossRef]
39. Cicero, A.F.G.; Baggioni, A. Berberine and Its Role in Chronic Disease. In *Anti-Inflammatory Nutraceuticals and Chronic Diseases*; Advances in Experimental Medicine and Biology; Springer: Cham, Switzerland, 2016; pp. 27–45. [CrossRef]
40. Bhatt, I.D.; Rawat, S.; Badhani, A.; Rawal, R.S. Nutraceutical Potential of Selected Wild Edible Fruits of the Indian Himalayan Region. *Food Chem.* **2017**, *215*, 84–91. [CrossRef]
41. Srivastava, S.; Kapoor, R.; Thathola, A.; Srivastava, R.P. Nutritional Quality of Leaves of Some Genotypes of Mulberry (*Morus alba*). *Int. J. Food Sci. Nutr.* **2006**, *57*, 305–313. [CrossRef]
42. Awasthi, A.K.; Nagaraja, G.; Naik, G.; Kanginakudru, S.; Thangavelu, K.; Nagaraju, J. Genetic Diversity and Relationships in Mulberry (Genus Morus) as Revealed by RAPD and ISSR Marker Assays. *BMC Genet.* **2004**, *5*, 1. [CrossRef]
43. Yadav, S.; Nair, N.; Biharee, A.; Prathap, V.M.; Majeed, J. Updated Ethnobotanical Notes, Phytochemistry and Phytopharmacology of Plants Belonging to the Genus Morus (Family: Moraceae). *Phytomed. Plus* **2022**, *2*, 100120. [CrossRef]
44. Chan, E.W.-C.; Lye, P.-Y.; Wong, S.-K. Phytochemistry, Pharmacology, and Clinical Trials of *Morus alba*. *Chin. J. Nat. Med.* **2016**, *14*, 17–30. [CrossRef] [PubMed]
45. Ercisli, S.; Orhan, E. Chemical Composition of White (*Morus alba*), Red (*Morus rubra*) and Black (*Morus nigra*) Mulberry Fruits. *Food Chem.* **2007**, *103*, 1380–1384. [CrossRef]
46. Kumar, R.; And, R.; Chauhan, S. Mulberry: Life Enhancer. *J. Med. Plants Res.* **2008**, *2*, 271–278.
47. Gryn-Rynko, A.; Bazylak, G.; Olszewska-Slonina, D. New Potential Phytotherapeutics Obtained from White Mulberry (*Morus alba* L.) Leaves. *Biomed. Pharmacother.* **2016**, *84*, 628–636. [CrossRef]
48. Kimura, T.; Nakagawa, K.; Kubota, H.; Kojima, Y.; Goto, Y.; Yamagishi, K.; Oita, S.; Oikawa, S.; Miyazawa, T. Food-Grade Mulberry Powder Enriched with 1-Deoxynojirimycin Suppresses the Elevation of Postprandial Blood Glucose in Humans. *J. Agric. Food Chem.* **2007**, *55*, 5869–5874. [CrossRef]
49. Kumari, K.; Sharma, S.; Kaushik, R. Wild Himalayan Fig: A Nutraceutical under exploited fruit of Western Himalayan region—A Review. *Int. J. Adv. Res.* **2017**, *5*, 833–839. [CrossRef]
50. Joshi, Y.; Joshi, A.K.; Prasad, N.; Juyal, D. A Review on *Ficus palmata* (Wild Himalayan Fig). *J. Phytopharm.* **2014**, *3*, 374–377. [CrossRef]
51. Kothiyal, S.C.; Saklani, S. Isolation, and Identification of *Ficus palmata* leaves and their antimicrobial activities. *J. Sci. Res.* **2017**, *9*, 193–200. [CrossRef]
52. Saklani, S.; Kothiyal, S. Phytochemical Screening of Garhwal Himalaya Wild Edible Fruit *Ficus palmata*. *Int. J. Pharm. Tech Res.* **2012**, *4*, 1185–1191.
53. Shabbir, A. *Berberis lycium* Royle: A Review of Its Traditional Uses, Phytochemistry and Pharmacology. *Afr. J. Pharm. Pharmacol.* **2012**, *6*, 2346–2353. [CrossRef]
54. Anjum, N.; Ridwan, Q.; Akhter, F.; Hanief, M. Phytochemistry and Therapeutic Potential of *Berberis lycium* Royle; an Endangered Species of Himalayan Region. *Acta Ecol. Sin.* **2022**, *43*, 577–584. [CrossRef]
55. Kaur, C.; Kapoor, H.C. Antioxidants in Fruits and Vegetables—The Millennium's Health. *Int. J. Food Sci. Technol.* **2008**, *36*, 703–725. [CrossRef]
56. Gupta, M.; Singh, A.; Joshi, H. *Berberis lycium* Multipotential Medicinal Application: An Overview. *Int. J. Chem. Stud.* **2015**, *10*, 10–13.
57. Bhardwaj, D.; Kaushik, N. Phytochemical and Pharmacological Studies in Genus Berberis. *Phytochem. Rev.* **2012**, *11*, 523–542. [CrossRef]
58. Gulfraz, M.; Mehmood, S.; Ahmad, A.; Fatima, N.; Praveen, Z.; Williamson, E.M. Comparison of the Antidiabetic Activity of *Berberis lycium* Root Extract and Berberine in Alloxan-Induced Diabetic Rats. *Phytother. Res.* **2008**, *22*, 1208–1212. [CrossRef] [PubMed]
59. Nazir, N.; Rahman, A.; Uddin, F.; Khan Khalil, A.A.; Zahoor, M.; Nisar, M.; Ullah, S.; Ullah, R.; Ezzeldin, E.; Mostafa, G.A.E. Quantitative Ethnomedicinal Status and Phytochemical Analysis of *Berberis lycium* Royle. *Agronomy* **2021**, *11*, 130. [CrossRef]
60. Sendri, N.; Bhandari, P. Polyphenolic Composition and Antioxidant Potential of Underutilized Himalayan Wild Edible Berries by High-performance Liquid Chromatography Coupled with Electrospray Ionization Quadrupole Time-of-flight Mass Spectrometry. *J. Sep. Sci.* **2021**, *44*, 4237–4254. [CrossRef]
61. Pradhan, P.C.; Saha, S. Anthocyanin Profiling of *Berberis lycium* Royle Berry and Its Bioactivity Evaluation for Its Nutraceutical Potential. *J. Food Sci. Technol.* **2016**, *53*, 1205–1213. [CrossRef]

62. Yanthan, M.; Biate, D.; Misra, A.K. Taxonomic Resolution of Actinorhizal Myrica Species from Meghalaya (India) through Nuclear RDNA Sequence Analyses. *Funct. Plant Biol.* **2011**, *38*, 738. [CrossRef]
63. Gusain, Y.S.; Khanduri, V.P. *Myrica esculenta* Wild Edible Fruit of Indian Himalaya: Need a Sustainable Approach for Indigenous Utilization. *Ecol. Environ. Conserv. J.* **2016**, *22*, S267–S270.
64. Rawat, S.; Jugran, A.; Giri, L.; Bhatt, I.D.; Rawal, R.S. Assessment of Antioxidant Properties in Fruits of *Myrica esculenta*: A Popular Wild Edible Species in Indian Himalayan Region. *Evid.-Based Complement. Altern. Med.* **2011**, *2011*, 51278. [CrossRef]
65. Kabra, A.; Martins, N.; Sharma, R.; Kabra, R.; Baghel, U.S. *Myrica esculenta* Buch.-Ham. Ex D. Don: A Natural Source for Health Promotion and Disease Prevention. *Plants* **2019**, *8*, 149. [CrossRef] [PubMed]
66. Nhiem, N.X.; Van Kiem, P.; Van Minh, C.; Tai, B.H.; Cuong, N.X.; Thu, V.K.; Anh, H.L.T.; Jo, S.-H.; Jang, H.-D.; Kwon, Y.-I.; et al. A New Monoterpenoid Glycoside from *Myrica esculenta* and the Inhibition of Angiotensin I-Converting Enzyme. *Chem. Pharm. Bull.* **2010**, *58*, 1408–1410. [CrossRef] [PubMed]
67. Bahuguna, D.P.; London, H.K.; Kharwal, H.; Joshi, D. Myrica Nagi: A Review on Active Constituents, Biological and Therapeutic Effects. *Int. J. Pharm. Pharm. Sci.* **2012**, *4*, 38–42.
68. Kumar, T.; Pande, K.K.; Sharma, H.; Koranga, M.; Pande, L. HRLC-ESI-MS Based Separation, and Identification of Anthocyanins Extracted from Popular Wild Edible Fruit of Himalaya: *Myrica esculenta* (Himalayan Bayberry). *J. Adv. Sci. Res.* **2020**, *11*, 269–275.
69. Kakar, M.; Kakar, I.; Akram, M. Antimicrobial and Phytochemical Exploration of *Duchesnea indica* Plant. *Plant Cell Biotechnol. Mol. Biol.* **2021**, *22*, 74–85.
70. Faghir, M.B.; Pourebrahim, S.; Shahi Shavvon, R. New Insight into the Molecular and Micromorphological Characteristics of *Potentilla indica* and *Potentilla reptans* (Rosaceae). *Iran. J. Bot.* **2022**, *28*, 77–95. [CrossRef]
71. Kar, T.; Nayak, A.K.N.A.K.; Dash, B.; Mandal, K.K.M.K.K. *Duchesnea indica* (Rosaceae): An Addition to the Flora of Odisha, India. *Biosci. Discov.* **2014**, *5*, 202–203.
72. Ahmad, I.; Ibrar, M.; Barkatullah; Ali, N. Ethnobotanical Study of Tehsil Kabal, Swat District, KPK, Pakistan. *J. Bot.* **2011**, *2011*, 368572. [CrossRef]
73. Peng, B.; Chang, Q.; Wang, L.; Hu, Q.; Wang, Y.; Tang, J.; Liu, X. Suppression of Human Ovarian SKOV-3 Cancer Cell Growth by Duchesnea Phenolic Fraction Is Associated with Cell Cycle Arrest and Apoptosis. *Gynecol. Oncol.* **2008**, *108*, 173–181. [CrossRef]
74. Zhu, M.; Dong, X.; Guo, M. Phenolic Profiling of *Duchesnea indica* Combining Macroporous Resin Chromatography (MRC) with HPLC-ESI-MS/MS and ESI-IT-MS. *Molecules* **2015**, *20*, 22463–22475. [CrossRef] [PubMed]
75. Qin, C.; Li, Y.; Zhang, R.; Niu, W.; Ding, Y. Separation and Elucidation of Anthocyanins in the Fruit of Mockstrawberry (*Duchesnea indica* Focke). *Nat. Prod. Res.* **2009**, *23*, 1589–1598. [CrossRef]
76. Wang, H.; Li, J.; Tao, W.; Zhang, X.; Gao, X.; Yong, J.; Duan, J.A. *Lycium ruthenicum* studies: Molecular biology, phytochemistry and pharmacology. *Food Chem.* **2018**, *240*, 759–766. [CrossRef]
77. Sharma, R.; Raghuvanshi, R.; Kumar, R.; Thakur, M.S.; Kumar, S.; Patel, M.K.; Chaurasia, O.P.; Saxena, S. Current Findings and Future Prospective of High-Value Trans Himalayan Medicinal Plant *Lycium ruthenicum* Murr: A Systematic Review. *Clin. Phytosci.* **2022**, *8*, 3. [CrossRef]
78. Chaurasia, O.; Ballabh, B. Herbal Formulations from Cold Desert Plants Used for Gynecological Disorders. *Ethnobot. Res. Appl.* **2011**, *9*. [CrossRef]
79. Proksch, P. Chinese Marine Materia Medica. By Huashi Guan and Shuguang Wang. Shanghai Scientific and Technical Publishers, China Ocean Press, and Chemical Industry Press: Shanghai, Beijing, China, 2009. *Mar. Drugs* **2014**, *12*, 193–195. [CrossRef]
80. Ballabh, B.; Chaurasia, O.P.; Ahmed, Z.; Singh, S.B. Traditional Medicinal Plants of Cold Desert Ladakh—Used against Kidney and Urinary Disorders. *J. Ethnopharmacol.* **2008**, *118*, 331–339. [CrossRef]
81. Chopra, R.N. *Glossary of Indian Medicinal Plants*; Council of Scientific & Industrial Research: New Delhi, India, 1956.
82. Gairola, S.; Sharma, J.; Bedi, Y.S. A Cross-Cultural Analysis of Jammu, Kashmir and Ladakh (India) Medicinal Plant Use. *J. Ethnopharmacol.* **2014**, *155*, 925–986. [CrossRef]
83. Yun, D.; Yan, Y.; Liu, J. Isolation, structure and biological activity of polysaccharides from the fruits of *Lycium ruthenicum* murr: A review. *Carbohydr. Polym.* **2022**, *291*, 119618. [CrossRef]
84. Liu, W.-R.; Qiao, W.-L.; Liu, Z.-Z.; Wang, X.-H.; Jiang, R.; Li, S.-Y.; Shi, R.-B.; She, G.-M. Gaultheria: Phytochemical and Pharmacological Characteristics. *Molecules* **2013**, *18*, 12071–12108. [CrossRef]
85. Alam, F.; Saqib, Q.N.; Ashraf, M. *Gaultheria trichophylla* (Royle): A Source of Minerals and Biologically Active Molecules, Its Antioxidant and Anti-Lipoxygenase Activities. *BMC Complement. Altern. Med.* **2017**, *17*, 3. [CrossRef] [PubMed]
86. Zhang, D.; Liu, R.; Sun, L.; Huang, C.; Wang, C.; Zhang, D.-M.; Zhang, T.-T.; Du, G.-H. Anti-Inflammatory Activity of Methyl Salicylate Glycosides Isolated from *Gaultheria yunnanensis* (Franch.) Rehder. *Molecules* **2011**, *16*, 3875–3884. [CrossRef] [PubMed]
87. Bahukh, A.; Aseesh, P.; Sekar, K.C.; Bhatt, I.D. Polyphenolics, Nutrients and Antioxidant Activity of *Gaultheria trichophylla* Royle: A High Value Wild Edible Plant of Trans Himalaya. *Hortic. Int. J.* **2017**, *1*, 39–43.
88. Moonlight, P.W.; Ardi, W.H.; Padilla, L.A.; Chung, K.-F.; Fuller, D.; Girmansyah, D.; Hollands, R.; Jara-Muñoz, A.; Kiew, R.; Leong, W.-C.; et al. Dividing and Conquering the Fastest–Growing Genus: Towards a Natural Sectional Classification of the Mega–Diverse Genus *Begonia* (Begoniaceae). *Taxon* **2018**, *67*, 267–323. [CrossRef]
89. Taram, M.; Borah, D.; Hughes, M. Two New Records of Begonia for the Flora of India from Arunachal Pradesh. *Phytotaxa* **2023**, *584*, 2. [CrossRef]

90. Bhattarai, B.; Rana, M. Diversified Morphological and Phytochemical Screening of Wild Begonia of Sikkim Himalaya. *Ecol. Environ. Conserv.* **2020**, *26*, S129–S138.
91. Roshan, R.; Ahmed, S.; ul Hassan, M.M. *Fragaria nubicola* (Rosaceae): A Review of Medicinal Uses, Phytochemistry and Pharmacology. *J. Pharmacogn. Phytochem.* **2019**, *8*, 3390–3393.
92. Staudt, G. Himalayan Species of Fragaria (Rosaceae). *Bot. JahrbÜCher Syst. Pflanzengesch. Pflanzengeogr.* **2006**, *126*, 483–508. [CrossRef]
93. Chakraborty, T.; Saha, S.; Bisht, N. First Report on the Ethnopharmacological Uses of Medicinal Plants by Monpa Tribe from the Zemithang Region of Arunachal Pradesh, Eastern Himalayas, India. *Plants* **2017**, *6*, 13. [CrossRef]
94. Antonio, R.L.; Kozasa, E.H.; Galduróz, J.C.F.; Dawa; Dorjee, Y.; Kalsang, T.; Norbu, T.; Tenzin, T.; Rodrigues, E. Formulas Used by Tibetan Doctors at Men-Tsee-Khang in India for the Treatment of Neuropsychiatric Disorders and Their Correlation with Pharmacological Data. *Phytother. Res.* **2013**, *27*, 552–563. [CrossRef]
95. Thakur, P. Sarika, Ethno-Medicinal Uses of Some Plants of Potter's Hill in Shimla (Himachal Pradesh, India). *Proc. Biol. Forum* **2016**, *8*, 417–422.
96. Rakhunde, P.B.; Ali, S.A. Antioxidant and Cytoprotective Effect of *Fragaria nubicola* on Ischemia Reperfusion Induced Brain Injury. *Ann. Exp. Biol.* **2014**, *2*, 33–38.
97. Bahukhandi, A.; Barola, A.; Sekar, K.C. Antioxidant Activity and Polyphenolics of *Fragaria nubicola*: A Wild Edible Fruit Species of Himalaya. *Proc. Natl. Acad. Sci. India Sect. B Biol. Sci.* **2020**, *90*, 761–767. [CrossRef]
98. Yuan, Q.; Zhao, L. The Mulberry (*Morus alba* L.) Fruit—A Review of Characteristic Components and Health Benefits. *J. Agric. Food Chem.* **2017**, *65*, 10383–10394. [CrossRef] [PubMed]
99. Miguel, M. Anthocyanins: Antioxidant and/or Anti-Inflammatory Activities. *J. Appl. Pharm. Sci.* **2011**, *1*, 7–15.
100. Tena, N.; Martín, J.; Asuero, A.G. State of the Art of Anthocyanins: Antioxidant Activity, Sources, Bioavailability, and Therapeutic Effect in Human Health. *Antioxidants* **2020**, *9*, 451. [CrossRef]
101. Liang, T.; Sun, G.; Cao, L.; Li, J.; Wang, L. A pH and NH_3 Sensing Intelligent Film Based on *Artemisia sphaerocephala* Krasch. Gum and Red Cabbage Anthocyanins Anchored by Carboxymethyl Cellulose Sodium Added as a Host Complex. *Food Hydrocoll.* **2019**, *87*, 858–868. [CrossRef]
102. Liu, S.; Fu, Y.; Nian, S. Buffering Colour Fluctuation of Purple Sweet Potato Anthocyanins to Acidity Variation by Surfactants. *Food Chem.* **2014**, *162*, 16–21. [CrossRef] [PubMed]
103. Tanaka, Y.; Sasaki, N.; Ohmiya, A. Biosynthesis of Plant Pigments: Anthocyanins, Betalains and Carotenoids. *Plant J.* **2008**, *54*, 733–749. [CrossRef]
104. Pietrini, F.; Iannelli, M.A.; Massacci, A. Anthocyanin Accumulation in the Illuminated Surface of Maize Leaves Enhances Protection from Photo-Inhibitory Risks at Low Temperature, without Further Limitation to Photosynthesis. *Plant Cell Environ.* **2002**, *25*, 1251–1259. [CrossRef]
105. Chalker-Scott, L. Environmental Significance of Anthocyanins in Plant Stress Responses. *Photochem. Photobiol.* **1999**, *70*, 1–9. [CrossRef]
106. Feild, T.S.; Lee, D.W.; Holbrook, N.M. Why Leaves Turn Red in Autumn. The Role of Anthocyanins in Senescing Leaves of Red-Osier Dogwood. *Plant Physiol.* **2001**, *127*, 566–574. [CrossRef]
107. Hoch, W.A.; Zeldin, E.L.; McCown, B.H. Physiological Significance of Anthocyanins during Autumnal Leaf Senescence. *Tree Physiol.* **2001**, *21*, 1–8. [CrossRef]
108. Ross, J.A.; Kasum, C.M. Dietary Flavonoids: Bioavailability, Metabolic Effects, and Safety. *Annu. Rev. Nutr.* **2002**, *22*, 19–34. [CrossRef]
109. Wu, H.; Oliveira, G.; Lila, M.A. Protein-binding Approaches for Improving Bioaccessibility and Bioavailability of Anthocyanins. *Compr. Rev. Food Sci. Food Saf.* **2023**, *22*, 333–354. [CrossRef]
110. Kamonpatana, K.; Giusti, M.M.; Chitchumroonchokchai, C.; MorenoCruz, M.; Riedl, K.M.; Kumar, P.; Failla, M.L. Susceptibility of Anthocyanins to Ex Vivo Degradation in Human Saliva. *Food Chem.* **2012**, *135*, 738–747. [CrossRef]
111. Fernandes, I.; Faria, A.; Calhau, C.; de Freitas, V.; Mateus, N. Bioavailability of Anthocyanins and Derivatives. *J. Fuctional Foods* **2014**, *7*, 54–66. [CrossRef]
112. Lila, M.A.; Burton-Freeman, B.; Grace, M.; Kalt, W. Unraveling Anthocyanin Bioavailability for Human Health. *Annu. Rev. Food Sci. Technol.* **2016**, *7*, 375–393. [CrossRef] [PubMed]
113. Shen, Y.; Zhang, N.; Tian, J.; Xin, G.; Liu, L.; Sun, X.; Li, B. Advanced Approaches for Improving Bioavailability and Controlled Release of Anthocyanins. *J. Control Release* **2022**, *341*, 285–299. [CrossRef] [PubMed]
114. Giusti, M.M.; Wrolstad, R.E. Acylated Anthocyanins from Edible Sources and Their Applications in Food Systems. *Biochem. Eng. J.* **2003**, *14*, 217–225. [CrossRef]
115. Oplatowska-Stachowiak, M.; Elliott, C.T. Food Colors: Existing and Emerging Food Safety Concerns. *Crit. Rev. Food Sci. Nutr.* **2017**, *57*, 524–548. [CrossRef]
116. McCann, D.; Barrett, A.; Cooper, A.; Crumpler, D.; Dalen, L.; Grimshaw, K.; Kitchin, E.; Lok, K.; Porteous, L.; Prince, E.; et al. Food Additives and Hyperactive Behaviour in 3-Year-Old and 8/9-Year-Old Children in the Community: A Randomised, Double-Blinded, Placebo-Controlled Trial. *Lancet* **2007**, *370*, 1560–1567. [CrossRef]
117. Cevallos-Casals, B.A.; Cisneros-Zevallos, L. Stability of Anthocyanin-Based Aqueous Extracts of Andean Purple Corn and Red-Fleshed Sweet Potato Compared to Synthetic and Natural Colorants. *Food Chem.* **2004**, *86*, 69–77. [CrossRef]

118. Amchova, P.; Kotolova, H.; Ruda-Kucerova, J. Health Safety Issues of Synthetic Food Colorants. *Regul. Toxicol. Pharmacol.* **2015**, *73*, 914–922. [CrossRef]
119. Bakowska-Barczak, A. Acylated Anthocyanins as Stable, Natural Food Colorants—A Review. *Pol. J. Food Nutr. Sci.* **2005**, *14*, 107–116.
120. Kim, I.; Lee, J. Variations in Anthocyanin Profiles and Antioxidant Activity of 12 Genotypes of Mulberry (*Morus* spp.) Fruits and Their Changes during Processing. *Antioxidants* **2020**, *9*, 242. [CrossRef]
121. Natić, M.M.; Dabić, D.Č.; Papetti, A.; Fotirić Akšić, M.M.; Ognjanov, V.; Ljubojević, M.; Tešić, Ž.L. Analysis and Characterisation of Phytochemicals in Mulberry (*Morus alba* L.) Fruits Grown in Vojvodina, North Serbia. *Food Chem.* **2015**, *171*, 128–136. [CrossRef] [PubMed]
122. Chen, C.; Mohamad Razali, U.H.; Saikim, F.H.; Mahyudin, A.; Mohd Noor, N.Q.I. *Morus alba* L. Plant: Bioactive Compounds and Potential as a Functional Food Ingredient. *Foods* **2021**, *10*, 689. [CrossRef]
123. Cortez, R.; Luna-Vital, D.A.; Margulis, D.; Gonzalez de Mejia, E. Natural Pigments: Stabilization Methods of Anthocyanins for Food Applications. *Compr. Rev. Food Sci. Food Saf.* **2017**, *16*, 180–198. [CrossRef]
124. Echegaray, N.; Munekata, P.E.S.; Gullón, P.; Dzuvor, C.K.O.; Gullón, B.; Kubi, F.; Lorenzo, J.M. Recent Advances in Food Products Fortification with Anthocyanins. *Crit. Rev. Food Sci. Nutr.* **2022**, *62*, 1553–1567. [CrossRef]
125. Ruta, L.L.; Farcasanu, I.C. Anthocyanins and Anthocyanin-Derived Products in Yeast-Fermented Beverages. *Antioxidants* **2019**, *8*, 182. [CrossRef] [PubMed]
126. Zhang, H.; Ma, Z.; Luo, X.; Li, X. Effects of Mulberry Fruit (*Morus alba* L.) Consumption on Health Outcomes: A Mini-Review. *Antioxidants* **2018**, *7*, 69. [CrossRef] [PubMed]
127. Salehi, B.; Selamoglu, Z.; Sener, B.; Kilic, M.; Kumar Jugran, A.; de Tommasi, N.; Sinisgalli, C.; Milella, L.; Rajkovic, J.; Flaviana, B.; et al. Berberis Plants—Drifting from Farm to Food Applications, Phytotherapy, and Phytopharmacology. *Foods* **2019**, *8*, 522. [CrossRef]
128. Sui, X.; Zhang, Y.; Zhou, W. Bread Fortified with Anthocyanin-Rich Extract from Black Rice as Nutraceutical Sources: Its Quality Attributes and in Vitro Digestibility. *Food Chem.* **2016**, *196*, 910–916. [CrossRef]
129. Albuquerque, B.R.; Pinela, J.; Barros, L.; Oliveira, M.B.P.P.; Ferreira, I.C.F.R. Anthocyanin-Rich Extract of Jabuticaba Epicarp as a Natural Colorant: Optimization of Heat- and Ultrasound-Assisted Extractions and Application in a Bakery Product. *Food Chem.* **2020**, *316*, 126364. [CrossRef] [PubMed]
130. López, C.J.; Caleja, C.; Prieto, M.A.; Sokovic, M.; Calhelha, R.C.; Barros, L.; Ferreira, I.C.F.R. Stability of a Cyanidin-3-O-Glucoside Extract Obtained from *Arbutus unedo* L. and Incorporation into Wafers for Colouring Purposes. *Food Chem.* **2019**, *275*, 426–438. [CrossRef] [PubMed]
131. Maner, S.; Sharma, A.K.; Banerjee, K. Wheat Flour Replacement by Wine Grape Pomace Powder Positively Affects Physical, Functional and Sensory Properties of Cookies. *Proc. Natl. Acad. Sci. India Sect. B Biol. Sci.* **2017**, *87*, 109–113. [CrossRef]
132. Papillo, V.A.; Locatelli, M.; Travaglia, F.; Bordiga, M.; Garino, C.; Arlorio, M.; Coïsson, J.D. Spray-Dried Polyphenolic Extract from Italian Black Rice (*Oryza sativa* L., Var. Artemide) as New Ingredient for Bakery Products. *Food Chem.* **2018**, *269*, 603–609. [CrossRef]
133. Karaaslan, M.; Ozden, M.; Vardin, H.; Turkoglu, H. Phenolic Fortification of Yogurt Using Grape, and Callus Extracts. *LWT-Food Sci. Technol.* **2011**, *44*, 1065–1072. [CrossRef]
134. Montibeller, M.J.; de Lima Monteiro, P.; Tupuna-Yerovi, D.S.; Rios, A.; de Oliveira Rios, V. Stability Assessment of Anthocyanins Obtained from Skin Grape Applied in Kefir and Carbonated Water as a Natural Colorant. *J. Food Preserv.* **2018**, *42*, e13698. [CrossRef]
135. Wallace, T.C.; Giusti, M.M. Determination of Color, Pigment, and Phenolic Stability in Yogurt Systems Colored with Nonacylated Anthocyanins from *Berberis boliviana* L. as Compared to Other Natural/Synthetic Colorants. *J. Food Sci.* **2008**, *73*, C241–C248. [CrossRef]
136. de Pascual-Teresa, S.; Sanchez-Ballesta, M.T. Anthocyanins: From Plant to Health. *Phytochem. Rev.* **2008**, *7*, 281–299. [CrossRef]
137. Gonçalves, A.C.; Nunes, A.R.; Falcão, A.; Alves, G.; Silva, L.R. Dietary Effects of Anthocyanins in Human Health: A Comprehensive Review. *Pharmaceuticals* **2021**, *14*, 690. [CrossRef]
138. He, X.; Fang, J.; Ruan, Y.; Wang, X.; Sun, Y.; Wu, N.; Huang, L. Structures, bioactivities and future prospective of polysaccharides from *Morus alba* (white mulberry): A review. *Food Chem.* **2018**, *245*, 899–910. [CrossRef]
139. Prashar, S.; Sharma, S.; Kumar, N.; Kaushik, R.; Chawla, P. Formulation, Characterization, and In Vitro Mineral Absorption of *Ficus palmata* Fruit Extract Nanoemulsion. *J. Am. Nutr. Assoc.* **2022**, *41*, 291–300. [CrossRef] [PubMed]
140. Garcia-Alonso, M.; Rimbach, G.; Sasai, M.; Nakahara, M.; Matsugo, S.; Uchida, Y.; Rivas-Gonzalo, J.C.; De Pascual-Teresa, S. Electron Spin Resonance Spectroscopy Studies on the Free Radical Scavenging Activity of Wine Anthocyanins and Pyranoanthocyanins. *Mol. Nutr. Food Res.* **2005**, *49*, 1112–1119. [CrossRef]
141. Derosa, G.; Romano, D.; D'Angelo, A.; Maffioli, P. *Berberis aristata* Combined with Silybum Marianum on Lipid Profile in Patients Not Tolerating Statins at High Doses. *Atherosclerosis* **2015**, *239*, 87–92. [CrossRef]
142. Derosa, G.; Romano, D.; D'Angelo, A.; Maffioli, P. *Berberis aristata/Silybum marianum* Fixed Combination (Berberol®) Effects on Lipid Profile in Dyslipidemic Patients Intolerant to Statins at High Dosages: A Randomized, Placebo-Controlled, Clinical Trial. *Phytomedicine* **2015**, *22*, 231–237. [CrossRef]

143. Derosa, G.; D'Angelo, A.; Maffioli, P. The Role of a Fixed *Berberis aristata/Silybum marianum* Combination in the Treatment of Type 1 Diabetes Mellitus. *Clin. Nutr.* **2016**, *35*, 1091–1095. [CrossRef] [PubMed]
144. Alam, M.A.; Islam, P.; Subhan, N.; Rahman, M.M.; Khan, F.; Burrows, G.E.; Nahar, L.; Sarker, S.D. Potential Health Benefits of Anthocyanins in Oxidative Stress Related Disorders. *Phytochem. Rev.* **2021**, *20*, 705–749. [CrossRef]
145. García-Alonso, M.; Rimbach, G.; Rivas-Gonzalo, J.C.; de Pascual-Teresa, S. Antioxidant and Cellular Activities of Anthocyanins and Their Corresponding Vitisins A-Studies in Platelets, Monocytes, and Human Endothelial Cells. *J. Agric. Food Chem.* **2004**, *52*, 3378–3384. [CrossRef]
146. Kamei, H.; Hashimoto, Y.; Koide, T.; Kojima, T.; Hasegawa, M. Anti-Tumor Effect of Methanol Extracts from Red and White Wines. *Cancer Biother. Radiopharm.* **1998**, *13*, 447–452. [CrossRef] [PubMed]
147. Shih, P.-H.; Yeh, C.-T.; Yen, G.-C. Effects of Anthocyanidin on the Inhibition of Proliferation, and Induction of Apoptosis in Human Gastric Adenocarcinoma Cells. *Food Chem. Toxicol.* **2005**, *43*, 1557–1566. [CrossRef] [PubMed]
148. Tsuda, T. Dietary Anthocyanin-Rich Plants: Biochemical Basis and Recent Progress in Health Benefits Studies. *Mol. Nutr. Food Res.* **2012**, *56*, 159–170. [CrossRef] [PubMed]
149. Nakaishi, H.; Matsumoto, H.; Tominaga, S.; Hirayama, M. Effects of Black Currant Anthocyanoside Intake on Dark Adaption and VDT Work-Induced Transient Refractive Alteration in Healthy Humans. *Altern. Med. Rev.* **2001**, *5*, 553–562.
150. Pojer, E.; Mattivi, F.; Johnson, D.; Stockley, C.S. The Case for Anthocyanin Consumption to Promote Human Health: A Review. *Compr. Rev. Food Sci. Food Saf.* **2013**, *12*, 483–508. [CrossRef]
151. Ghosh, D.; Konishi, T. Anthocyanins and Anthocyanin-Rich Extracts: Role in Diabetes and Eye Function. *Asia Pac. J. Clin. Nutr.* **2007**, *16*, 200–208.
152. Smeriglio, A.; Barreca, D.; Bellocco, E.; Trombetta, D. Chemistry, Pharmacology and Health Benefits of Anthocyanins. *Phytother. Res.* **2016**, *30*, 1265–1286. [CrossRef]
153. Roy, S.; Rhim, J.-W. Anthocyanin food colorant and its application in pH-responsive color change indicator films. *Crit. Rev. Food Sci. Nutr.* **2021**, *61*, 2297–2325. [CrossRef]
154. Kim, H.J.; Roy, S.; Rhim, J.-W. Gelatin/agar-based color-indicator film integrated with *Clitoria ternatea* flower anthocyanin and zinc oxide nanoparticles for monitoring freshness of shrimp. *Food Hydrocoll.* **2022**, *124*, 107294. [CrossRef]
155. Gupta, V.; Biswas, D.; Roy, S. A comprehensive review of biodegradable polymer-based films and coatings and their food packaging applications. *Materials* **2022**, *15*, 5899. [CrossRef]
156. Roy, S.; Priyadarshi, R.; Ezati, P.; Rhim, J.-W. Curcumin and its uses in active and smart food packaging applications-A comprehensive review. *Food Chem.* **2022**, *375*, 131885. [CrossRef] [PubMed]
157. Oun, A.A.; Roy, S.; Shin, G.H.; Yoo, S.; Kim, J.T. pH-sensitive smart indicators based on cellulose and different natural pigments for tracing kimchi ripening stages. *Int. J. Biol. Macromol.* **2023**, *242*, 124905. [CrossRef] [PubMed]
158. Zhang, W.; Roy, S.; Ezati, P.; Yang, D.P.; Rhim, J.-W. Tannic acid: A green crosslinker for biopolymer-based food packaging films. *Trends Food Sci. Technol.* **2023**, *136*, 11–23. [CrossRef]
159. Santhosh, R.; Nath, D.; Sarkar, P. Novel Food Packaging Materials Including Plant-Based Byproducts: A Review. *Trends Food Sci. Technol.* **2021**, *118*, 471–489. [CrossRef]
160. Roy, S.; Kim, H.J.; Rhim, J.-W. Effect of blended colorants of anthocyanin and shikonin on carboxymethyl cellulose/agar-based smart packaging film. *Int. J. Biol. Macromol.* **2021**, *183*, 305–315. [CrossRef]
161. Ma, Q.; Wang, L. Preparation of a Visual PH-Sensing Film Based on Tara Gum Incorporating Cellulose and Extracts from Grape Skins. *Sens. Actuators B Chem.* **2016**, *235*, 401–407. [CrossRef]
162. Mushtaq, M.; Gani, A.; Gani, A.; Punoo, H.A.; Masoodi, F.A. Use of Pomegranate Peel Extract Incorporated Zein Film with Improved Properties for Prolonged Shelf Life of Fresh Himalayan Cheese (Kalari/Kradi). *Innov. Food Sci. Emerg. Technol.* **2018**, *48*, 25–32. [CrossRef]
163. Halliwell, B.; Murcia, M.A.; Chirico, S.; Aruoma, O.I. Free Radicals and Antioxidants in Food and *in Vivo:* What They Do and How They Work. *Crit. Rev. Food Sci. Nutr.* **1995**, *35*, 7–20. [CrossRef]
164. Ebrahimzadeh, S.; Biswas, D.; Roy, S.; McClements, D.J. Incorporation of essential oils in edible seaweed-based films: A comprehensive review. *Trends Food Sci. Technol.* **2023**, *135*, 43–56. [CrossRef]
165. Yan, J.; Zhang, H.; Yuan, M.; Qin, Y.; Chen, H. Effects of Anthocyanin-Rich *Kadsura coccinea* Extract on the Physical, Antioxidant, and pH-Sensitive Properties of Biodegradable Film. *Food Biophys.* **2022**, *17*, 375–385. [CrossRef]
166. Kaya, M.; Ravikumar, P.; Ilk, S.; Mujtaba, M.; Akyuz, L.; Labidi, J.; Salaberria, A.M.; Cakmak, Y.S.; Erkul, S.K. Production and Characterization of Chitosan Based Edible Films from *Berberis crataegina*'s Fruit Extract and Seed Oil. *Innov. Food Sci. Emerg. Technol.* **2018**, *45*, 287–297. [CrossRef]
167. Ma, Y.; Ding, S.; Fei, Y.; Liu, G.; Jang, H.; Fang, J. Antimicrobial Activity of Anthocyanins and Catechins against Foodborne Pathogens *Escherichia coli* and Salmonella. *Food Control* **2019**, *106*, 106712. [CrossRef]
168. Campos, C.A.; Gerschenson, L.N.; Flores, S.K. Development of Edible Films and Coatings with Antimicrobial Activity. *Food Bioprocess Technol.* **2011**, *4*, 849–875. [CrossRef]
169. Qin, Y.; Xu, F.; Yuan, L.; Hu, H.; Yao, X.; Liu, J. Comparison of the Physical and Functional Properties of Starch/Polyvinyl Alcohol Films Containing Anthocyanins and/or Betacyanins. *Int. J. Biol. Macromol.* **2020**, *163*, 898–909. [CrossRef]
170. Choi, H.-J.; Choi, S.-W.; Lee, N.; Chang, H.-J. Antimicrobial Activity of Chitosan/Gelatin/Poly(Vinyl Alcohol) Ternary Blend Film Incorporated with *Duchesnea indica* Extract in Strawberry Applications. *Foods* **2022**, *11*, 3963. [CrossRef] [PubMed]

171. Santos, L.G.; Alves-Silva, G.F.; Martins, V.G. Active-Intelligent and Biodegradable Sodium Alginate Films Loaded with *Clitoria ternatea* Anthocyanin-Rich Extract to Preserve and Monitor Food Freshness. *Int. J. Biol. Macromol.* **2022**, *220*, 866–877. [CrossRef] [PubMed]
172. Ansari, S.; Chauhan, B.; Kalam, N.; Kumar, G. Current Concepts and Prospects of Herbal Nutraceutical: A Review. *J. Adv. Pharm. Technol. Res.* **2013**, *4*, 4. [CrossRef]
173. Alappat, B.; Alappat, J. Anthocyanin Pigments: Beyond Aesthetics. *Molecules* **2020**, *25*, 5500. [CrossRef]
174. Mattioli, R.; Francioso, A.; Mosca, L.; Silva, P. Anthocyanins: A Comprehensive Review of Their Chemical Properties and Health Effects on Cardiovascular and Neurodegenerative Diseases. *Molecules* **2020**, *25*, 3809. [CrossRef]
175. Garzón, G.A.; Wrolstad, R.E. Major Anthocyanins and Antioxidant Activity of Nasturtium Flowers (*Tropaeolum majus*). *Food Chem.* **2009**, *114*, 44–49. [CrossRef]
176. Azuma, K.; Ohyama, A.; Ippoushi, K.; Ichiyanagi, T.; Takeuchi, A.; Saito, T.; Fukuoka, H. Structures and Antioxidant Activity of Anthocyanins in Many Accessions of Eggplant and Its Related Species. *J. Agric. Food Chem.* **2008**, *56*, 10154–10159. [CrossRef] [PubMed]
177. Szymanowska, U.; Złotek, U.; Karaś, M.; Baraniak, B. Anti-Inflammatory and Antioxidative Activity of Anthocyanins from Purple Basil Leaves Induced by Selected Abiotic Elicitors. *Food Chem.* **2015**, *172*, 71–77. [CrossRef] [PubMed]
178. Sapian, S.; Taib, I.S.; Katas, H.; Latip, J.; Zainalabidin, S.; Hamid, Z.A.; Anuar, N.N.M.; Budin, S.B. The Role of Anthocyanin in Modulating Diabetic Cardiovascular Disease and Its Potential to Be Developed as a Nutraceutical. *Pharmaceuticals* **2022**, *15*, 1344. [CrossRef]
179. Strugała, P.; Dzydzan, O.; Brodyak, I.; Kucharska, A.Z.; Kuropka, P.; Liuta, M.; Kaleta-Kuratewicz, K.; Przewodowska, A.; Michałowska, D.; Gabrielska, J.; et al. Antidiabetic and Antioxidative Potential of the Blue Congo Variety of Purple Potato Extract in Streptozotocin-Induced Diabetic Rats. *Molecules* **2019**, *24*, 3126. [CrossRef]
180. Sarikaphuti, A.; Nararatwanchal, T.; Hashiguchi, T.; Ito, T.; Thaworanunta, S.; Kikuchi, K.; Omaya, Y.; Maruyama, I.; Tancharoen, S. Preventive Effects of *Morus alba* L. Anthocyanins on Diabetes in Zucker Diabetic Fatty Rats. *Exp. Ther. Med.* **2013**, *6*, 689–695. [CrossRef]
181. Lin, B.-W.; Gong, C.-C.; Song, H.-F.; Cui, Y.-Y. Effects of Anthocyanins on the Prevention and Treatment of Cancer. *Br. J. Pharmacol.* **2017**, *174*, 1226–1243. [CrossRef]
182. Bontempo, P.; De Masi, L.; Carafa, V.; Rigano, D.; Scisciola, L.; Iside, C.; Grassi, R.; Molinari, A.M.; Aversano, R.; Nebbioso, A.; et al. Anticancer Activities of Anthocyanin Extract from Genotyped *Solanum tuberosum* L. "Vitelotte". *J. Funct. Foods* **2015**, *19*, 584–593. [CrossRef]
183. Joshi, R.; Rana, A.; Kumar, V.; Kumar, D.; Padwad, Y.S.; Yadav, S.K.; Gulati, A. Anthocyanins Enriched Purple Tea Exhibits Antioxidant, Immunostimulatory and Anticancer Activities. *J. Food Sci. Technol.* **2017**, *54*, 1953–1963. [CrossRef]
184. Kapoor, B.B.S.; Sood, P.; Modgil, R.; Sood, M. Berberis lycium a Medicinal Plant with Immense Value. *Indian J. Pharm. Biol. Res.* **2013**, *1*, 27. [CrossRef]
185. Gaur, P.; Bhatia, S.; Andola, H.; Gupta, R. In Vitro Radical Scavenging Activity and Antimicrobial Potential of *Berberis asiatica* Roxb. Ex DC. Fruit Extracts in Four Different Processed Forms. *Indian J. Tradit. Knowl.* **2017**, *16*, 706–713.
186. Tewari, D.; Zengin, G.; Ak, G.; Sinan, K.I.; Cziáky, Z.; Mishra, S.T.; Jekő, J. Phenolic Profiling, Antioxidants, Multivariate, and Enzyme Inhibitory Properties of Wild Himalayan Fig (*Ficus palmata* Forssk.): A Potential Candidate for Designing Innovative Nutraceuticals and Related Products. *Anal. Lett.* **2021**, *54*, 1439–1456. [CrossRef]

foods

MDPI

Article

Lotus Root Polysaccharide-Phenol Complexes: Interaction, Structure, Antioxidant, and Anti-Inflammatory Activities

Kaidi Peng [1,2,†,‡], Yin Li [1,‡], Ying Sun [1,2], Wei Xu [1,2], Hongxun Wang [1,2], Rui Zhang [1,3] and Yang Yi [1,2,*]

1 Hubei Key Laboratory for Processing and Transformation of Agricultural Products, College of Food Science and Engineering, Wuhan Polytechnic University, Wuhan 430023, China
2 Hubei Industrial Technology Research Institute of Jingchu Special Foods, Jingzhou 434000, China
3 National R & D Center for Se-rich Agricultural Products Processing, School of Modern Industry for Se-lenium Science and Engineering, Wuhan Polytechnic University, Wuhan 430023, China
* Correspondence: yiy86@whpu.edu.cn; Tel.: +86-138-8615-2207
† Current address: No. 68, Xuefu South Road, Changqing Garden Street, Dongxihu District, Wuhan 430023, China.
‡ These authors contributed equally to this work.

Abstract: This research aimed to explore the interaction between lotus root polysaccharides (LRPs) and phenolic compounds, and to study the effects of phenolic binding on the structural and functional properties of LRPs. The influences of pH, temperature, and NaCl and phenol concentration on the binding ratio of gallic acid (GA)/epigallocatechin (EGC) to LRPs were evaluated. LRP-GA/EGC complexes with different phenolic binding amounts were then prepared and characterized via ultraviolet–visible (UV–Vis) and Fourier-transform infrared (FTIR) spectroscopy, and average molecular weight (MW) measurements. The results suggest that hydrogen bonds contributed to the binding of GA/EGC and LRPs. The phenolic binding led to significant changes in the structure and MW of LRPs. Moreover, antioxidant activity and the macrophage-stimulating effect of LRPs were improved after binding with GA/EGC, depending on the binding amount and type of polyphenol. Interestingly, LRP-GA/EGC complexes with polyphenol binding amounts of 105.4 mg/g and 50.71 mg/g, respectively, showed better stimulation effects on the anti-inflammatory cytokine IL10 secretion of macrophages when compared to LRPs. These results show the great potential of phenolic binding to be applied to improve the structure and functional activity of LRPs.

Keywords: lotus root polysaccharides; gallic acid; epigallocatechin; interaction; antioxidant; anti-inflammatory

Citation: Peng, K.; Li, Y.; Sun, Y.; Xu, W.; Wang, H.; Zhang, R.; Yi, Y. Lotus Root Polysaccharide-Phenol Complexes: Interaction, Structure, Antioxidant, and Anti-Inflammatory Activities. *Foods* **2023**, *12*, 577. https://doi.org/10.3390/foods12030577

Academic Editor: Philippe Michaud

Received: 28 December 2022
Revised: 23 January 2023
Accepted: 24 January 2023
Published: 28 January 2023

1. Introduction

Lotus (*Nelumbo nucifera* Gaertn.) root is one of the most popular aquatic vegetables cultivated and consumed in Southeast Asia. It has attracted considerable attention due to its edible properties and therapeutic potential [1]. Polysaccharides are suggested to be important functional factors that contribute to its potential health benefits, e.g., antioxidant, immunomodulatory, and anti-obesity activities [2,3]. Lotus root also contains phenolic compounds that are related to the benefits. Recently, interesting results have been reported that polysaccharides could interact with polyphenols spontaneously when they are physically mixed, leading to the improved bioaccessibility, stability, and flavor of end products [4,5]. For example, Tudorache and Bordenave [6] observed that phenolic compounds mediated the colloidal aggregation and decreased the pseudo-first behavior of polysaccharide solutions. Jakobek and Matić [7] reported that associating with dietary fiber might control the amount of bioaccessible polyphenols in the upper or lower parts of the digestive tract. Concerning lotus root polysaccharides (LRPs), these have a great opportunity to contact and interact with polyphenols whenever they are released from the cell during lotus root harvesting, processing, and eating.

Polysaccharides may interact with polyphenols by covalent and non-covalent interactions [8]. Non-covalent interactions occur more easily during common processing conditions and have garnered more interest from researchers [9]. It was proposed that non-covalent interactions are mediated by hydrogen bonds, hydrophobic interactions, and electrostatic interactions. The intensity of non-covalent interactions is likely to be dependent on the physical–chemical properties of substrates as well as environmental factors (pH, temperature, and ionic strength). Studies have reported that hydrogen bonds are the main driving force for the complex formation between epigallocatechin gallate and oat β-glucan [10], whereas both hydrogen bonding and electrostatic forces are responsible for the interactions between ferulic acid and arabinan-rich pectic polysaccharides from rapeseed meal [9].

However, the nature of the binding mechanism between LRPs and polyphenols, as well as the positive effects on the nutritional value of LRPs, have rarely been explored. In this study, we aimed to reveal the interaction mechanism of LRPs and polyphenols, study the effects of phenolic binding on the structure, antioxidant, and macrophage-stimulating activities of LRPs. Polysaccharides were extracted from lotus root and GA/EGC were selected as the representatives of phenolic compounds. The effects of environment factors (pH, temperature, and NaCl concentration) on the interactions between LRPs and GA/EGC were evaluated. In addition, the complexes of LRP-GA/EGC were prepared with different binding ratios of GA/EGC and instrumentally characterized, and the effects of binding GA/EGC on the antioxidant and anti-immunomodulatory effects of LRPs were examined.

2. Materials and Methods

2.1. Materials and Chemicals

Fresh lotus roots (cultivar Elian No. 5) used in this work were purchased from Wuhan Jinshui-qiliang Agricultural Products Co., Ltd (Wuhan, China). Lotus roots (Figure 1) were washed with tap-water to remove mud, and knots were discarded. Roots were then peeled, sliced, and dried in an electric thermostatic drying oven (DHG-9140A, Shanghai Yiheng Scientific Instrument Co., Ltd, Shanghai, China) at 65 °C for 12 h. The dried slices were ground into fine powder and stored in a desiccator at room temperature for further use.

Figure 1. Image of lotus roots.

Gallic acid (≥99%) and epigallocatechin (≥98%) were purchased from Shanghai Macklin Biochemical Co., Ltd (Shanghai, China). Coomassie brilliant blue assay kit and ferric reducing antioxidant power (FRAP) kit were obtained from Shanghai Beyotime Technology (Shanghai, China). Gibco Dulbecco's Modified Eagle Medium (DMEM) and phosphate-buffered saline (PBS) were purchased from Hyclone Company (Logan, American). Lipopolysaccharides were obtained from Beijing Biotopped Co., Ltd (Beijing, China). RNA extraction solution, primers, and first strand cNDA synthesis kit were purchased from Wuhan Servicebio Technology Co., Ltd (Wuhan, China). Mouse RAW264.7 macrophages were provided by Biosea Biotechnology Co. Ltd (Wuhan, China). Folin–Ciocalteu reagent, 2,2-diphenyl-1-picrylhydrazyl (DPPH), and other analytic reagents were purchased from Sinopharm Chemical Reagent Co., Ltd (Shanghai, China).

2.2. Isolation and Proximate Composition of Lotus Root Polysaccharides

Lotus root polysaccharides (LRPs) were prepared according to a previous study with minor modifications [11]. Briefly, dried lotus root powder was extracted with distilled water (solid:liquid ratio of 1:8, *w/v*) in a water bath with ultrasound treatment at 200 W for 5 min and then placed in a 90 °C water bath with stirring for 3 h. The mixture was centrifuged at 4500 rpm for 10 min. The supernatant was mixed with 30% (*v/v*) ethanol distilled water solution, and incubated at 4 °C for 3 h to precipitate starches. The non-starch solution was obtained by centrifugation at 4500 rpm for 6 min to remove the sediment and confirmed using iodide/iodine reagent. Then, the non-starch solution was concentrated using a rotary evaporator under vacuum at 55 °C and subjected to the Sevage method for the removal of proteins. The concentrated solution was precipitated with three-fold volumes of 95% (*v/v*) ethanol distilled water solution at 4 °C for 12 h. The precipitates were collected by centrifugation at 4500 rpm for 6 min, re-dissolved in distilled water, concentrated again as mentioned above, and lyophilized to obtain LRPs.

The total sugar content was measured according to the phenol sulfuric acid method [12] and presented as glucose equivalents [13]. The reducing sugar content was determined by the 3,5-dinitrosalicylic acid (DNS) method [14]. The protein content was evaluated using a Coomassie Brilliant Blue staining kit according to the manufacturer's instructions. The total phenolic content was determined by the Folin–Ciocalteu method [15]. Briefly, 0.125 mL sample solution, 0.625 mL distilled water, and 0.125 mL Folin–Ciocalteu reagent were mixed well and incubated in the dark at room temperature for 10 min. Then, 1.25 mL 7% $NaCO_3$ was added, and the mixture was kept in the dark for 90 min, followed by absorbance measurement using a spectrometer (Shimadzu, UV-1800, Kyoto, Japan) at 760 nm. A calibration curve was established using GA (20–100 µg/mL) as the standard. Each sample measurement was performed in three replicates.

2.3. Preparation of LRP-GA and LRP-EGC Complexes

Polysaccharide-phenol complexes were prepared following the method of Li et al. [16] with slight modifications. LRP (8 mL, 1 mg/mL), GA, and EGC solutions (4 mL, varying concentrations) were prepared. LRP solutions were made at low concentration to allow complete dissolving in distilled water and avoid self-aggregation or phenolic-induced aggregation [17]. The LRP & GA and LRP & EGC mixtures were obtained by mixing corresponding solutions in a 30 mL screw-capped glass bottle, and then stirring for 30 min at different pH, temperatures, or NaCl concentrations. Mixture solutions of 10 mL were transferred to pre-treated dialysis bags (MWCO1000, Spectrum Laboratories, Compton, CA, USA), followed by immersing in 190 mL distilled water (equal to 20-fold dilution of mixture solution) and incubating for 4 h at the same temperature as before. The reserved solution in dialysis bags was collected and freeze-dried as polysaccharide-phenol complexes. The varied preparation conditions were concentration of GA/EGC (0.5, 1, 2, 4, and 8 mg/mL), pH (3, 4, 5, 6, and 7, adjusted by 0.1 M HCl or 0.1 M NaOH), temperature (0, 15, 30, 45, and 60 °C) and NaCl concentration (0, 20, 40, 60, and 80 mmol/L).

After the dialysis bag was removed, the phenol concentration in the dialysate was determined by the Folin–Ciocalteu method. The binding ratio (Y), in terms of the amount of GA/EGC (mg) bound per gram of LRPs, was calculated by the following equation:

$$Y = (M_0 - CV)/M_{LRPS} \tag{1}$$

where M_0 is the mass of the GA/EGC applied (mg), C is the phenol concentration in the dialysate (mg/mL), V is the total volume of the dialysis system (200 mL), and M_{LRPs} is the mass of LRPs applied (8 mg).

2.4. *Characterization of LRP-GA and LRP-EGC Complexes*

The LRP-GA (LRP-GA_1, LRP-GA_2, LRP-GA_3) and LRP-EGC (LRP-EGC_1, LRP-EGC_2, LRP-EGC_3) complexes with different binding ratios of GA/EGC to LRPs were prepared at optimum conditions and named as shown in Table 1.

Table 1. Preparation conditions of LRP-GA and LRP-EGC complexes with different binding amounts of GA/EGC: concentration of GA/EGC (mg/mL), mass ratio of LRPs to GA/EGC, pH, temperature.

Complex	Name	Preparation Conditions			
		Concentration of GA/EGC (mg/mL)	Mass Ratio of LRPs to GA/EGC	pH	Temperature (°C)
LRP-GA	LRP-GA_1	8.0	1:4	7	0
	LRP-GA_2	2.0	1:1	7	0
	LRP-GA_3	0.5	4:1	7	0
LRP-EGC	LRP-EGC_1	8.0	1:4	5	0
	LRP-EGC_2	2.0	1:1	5	0
	LRP-EGC_3	0.5	4:1	5	0

2.4.1. Ultraviolet–Visible (UV–Vis) Spectrum Analysis

The UV–Vis spectra of fifteen samples (LRPs, GA, EGC, complexes, and their physical mixtures) were determined at a wavelength range of 190–500 nm using a UV–Vis spectrophotometer (A360, Aoyi Instruments Shanghai CO., Ltd, Shanghai, China). The aqueous solutions of LRPs and phenolic compounds were prepared with distilled water at the concentration of 0.02 mg/mL. The LRP-GA_1, LRP-GA_2, and LRP-GA_3 were dissolved in distilled water at 0.02 mg/mL, 0.05 mg/mL, and 0.1 mg/L, respectively (same preparation concentrations for LRP-EGC complexes). The physical mixtures of LRP & GA and LRP & EGC were prepared by mixing the LRPs and phenol solution with the same molar ratio as the complexes. All the samples were prepared and then scanned immediately.

2.4.2. Fourier-Transform Infrared (FTIR) Spectrum Analysis

The FTIR spectra of samples were measured using an FTIR spectrophotometer (Nexus 5DXC FT-IR, Thermo Nicolet, Madison, USA). The samples (2 mg) were ground with KBr and pressed into pellets for FTIR measurement in the frequency range 4000–400 cm^{-1}, with a scanning number of 32 and a resolution of 4 cm^{-1} [18].

2.4.3. High-Performance Size-Exclusion Chromatography

The molecular weight (MW) distributions of LRPs and LRP-phenol complexes were evaluated by the method of Yi et al. [11] and using a high-performance size-exclusion chromatograph (HPSEC) equipped with a column of Aglient PL aquagel OH. Samples were dissolved in the mobile phase with a concentration of 0.2 mol/L, and then centrifuged at 13,000 rpm for 5 min. The supernatant was filtered through 0.22 μm film and used for injection. It was eluted by the mobile phase of 0.2 mol/L $CH_3COOHNH_4$ solution at a flow rate of 0.7 mL/min and column temperature of 30 °C. The peaks were determined using a multi-angle laser-light scattering detector (DAWN HELEOS-II 18, Wyatt Technology Co., Santa Barbara, CA, USA) and refractive index detector (Optilab rEX, Wyatt Technology Co.).

2.5. *Antioxidant Activities of LRP-GA and LRP-EGC Complexes*

The 2,2-diphenyl-1-picrylhydrazyl (DPPH) radical scavenging ability was assessed based on the method reported by Rodríguez et al. [19]. In brief, 100 µL aqueous solution of sample (varying concentrations ranging from 0.1 mg/mL to 1.0 mg/mL), 1 mL DPPH (100 µmol/L) solution, and 1 mL distilled water were mixed well and then incubated for

30 min in the dark. After reaction, the absorbance of the mixture was determined at 517 nm. Similarly, a control was prepared with methanol instead of sample. Blank sample was obtained using methanol instead of DPPH solution. The DPPH radical scavenging ability was expressed as the scavenging ratio (S, %) and calculated by the equation:

$$S = [1 - (A_s - A_b)/A_c] \times 100 \quad (2)$$

where A_c, A_s, and A_b are the absorbance of the control, sample, and blank sample, respectively. The sample concentration required for a 50% (IC_{50}, mg/mL) scavenging ratio against DPPH radical was calculated according to the linear equation of concentration vs. scavenging rate.

The evaluation of ferric reducing antioxidant power (FRAP) was performed using the FRAP kit according to its instruction. The results were expressed as millimoles of Fe^{2+} equivalents per 1 g of sample.

2.6. Immunomodulatory Activity of LRP-GA and LRP-EGC Complexes

The immunomodulatory activity was measured by the NO production and the mRNA expression levels of tumor necrosis factor-α (TNF-α) and interleukin-10 (IL-10). In detail, RAW264.7 macrophage suspension was prepared with DMEM medium at a density of 5×10^5 cells/mL and then placed in 48-well culture plates (500 µL/well). Cells were incubated at 37 °C in a 5% CO_2 incubator for 3 h, followed by removing the supernatant. Adherent cells were obtained and re-incubated in medium alone (control group) or medium containing samples and/or LPS. All aqueous solutions of samples and LPS were prepared with DMEM medium. Aqueous solutions of LRPs and LRP-phenol complexes were prepared with the same LRP concentration of 200 μg/mL; aqueous solution of LPS was prepared with a concentration of 500 ng/mL and used as a positive control. After incubation for 24 h and 48 h, the NO concentration in the macrophage culture supernatant was determined by the Greiss method [12,13].

Cells incubated for 24 h with the same stimulation procedures mentioned above were collected, washed with phosphate buffered saline (PBS) twice, and used for extraction of total cellular RNA. Total cellular RNA was isolated using the RNA extraction solution (Servicebio, Wuhan, China) and treated following the manufacturer's protocol. cDNA was synthesized with 2 μg total cellular RNA using hexamers for priming and a first-strand cNDA synthesis kit (Servicebio, Wuhan, China) according to the manufacturer's instructions. Quantitative real-time PCR was performed using a Real-time PCR System (Thermo fisher scientific, USA) with 2 × SYBR Green qPCR Master Mix (None ROX, Servicebio, Wuhan, China) following the manufacturer's recommendations. The primers used in the reaction and the glyceraldehyde-3-phosphate dehydrogenase (GAPDH) gene serving as endogenous references were as follows: 5′-CCTCGTCCCGTAGACAAAATG-3′ (sense primer) and 5′ TGA GGTCAATGAAGGGGTCGT-3′ (antisense primer) for GAPDH, 5′-CTC TTCTGTCTACTGAACTTCGGG -3′ (sense primer) and 5′-GGTGGTTTGTGAGTGTGAGGGT-3′ (antisense primer) for TNF-α, 5′-TGCCAAGCCTTATCGG AAATG-3′ (sense primer) and 5′-AAATCACTCTTCACCTGCTCCAC-3′ (antisense primer) for IL-10 [20].

2.7. Statistical Analysis

All the experiments were performed in triplicate, and the data were recorded as mean ± standard deviation. Statistical analysis was performed with SPSS 19.0 software (IBM, Armonk, NY, USA), and the significant differences ($p < 0.05$) between groups were estimated with one-way ANOVA followed by the Student–Newman–Keuls test.

3. Results and Discussion

3.1. Physicochemical Properties of LRPs

The physicochemical properties of LRP powders were measured after water extraction, 30% alcohol precipitation, deproteinization, 75% alcohol precipitation, and lyophilization.

The total sugar content of LRPs was 87.48%, reducing sugar content was 2.79%, polyphenol content was 1.13%, and protein content was 0.20%.

3.2. Factors that Influence the Binding Ratio of GA/EGC to LRPs

The binding interactions between polyphenols and polysaccharides were mostly non-covalent under normal processing conditions [18], and were mainly mediated by hydrogen bonds, hydrophobic interactions, and ion interactions. These could be influenced by environmental conditions like compounds, pH, temperature, and ionic strength [16]. Thus, the individual effects of pH, temperature, concentration of NaCl, and concentration of GA/EGC on the binding ratio were investigated to determine the complexation mechanism between GA/EGC and LRPs.

3.2.1. Effect of pH

The binding ratios of GA/EGC to LRPs at different pH (2–7) are shown in Figure 2A. The binding ratio of GA to LRPs slightly decreased from pH 3 to 4, and then significantly increased with pH increasing from 4 to 7. The highest binding ratio value of 567.25 ± 34.89 mg/g was observed at pH 7. Badhani and Kakkar [21] reported that gallic acid had a pKa of 5 and existed in the anionic form in the pH range 5 to 9 with the deprotonation of the carboxy-group or hydroxy-group [22].

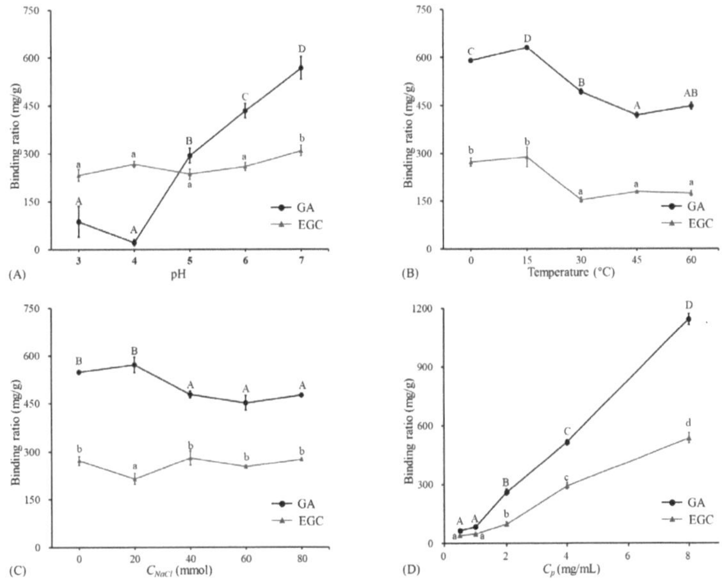

Figure 2. The binding ratios of GA and EGC to LRPs at different pH, temperatures, NaCl concentrations, and phenol concentrations. (**A**) The binding ratios of GA and EGC to LRPs at different pH, temperature of 15 °C, and mass ratio of GA/EGC to LRPs of 2:1; (**B**) the binding ratios of GA and EGC to LRPs at different temperatures, pH 7, and mass ratio of GA/EGC to LRPs of 2:1; (**C**) the binding ratios of GA and EGC to LRPs at different NaCl concentrations, pH 7, temperature of 15 °C, and mass ratio of GA/EGC to LRPs of 2:1; (**D**) the binding ratios of GA and EGC to LRPs at different C_p phenol concentrations (different mass ratios of GA/EGC to LRPs of 1:4, 1:2, 1:1, 2:1, 4:1), pH 7, and temperature of 15 °C. Means of the samples with different letters (A–D, a–d) differ significantly ($p < 0.05$).

According to previous studies, if polyphenols are absorbed onto polysaccharides via electrostatic interaction, the electrostatic repulsion increases with pH increasing and the polyphenols absorbed can be reduced [9,23]. However, results in this study were not shown in this way. This implies that electrostatic interaction may not play an important role in the binding of GA and LRPs. Phan et al. [23] also obtained similar results regarding the complexation of ferulic acid and cellulose. Studies reported that the higher the ratio of deprotonated/protonated carboxyl groups (COO^-/COOH), the higher the repulsion between dietary fiber molecules, which stabilized those molecules as micelles and led to better entrapment of hydrophobic polyphenol curcumin [7,24]. For EGC, the binding ratio remained stable from pH 3 to 6, and showed a significant increase from pH 6 to 7. EGC (pKa is 7.87) was protonated in the pH range 2–7 [25]. With increasing pH, the decreased H^+ had little effect on the binding, and then due to the increased neutral fraction, the binding amount had the highest value (308.27 ± 15.96 mg/g) at pH 7. This indicated that electrostatic interactions made little sense in the association between EGC and LRPs. Similarly, Le Bourvellec et al. [26]reported that the binding of procyanidins and apple cell-wall material had no significant variation within the pH range 2–7. The pH had a different effect on the binding of GA and EGC to LRPs. This is consistent with previous literature reporting that the pH effect on the combination of polysaccharides and polyphenols is greatly dependent on the type of polyphenol [27].

3.2.2. Effect of Temperature

The binding ratios of GA/EGC to LRPs at different temperatures are shown in Figure 2B. The binding amount of GA significantly increased with temperature increasing from 0 °C to 15 °C, and then significantly decreased with temperature increasing from 15 °C to 45 °C, and showed a slight increase from 45 °C to 60 °C. The binding amounts of EGC to LRPs had no change with temperature increasing from 0 °C to 15 °C, and then sharply decreased when the temperature increased from 15 °C to 30 °C, and remained stable with the temperature continually increasing from 30 °C to 60 °C. Both GA and EGC had larger binding ratios (630.96 ± 6.77 mg/g and 288.13 ± 30.13 mg/g, respectively) to LRPs at 15 °C than at other temperatures. It was inferred that the binding of GA/EGC to LRPs was not an entire endothermic process or physical adsorption process [28]. When the temperature rose to 15 °C, the notable decrease of the binding amount with increasing temperature implied the hydrogen bonding would drive the sorption of GA/EGC to LRPs [9]. A similar observation was reported by Wu et al. [28] that increasing temperature from 20 °C to 30 °C increased the adsorption of oat to β-glucan, while the increase in temperature from 30 °C to 60 °C adversely affected the binding.

3.2.3. Effect of NaCl Concentration

The data in Figure 2C show the binding ratio of GA and EGC to LRPs at different NaCl concentrations. In general, by increasing the NaCl concentration, the binding ratio of GA to LRPs first increased slightly ($p > 0.05$), reached its maximum point (572.31 ± 24.77 mg/g) at an NaCl concentration of 20 mmol, and then kept stable. Conversely, the binding ratio of EGC at an NaCl concentration of 20 mmol appeared to have a lower binding amount than at lower or higher NaCl concentration values. Theoretically, if hydrophobic interactions are formed between the hydrophobic regions of LRPs and the hydrophobic groups (e.g., aromatic rings) of GA/EGC, one of their characteristics is that their strength increases with an increase in ionic strength [23,29]. However, this phenomenon was not observed in this study. This attested that the hydrophobic interactions between GA/EGC and LRPs were weak. Previous research showed similar results that hydrophobic interactions were not a major contributor between ferulic acid and arabinan-rich pectic polysaccharide from rapeseed meal [9].

3.2.4. Effect of Phenol Concentration

As shown in Figure 2D, the binding amount of GA/EGC to LRPs increased approximately linearly with increasing the initial GA/EGC concentration from 0.5 mg/mL to 8 mg/mL. This suggests there was equilibrium between free and bound phenol, whereas the increase in concentration could result in the increase in binding capacity [8]. Because of the limited solubility of GA/EGC in distilled water, the experiment could not be conducted at higher phenol concentrations. GA/EGC had maximum bound ratio values of 1144.18 ± 27.50 and 535.98 ± 26.16 mg/g, respectively. At the same preparation conditions, GA showed better binding capacity to LRPs compared to EGC. Both GA and EGC had hydroxyl groups as well as aromatic rings, whereas the differing molecular weights, degrees of galloylation, and number and position of various functional groups could lead to their varied adsorption amounts to LRPs.

3.3. Characterization of LRP-GA and LRP-EGC Complexes

The interactions between LRPs and phenolic compounds (GA and EGC) as well as the structural properties of LRP-GA and LRP-EGC complexes with different binding amounts of GA/EGC were evaluated. LRP-GA_1, LRP-GA_2, LRP-GA_3, LRP-EGC_1, LRP-EGC_2, and LRP-EGC_3 complexes were prepared, respectively, with binding ratios of 1115.05, 336.14, 105.42, 698.71, 169.09, and 50.71 mg/g. The corresponding mass ratios of bound GA and EGC to LRP-GA/EGC complexes were 52.72%, 25.16%, 9.62%, 41.13%, 14.46%, and 4.83%, respectively.

3.3.1. UV–Vis Spectrum Analysis

The UV–Vis spectrum of LRPs had the characteristic absorption of unsaturated carbonyl and carboxyl groups in the wavelength range 190–220 nm, and weak absorption of a few proteins incorporated in LRPs at 280 nm (Figure 3A) [13]. Both GA and LRP & GA mixtures had strong absorption at around 210 nm and 260 nm (Figure 3A), due to the bathochromic effect of hydroxyl groups on the aromatic ring [30]. For LRP-GA complexes, all showed the similar strong absorption peak at 210 nm, but a weaker peak at 260 nm compared to their corresponding physical mixtures. Particularly, the peak at 260 nm was not obvious for the LRP-GA_3 complex. Similarly, the absorption peak was observed at 270 nm on the UV–Vis spectrum of EGC and LRP & EGC mixtures (Figure 3B), and it disappeared in the LRP-EGC complexes. It was implied that the hydroxyl group of GA/EGC was involved in the interaction with LRPs. Similarly, Li et al. [16] found that the chromatographic peaks of GA/catechin disappeared in the polysaccharide-GA/catechin complexes.

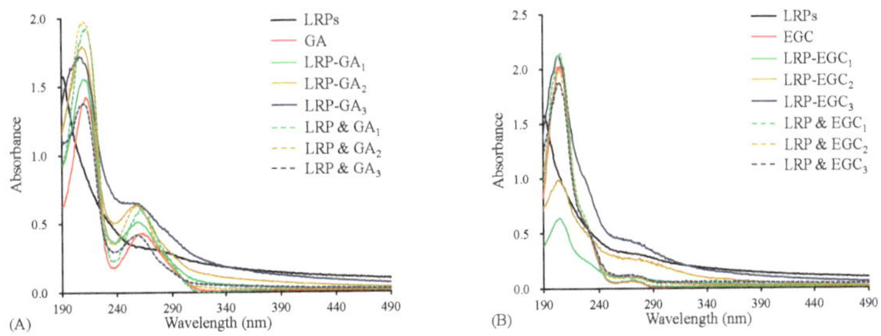

Figure 3. The UV–Vis spectra of LRPs, polyphenols, LRP-phenol complexes, and their physical mixtures. LRPs, GA, LRP-GA complexes, and their physical mixture (**A**); LRPs, EGC, LRP-EGC complexes, and their physical mixture (**B**).

3.3.2. FTIR Spectrum Analysis

FTIR spectrum analysis was carried out to characterize the binding of GA and EGC on LRPs. Figure 4A shows the FTIR spectrum of LRPs, GA, their physical mixture, LRP-GA_1, LRP-GA_2, and LRP-GA_3 complexes. The GA spectrum exhibited typical broad absorption bands of O-H stretching vibrations (3316 cm^{-1} and 3288 cm^{-1}), C=O stretching vibration (1709 cm^{-1}), aromatic ring with substitutions (1620 cm^{-1}, 1541 cm^{-1}), C-C and C-O stretching vibrations (1446 cm^{-1}, 1200–1300 cm^{-1}, 1024 cm^{-1}), and C-H bending vibration (867 cm^{-1}) [21,30,31] . The LRP & GA physical mixture had a similar FTIR spectrum to GA, except for the absorption peak at 2925 cm^{-1}. The broad peak at 2925 cm^{-1} was attributed to the C-H stretching vibration of LRPs, which was also observed on the FTIR spectra of LRPs, LRP-GA$_1$, LRP-GA$_2$, and LRP-GA$_3$ [11]. The O-H stretching vibration bands in LRP-GA complex spectrum (at 3397 cm^{-1}, 3294 cm^{-1}, and 3391 cm^{-1} for LRP-GA$_1$, LRP-GA$_2$, and LRP-GA$_3$, respectively) were broader than that in the GA and LRP spectrum. It was reported that the O-H stretching band is a sensitive indicator of the strength of the hydrogen bond, and broadening of the O-H stretching band is often observed in strong hydrogen bonds [32]. Here, it was suggested that the hydrogen bond was formed between LRPs and GA. Compared to GA, the bands located at 1709 cm^{-1}, 1541 cm^{-1}, 1446 cm^{-1}, 1244 cm^{-1}, and 1203 cm^{-1} disappeared for LRP-GA complexes; the absorption peaks at 1340 cm^{-1} and 1308 cm^{-1} were attenuated and shifted, leading to absorption peaks at 1398 cm^{-1} and 1406 cm^{-1} on the spectra of LRP-GA$_2$ and LRP-GA$_3$, respectively. LRP-GA$_1$, LRP-GA$_2$, and LRP-GA$_3$ spectra had characteristic peaks at 1041 cm^{-1}, 1078 cm^{-1}, and 1078 cm^{-1}, respectively, which was likely related to the shift of the GA band at 1024 cm^{-1}. The absorption intensity of bands at 897 cm^{-1} decreased.

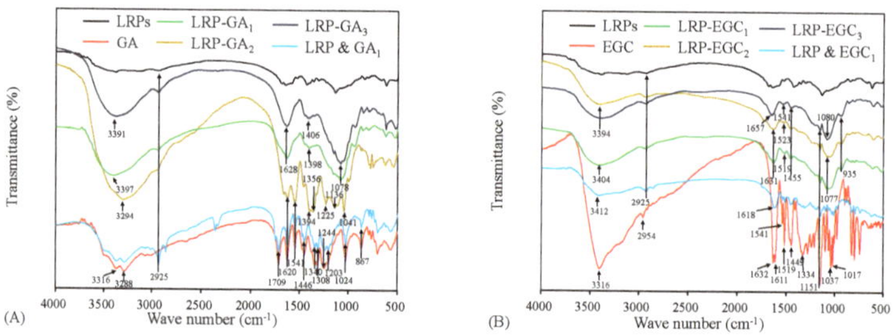

Figure 4. FTIR spectra of LRPs, polyphenols, LRP-phenol complexes, and their physical mixtures. LRPs, GA, LRP-GA complexes, and their physical mixture (**A**); LRPs, EGC, LRP-EGC complexes, and their physical mixture (**B**).

The FTIR spectrum of EGC, physical mixture of LRPs and EGC, and LRP-EGC complexes are shown in Figure 4B. For EGC, the prominent peaks at 3316 cm^{-1} and 2954 cm^{-1} represent the stretching vibrations of O-H and C-H, respectively. The bands within 1400–1632 cm^{-1} and 1200–1334 cm^{-1} are related to aromatic rings [31,33,34]. Peaks observed within 1017–1151 cm^{-1} correspond to C-O stretching vibrations [35,36]. After binding to LRPs, the peak at 3316 cm^{-1} (O-H vibrational frequency) was shifted to a higher frequency, which was mainly due to the migration of the electron cloud in oxygen atoms on the phenol to a higher wavenumber [37]. The absorption peaks at 2954 cm^{-1} (C-H stretching vibrations) and 1200–1334 cm^{-1} (aromatic ring) disappeared. The intensity of the broad peaks at 1632, 1611, 1519, 1449, 1151, 1037, and 1017 cm^{-1} were shifted and diminished. The results imply that the interaction of LRPs and GA/EGC exists, which obviously altered the FTIR spectrum properties of LRPs and GA/EGC [16].

3.4. Molecular Weight (MW) Distributions of LRP-Phenol Complexes

The elution profiles of LRPs and LRP-phenol complexes as a function of the elution time obtained by the HPSEC-MALL-RI method are presented in Figure 5, the retention time and peak area percentages of their fractions in different MW ranges are summarized in Table 2 and visually presented in Figure 6. The peak area percentage is proportional to the concentration of fractions, which can be used to calculate the weight-average MW. In general, the weight-average MWs of LRP-phenol complexes were significantly higher than those of LRPs. The peak area percentages of LRP fractions in the MW ranges of $1 \times 10^{-3} \times 10^4$, $1 \times 10^{-4} \times 10^5$, and $1 \times 10^{-5} \times 10^6$ were 84.9%, 2.5%, and 12.5%, respectively. Compared to LRPs, fractions in a higher MW range of $1 \times 10^{-6} \times 10^7$ were observed for LRP-GA complexes; fractions in the MW range of $1 \times 10^{-4} \times 10^5$ increased and in $1 \times 10^{-3} \times 10^4$ decreased for LRP-GA$_1$ and LRP-GA$_2$ complexes. Particularly, both had the lowest MW of 17.36 kDa and 6.79 kDa, respectively. This indicated that free GA was not found in LRP-GA_1/GA_2 complexes (the MW of GA is 306 Da). These results imply that GA was successfully combined with LRPs. The fraction of LRP-GA_3 in the lower MW range of $1 \times 10^{-2} \times 10^3$ was observed, which could be due to the low binding ratio and aggregation of GA. LRP-EGC complexes also had fractions in the high MW range of $1 \times 10^{-6} \times 10^7$. By contrasting to LRPs, fractions in the ranges $1 \times 10^{-5} \times 10^6$ and $1 \times 10^{-3} \times 10^4$ increased for LRP-EGC_1 and LRP-EGC_3, respectively. Several studies [13,16] had similar results showing that polysaccharide-phenol complexes had different MW distributions with polysaccharides.

Table 2. The molecular weights of LRPs and LRP-phenol complexes detected by the HPSEC−MALL−RI method.

Sample	Retention Time (min)	Molecular Weight (Da)	Peak Area Percentage (%)	Average Molecular Weight (Da)
LRP	7.237–11.370	2.561×10^5 (±1.031%)	12.5	3.618×10^4
	11.370–12.244	8.771×10^3 (±8.120%)	8.3	
	12.244–13.934	4.055×10^3 (±10.646%)	26.8	
	13.934–16.950	2.782×10^3 (±16.105%)	47.1	
	16.950–20.058	3.320×10^4 (±29.285%)	2.5	
	21.204–23.165	8.040×10^3 (±44.484%)	2.7	
LRP-GA_1	7.156–8.877	1.540×10^6 (±10.451%)	0.9	1.106×10^5
	8.877–11.089	3.246×10^5 (±40.231%)	5.8	
	11.089–12.283	1.736×10^4 (±27.878%)	18.9	
	12.283–14.285	1.065×10^5 (±39.474%)	28.1	
	14.285–17.411	2.357×10^4 (±28.070%)	46.3	
LRP-GA_2	6.842–8.870	1.350×10^6 (±4.031%)	2.5	6.501×10^4
	8.870–11.177	1.786×10^5 (±19.428%)	9.6	
	11.177–12.296	6.790×10^3 (±21.058%)	20.4	
	12.296–14.078	2.497×10^4 (±31.497%)	25.3	
	14.078–16.980	1.505×10^4 (±24.199%)	34.7	
	17.015–19.601	1.584×10^4 (±11.543%)	7.5	
LRP-GA_3	6.815–8.896	1.617×10^6 (±1.334%)	2.9	6.663×10^4
	8.926–11.038	1.749×10^5 (±0.784%)	9.5	
	11.038–12.727	3.523×10^3 (±11.649%)	35.5	
	12.727–14.024	1.291×10^3 (±27.174%)	18.2	
	14.024–14.959	9.475×10^3 (±17.222%)	17.3	
	14.990–17.493	8.874×10^3 (±25.312%)	16.6	
LRP-EGC_1	6.875–8.866	1.102×10^6 (±1.355%)	2.5	4.691×10^4
	8.866–11.279	1.016×10^5 (±0.856%)	16.4	
	11.279–12.516	3.479×10^3 (±15.877%)	37.9	
	12.516–13.964	3.221×10^3 (±17.625%)	20.6	
	13.964–14.899	2.195×10^3 (±12.611%)	13.7	
	14.899–16.317	4.688×10^3 (±14.529%)	8.9	

Table 2. *Cont.*

LRP-EGC$_2$	6.905–8.926	1.066×10^6 (±1.419%)	2.4	7.895×10^4
	8.926–11.128	1.213×10^5 (±0.931%)	10.8	
	11.128–13.994	1.446×10^3 (±17.875%)	62.7	
	13.994–14.959	9.786×10^2 (±22.227%)	15.5	
	14.959–17.041	1.886×10^3 (±31.802%)	8.6	
LRP-EGC$_3$	7.026–8.806	2.169×10^6 (±1.727%)	1.8	5.799×10^4
	8.836–10.978	1.960×10^5 (±2.232%)	8.3	
	10.978–13.964	3.035×10^3 (±12.892%)	63.1	
	13.994–16.166	2.196×10^3 (±16.236%)	26.1	
	16.166–17.403	2.424×10^4 (±25.006%)	0.8	

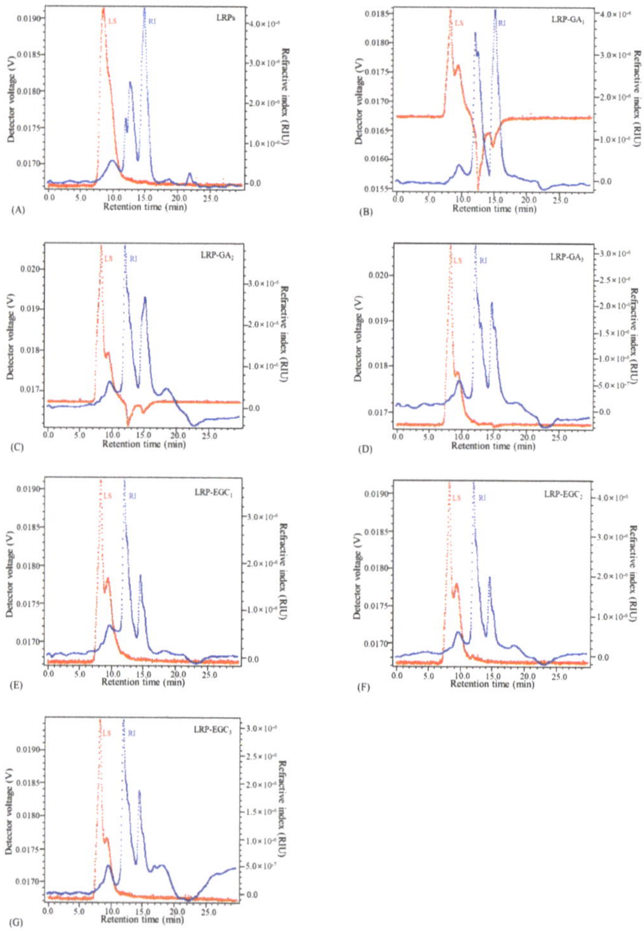

Figure 5. Elution profiles of LRPs and LRP-phenol complexes obtained by the HPSEC−MALL−RI method. LRPs(A);LRP-GA$_1$(B); LRP-GA$_2$(C); LRP-GA$_3$(D); LRP-EGC$_1$(E); LRP-EGC$_2$(F); LRP-EGC$_3$(G). Dotted line in blue presents the refractive index (RI) intensity, dotted line in red presents the light scattering (LS) intensity.

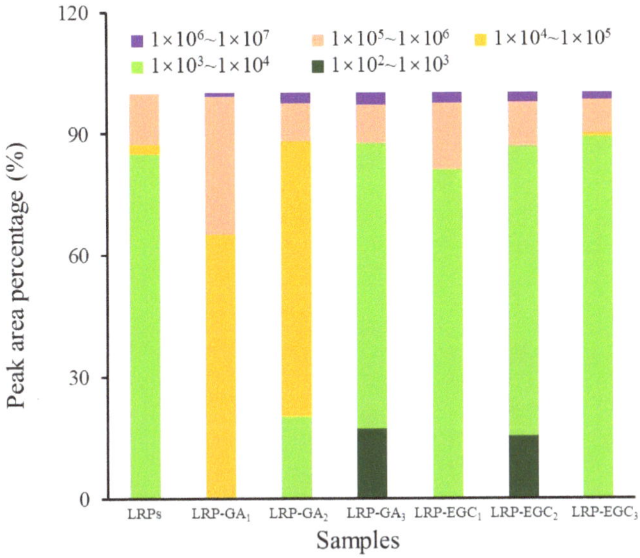

Figure 6. The peak area percentage of fractions with different molecular weights from LRPs and LRP-phenol complexes.

3.5. *Antioxidant Activities of LRPs and LRPs-GA/EGC Complexes*

As shown in Figure 7A, the DPPH radical scavenging ratio of LRPs linearly increased with the increase in concentration and reached 20.6% at 1 mg/mL. The results are similar to those obtained by Su & Li. [38]. In general, LRP-GA/EGC complexes and their mixtures had significantly stronger DPPH radical scavenging activity than LRPs, which was due to the hydrogen-donating ability and redox property of GA/EGC (Figure 7A,B) [39]. Yong et al. [36] also reported that dialdehyde starch displayed enhanced antioxidant activity after grating with catechins. The IC_{50} values of LRP-GA_1, LRP-GA_2, LRP-GA_3, LRPs & GA_1, LRPs & GA_2, and LRPs & GA_3 were 0.02, 0.06, 0.22, 0.02, 0.04, and 0.12 mg/mL, respectively (Figure 7E). The higher binding ratio/mass ratio of GA in samples contributes to better DPPH radical scavenging activity. Moreover, LRP-GA_2 and LRP-GA_3 complexes showed less free-scavenging activity than their corresponding mixtures; this may be explained by the fact that the hydrogen bonds combine between LRPs with GA and consumption of hydroxyl groups. Similarly, LRP-EGC complexes exhibited lower DPPH radical scavenging ratios as well as higher IC_{50} values compared to their mixtures. Notably, there was a difference in DPPH radical scavenging activities between LRP-GA and LRP-EGC complexes, which was due to the different type and total content of phenol.

The antioxidant activities of LRPs, LRP-GA/EGC complexes, and their mixtures evaluated by FRAP are presented in Figure 7C,D,F. LRPs had the lowest FRAP value of 0.189 mmol Fe^{2+}/g sample, showing their weak electron donation ability. Regarding LRP-GA/EGC complexes and their corresponding mixtures, the FRAP values increased with the molar ratio of GA/EGC increasing, suggesting that the phenolic compounds were mainly responsible for the electron donation even when complexed to the LRPs [40]. The FRAP values of LRP-GA_3 and LRP-EGC complexes were lower than those of their corresponding mixtures, whereas LRP-GA_1 and LRP-GA_2 showed higher FRAP values than LRP&GA_1 and LRP&GA_2 mixtures, respectively.

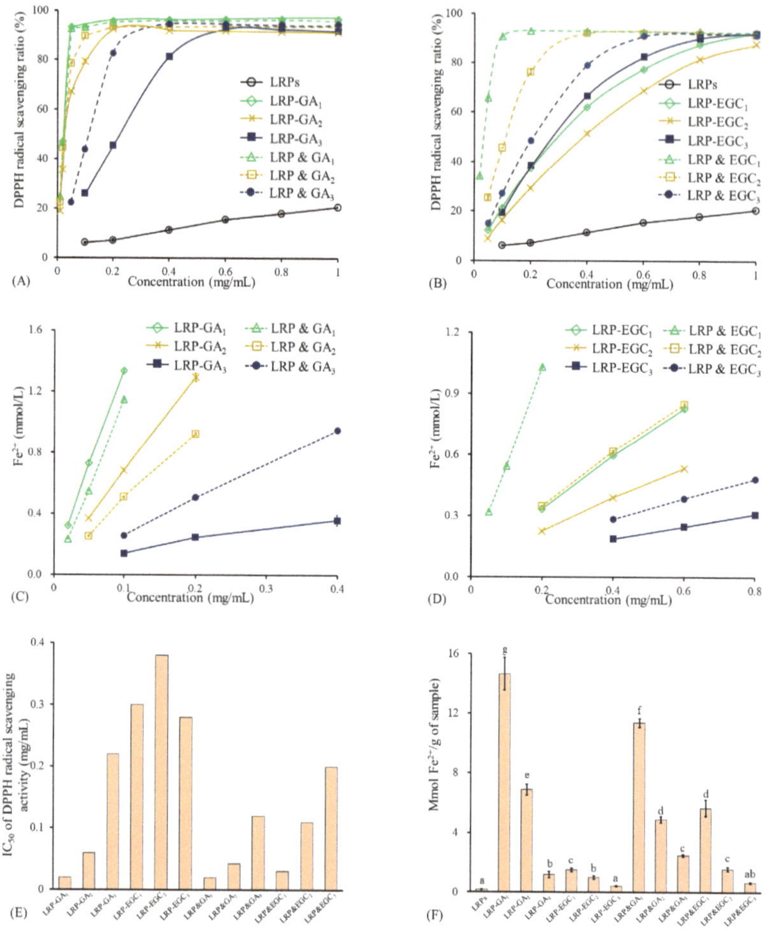

Figure 7. Antioxidant activities of LRPs, LRP-phenol complexes, and their physical mixtures. (**A**) DPPH radical scavenging ratio of LRPs, LRP-GA complexes, and their physical mixtures at different concentrations; (**B**) DPPH radical scavenging ratio of LRPs, LRP-EGC complexes, and their physical mixtures at different concentrations; (**C**) FRAP activities of LRPs, LRP-GA complexes, and their physical mixtures at different concentrations; (**D**) FRAP activities of LRPs, LRP-EGC complexes, and their physical mixtures at different concentrations. (**E**) IC_{50} values for DPPH radical scavenging activity of LRP-phenol complexes and their physical mixtures; (**F**) FRAP activities of LRPs, LRP-phenol complexes, and their physical mixtures expressed as mmol Fe^{2+}/g sample. Means of the samples with different letters (a–g) differ significantly (p <0.05).

3.6. Immunomodulatory Activity of LRPs-GA/EGC Complexes on RAW264.7 Cells

NO is an important immune-regulatory signaling molecule whose release from macrophages can suppress the growth of bacterial infection or cancer cells [41]. Further, excessive production of NO can induce oxidative stress and inflammatory disease [42]. The levels of NO and LRP-induced NO production from RAW 264.7 cells treated with different samples for 24 h and 48 h are shown in Figure 8A,B. LRPs and LRP-GA/EGC complexes were used at same LRP concentration of 200 μg/mL. Cells treated with LPS at 500 ng/mL were applied as a positive group. Regarding cells incubated for 24 h, the level of NO produced from LRP-treated cells had no significant difference to that from the control; values were 8.58 and 8.01 μmol/L, respectively. The LRP-GA/EGC complexes and LPS all stimulated NO production as compared to the control, showing increasing NO production in the order: LRP-EGC_2 < LRP-EGC_1 < LRP-EGC_3 ≤ LPS < LRP-GA_1 ≤ LRP-GA_2 < LRP-GA_3 at 13.05, 14.7, 15.59, 15.72, 17.0, 17.7, and 18.95 μmol/L, respectively. Combination with GA/EGC significantly increased the stimulation of LRPs, whereas the increase in the GA/EGC compound in complexes had no positive effect on NO levels, suggesting that LRPs in complexes play the major role in the stimulation mechanism. A similar phenomenon was observed for the cells incubated after 48 h. For cells stimulated by LPS, LRP-EGC complex showed an inhibitory effect on the LPS-induced NO production. Particularly, LRP-EGC_1 and LRP-EGC_3 significantly decreased the LPS-induced NO production for cells incubated after 24 h, and LRP-EGC_1 showed a significantly inhibitory effect on LPS-induced cells after 48 h.

As mentioned above, LRP-GA_3 and LRP-EGC_3 showed relatively better immunostimulatory effects with NO production compared with other complexes. Here, the effects of LRPs, LRP-GA_3/EGC_3 complexes on mRNA expression of cytokines (TNF-α, IL10) in RAW264.7 cells were investigated. Compared with the control untreated cells, the mRNA expression of TNF-α increased by 1.65, 1.89, 2.54, and 1.81-fold when cells were treated with LPS, LRPs, LRP-GA_3, and LRP-EGC_3, respectively (Figure 8C), but the expression levels of anti-inflammatory cytokine IL-10 were elevated. Especially, LRP-GA_3 and LRP-EGC_3 increased IL-10 expression by 6.6- and 20.1-fold compared with the control. The IL-10 expression levels induced by LRP-GA_3 and LRP-EGC_3 were significantly higher than those stimulated by LRPs. The results show that binding with GA/EGC has great potential in improving the beneficial immunomodulation of LRPs on macrophages.

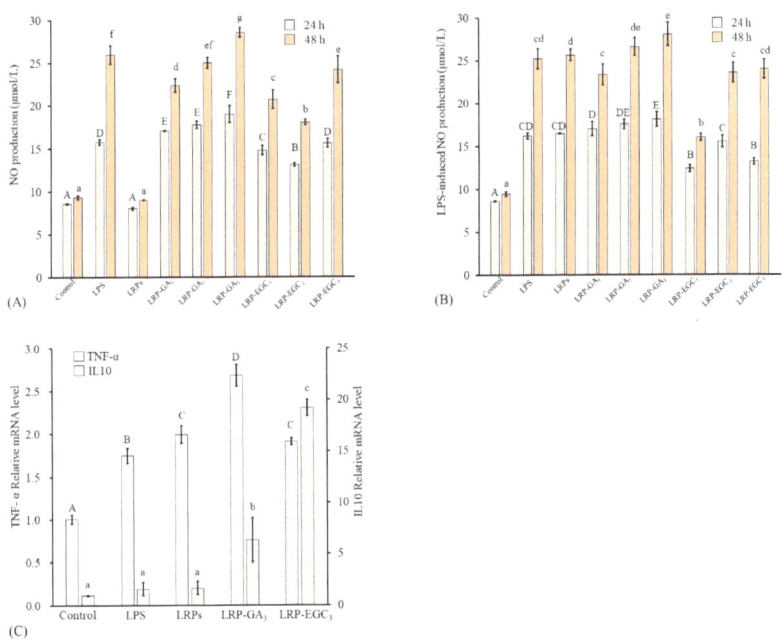

Figure 8. Macrophage-stimulating activities of LRPs and LRP-phenol complexes. (**A**) NO production of RAW264.7 cells stimulated by LRPs and LRP-phenol complexes; (**B**) NO production of LPS-induced RAW264.7 cells suppressed by LRPs and LRP-phenol complexes; (**C**) mRNA expression of cytokines (TNF-α, IL10) in RAW264.7 cells stimulated by LRPs, LRP-GA_3, and LRP-EGC_3. Means of the samples with different letters (A–F, a–g) differ significantly ($p < 0.05$).

4. Conclusions

In this work, the binding interactions between GA/EGC and LRPs were studied by evaluating the effect of environmental factors (pH, temperature, and NaCl concentration) on the binding amount of GA/EGC to LRPs, and measuring the UV–Vis spectra, FTIR spectra, and MW of LRPs-GA/EGC complexes. It was demonstrated that GA/EGCs successfully bond to LRPs by hydrogen bond interactions, leading to the changes in structure and MW distribution of LRPs. The binding of phenolic compounds significantly increased the DPPH radical scavenging activity and FRAP capacity of LRPs, and the increase degree was dependent not only on the phenol type but also on the mass ratio of the phenols in complexes. Moreover, GA and EGC combined to LRPs promoted the NO production of macrophages in normal conditions; EGC with a mass ratio of 698.71 mg/g bound to LRPs showed inhibitory effects on the NO over-production stimulated by LPS. Binding with GA/EGC did not enhance the TNF-α production induced by LRPs, but LRP-GA/EGC (with binding ratios of 105.42 and 50.71 mg/g, respectively) significantly improved the expression of anti-inflammatory cytokine IL10 as compared with LRPs. These results contribute to knowledge on the interaction mechanism of LRPs and phenolic compounds, and provide evidence that binding with phenol is a promising way to improve the structure, antioxidant, and anti-inflammatory properties of LRPs. Further research into the bioaccessibility and stability of phenolic compounds bound to LRPs under different conditions (for example, gastrointestinal digestion, light, humidity, and so on) will be carried out to obtain more knowledge on the nutritional aspects of LRP-phenol complexes.

Author Contributions: K.P.: investigation, methodology, validation, writing—original draft. Y.L.: investigation, methodology, writing—original draft. Y.S.: methodology, data analysis. Y.Y.: project administration, funding acquisition, writing—review and editing. W.X.: visualization, writing—review and editing. H.W.: funding acquisition, resources, editing. R.Z.: writing—review and editing. All authors have read and agreed to the published version of the manuscript.

Funding: This work was supported by the Science and Technology Department of Hubei Province (grant number 2022BBA0023), the Young Top-notch Talent Cultivation Program of Hubei Province, and the Wuhan Science and Technology Bureau (grant number 2022020801010389).

Data Availability Statement: Data is contained within the article.

Conflicts of Interest: The authors declare that they have no known competing financial interests or personal relationships that could have appeared to influence the work reported in this paper.

Abbreviations

The following abbreviations are used in this manuscript:

GA	gallic acid
EGC	epigallocatechin
UV–Vis	ultraviolet–visible
FTIR	Fourier-transform infrared
MW	molecular weight
LPS	liposaccharide
DMEM	Gibco Dulbecco's Modified Eagle Medium
PBS	phosphate-buffered saline
DPPH	2,2-diphenyl-1-picrylhydrazyl
DNS	3,5-dinitrosalicylic acid
HPSEC	high-performance size-exclusion chromatography
DPPH	2,2-diphenyl-1-picrylhydrazyl
FRAP	ferric-reducing antioxidant power
TNF-α	tumor necrosis factor-α
IL-10	interleukin-10

References

1. Yi, Y.; Sun, J.; Xie, J.; Min, T.; Wang, L.-M.; Wang, H.-X.Phenolic profiles and antioxidant activity of lotus root varieties. *Molecules* **2016**, *21*, 863. [CrossRef] [PubMed]
2. Wang, H.-X.; Yi, Y.; Sun, J.; Lamikanra, O.; Min, T. Fingerprint Profiling of Polysaccharides from Different Parts of Lotus Root Varieties. *RSC Adv.* **2018**, *8*, 16574–16584. [CrossRef] [PubMed]
3. Yi, Y.; Lamikanra, O.; Sun, J.; Wang, L.-M.; Min, T.; Wang, H.-X. Activity Diversity Structure-Activity Relationship of Polysaccharides from Lotus Root Varieties. *Carbohydr. Polym.* **2018**, *190*, 67–76. [CrossRef]
4. Bermúdez-Oria, A.; Rodríguez-Gutiérrez, G.; Fernández-Prior, Á.; Vioque, B.; Fernández-Bolaños, J. Strawberry Dietary Fiber Functionalized with Phenolic Antioxidants from Olives. Interactions between Polysaccharides and Phenolic Compounds. *Food Chem.* **2019**, *280*, 310–320. [CrossRef] [PubMed]
5. Hayashi, N.; Ujihara, T.; Kohata, K. Reduction of Catechin Astringency by the Complexation of Gallate-Type Catechins with Pectin. *Biosci. Biotechnol. Biochem.* **2005**, *69*, 1306–1310. [CrossRef]
6. Tudorache, M.; Bordenave, N. Phenolic Compounds Mediate Aggregation of Water-Soluble Polysaccharides and Change Their Rheological Properties: Effect of Different Phenolic Compounds. *Food Hydrocoll.* **2019**, *97*, 105193. [CrossRef]
7. Jakobek, L.; Matić, P. Non-Covalent Dietary Fiber—Polyphenol Interactions and Their Influence on Polyphenol Bioaccessibility. *Trends Food Sci. Technol.* **2019**, *83*, 235–247. [CrossRef]
8. Guo, Q.; Ma, Q.; Xue, Z.; Gao, X.; Chen, H. Studies on the Binding Characteristics of Three Polysaccharides with Different Molecular Weight and Flavonoids from Corn Silk (Maydis Stigma). *Carbohydr. Polym.* **2018**, *198*, 581–588. [CrossRef]
9. Zhang, D.; Zhu, J.; Ye, F.; Zhao, G. Non-Covalent Interaction between Ferulic Acid and Arabinan-Rich Pectic Polysaccharide from Rapeseed Meal. *Int. J. Biol. Macromol.* **2017**, *10*, 307–315. [CrossRef]
10. Gao, R.; Liu, H.; Peng, Z.; Wu, Z.; Wang, Y.; Zhao, G. Adsorption of (-)-Epigallocatechin-3-Gallate (EGCG) onto Oat β-Glucan. *Food Chem.* **2012**, *132*, 1936–1943. [CrossRef]
11. Yi, Y.; Huang, X.-Y.; Zhong, Z.-T.; Huang, F.; Li, S.-Y.; Wang, L.-M.; Min, T.; Wang, H.-X. Structural and Biological Properties of Polysaccharides from Lotus Root. *Int. J. Biol. Macromol.* **2019**, *130*, 454–461. [CrossRef] [PubMed]
12. Dubois, M.; Gilles, K.A.; Hamilton, J.K.; Rebers, P. t; Smith, F. Colorimetric Method for Determination of Sugars and Related Substances. *Anal. Chem.* **1956**, *28*, 350–356. [CrossRef]
13. Yi, Y.; Tang, H.-S.; Sun, Y.; Xu, W.; Min, T.; Wang, H.-X. Comprehensive Characterization of Lotus Root Polysaccharide-Phenol Complexes. *Food Chem.* **2022**, *366*, 130693. [CrossRef]
14. Zhao, K.; Xue, P.; Gu, G. Study on Determination of Reducing Sugar Content Using 3, 5-Dinitrosalicylic Acid Method. *Food Sci.* **2008**, *29*, 534–536.
15. Guo, X.-D.; Wu, C.-S.; Ma, Y.-J.; Parry, J.; Xu, Y.-Y.; Liu, H.; Wang, M. Comparison of Milling Fractions of Tartary Buckwheat for Their Phenolics and Antioxidant Properties. *Food Res. Int.* **2012**, *49*, 53–59. [CrossRef]
16. Li, S.; Li, J.; Zhu, Z.; Cheng, S.; He, J.; Lamikanra, O. Soluble Dietary Fiber and Polyphenol Complex in Lotus Root: Preparation, Interaction and Identification. *Food Chem.* **2020**, *314*, 126219. [CrossRef]
17. Dridi, W.; Bordenave, N. Influence of Polysaccharide Concentration on Polyphenol-Polysaccharide Interactions. *Carbohydr. Polym.* **2021**, *274*, 118670. [CrossRef] [PubMed]
18. Tsirigotis-Maniecka, M.; Pawlaczyk-Graja, I.; Ziewiecki, R.; Balicki, S.; Matulová, M.; Capek, P.; Czechowski, F.; Gancarz, R. The Polyphenolic-Polysaccharide Complex of *Agrimonia Eupatoria* L. as an Indirect Thrombin Inhibitor-Isolation and Chemical Characterization. *Int. J. Biol. Macromol.* **2019**, *125*, 124–132. [CrossRef]
19. Rodríguez, R.; Jaramillo, S.; Rodríguez, G.; Espejo, J.A.; Guillén, R.; Fernández-Bolaños, J.; Heredia, A.; Jiménez, A. Antioxidant Activity of Ethanolic Extracts from Several Asparagus Cultivars. *J. Agric. Food Chem.* **2005**, *53*, 5212–5217. [CrossRef]
20. Shao, P.; Chen, X.; Sun, P. Chemical Characterization, Antioxidant and Antitumor Activity of Sulfated Polysaccharide from Sargassum Horneri. *Carbohydr. Polym.* **2014**, *105*, 260–269. [CrossRef]
21. Badhani, B.; Kakkar, R. DFT Study of Structural and Electronic Properties of Gallic Acid and Its Anions in Gas Phase and in Aqueous Solution. *Struct. Chem.* **2017**, *28*, 1789–1802. [CrossRef]
22. Karimova, N.V.; Luo, M.; Sit, I.; Grassian, V.H.; Gerber, R.B. Absorption Spectra and the Electronic Structure of Gallic Acid in Water at Different PH: Experimental Data and Theoretical Cluster Models. *J. Phys. Chem. A* **2022**, *126*, 190–197. [CrossRef] [PubMed]
23. Phan, A.D.T.; D'Arcy, B.R.; Gidley, M.J. Polyphenol–Cellulose Interactions: Effects of PH, Temperature and Salt. *Int. J. Food Sci. Technol.* **2016**, *51*, 203–211. [CrossRef]
24. Liu, J.; Li, J.; Ma, Y.; Chen, F.; Zhao, G. Synthesis, Characterization, and Aqueous Self-Assembly of Octenylsuccinate Oat β-Glucan. *J. Agric. Food Chem.* **2013**, *61*, 12683–12691. [CrossRef] [PubMed]
25. Kumamoto, M.; Sonda, T.; Nagayama, K.; Tabata, M. Effects of PH and Metal Ions on Antioxidative Activities of Catechins. *Biosci. Biotechnol. Biochem.* **2001**, *65*, 126–132. [CrossRef]
26. Le Bourvellec, C.; Guyot, S.; Renard, C. Non-Covalent Interaction between Procyanidins and Apple Cell Wall Material: Part I. Effect of Some Environmental Parameters. *Biochim. Biophys. Acta-BBA-Gen. Subj.* **2004**, *1672*, 192–202. [CrossRef]
27. Zhu, F. Interactions between Cell Wall Polysaccharides and Polyphenols. *Crit. Rev. Food Sci. Nutr.* **2018**, *58*, 1808–1831. [CrossRef]
28. Wu, Y.; Lin, Q.; Chen, Z.; Xiao, H. The Interaction between Tea Polyphenols and Rice Starch during Gelatinization. *Food Sci. Technol. Int.* **2011**, *17*, 569–577. [CrossRef]

29. Le Bourvellec, C.; Renard, C. Interactions between Polyphenols and Macromolecules: Quantification Methods and Mechanisms. *Crit. Rev. Food Sci. Nutr.* **2012**, *52*, 213–248. [CrossRef]
30. Arizmendi-Cotero, D.; Villanueva-Carvajal, A.; Gómez-Espinoza, R.M.; Dublán-García, O.; Dominguez-Lopez, A. Radical Scavenging Activity of an Inulin-Gallic Acid Graft and Its Prebiotic Effect on *Lactobacillus Acidophilus* in Vitro Growth. *J. Funct. Foods* **2017**, *29*, 135–142. [CrossRef]
31. Liu, J.; Lu, J.; Kan, J.; Wen, X.; Jin, C. Synthesis, Characterization and in Vitro Anti-Diabetic Activity of Catechin Grafted Inulin. *Int. J. Biol. Macromol.* **2014**, *64*, 76–83. [CrossRef] [PubMed]
32. Wu, Z.; Li, H.; Ming, J.; Zhao, G. Optimization of Adsorption of Tea Polyphenols into Oat β-Glucan Using Response Surface Methodology. *J. Agric. Food Chem.* **2011**, *59*, 378–385. [CrossRef]
33. Zhang, M.; Huang, C.; Xie, J.; Shao, Z.; Li, X.; Bian, X.; Xue, B.; Gan, J.; Sun, T. Physical, Mechanical and Biological Properties of Phenolic Acid-Grafted Soluble Soybean Polysaccharide Films. *Foods* **2022**, *11*, 3747. [CrossRef] [PubMed]
34. Xia, J.; Wang, D.; Liang, P.; Zhang, D.; Du, X.; Ni, D.; Yu, Z. Vibrational (FT-IR, Raman) Analysis of Tea Catechins Based on Both Theoretical Calculations and Experiments. *Biophys. Chem.* **2020**, *256*, 106282. [CrossRef] [PubMed]
35. Poljansek, I.; Krajnc, M. Characterization of Phenol-Formaldehyde Prepolymer Resins by in Line FT-IR Spectroscopy. *Acta Chim. Slov.* **2005**, *52*, 238.
36. Yong, H.; Hu, H.; Wang, Z.; Yun, D.; Kan, J.; Liu, J. Structure, Stability and Antioxidant Activity of Dialdehyde Starch Grafted with Epicatechin, Epicatechin Gallate, Epigallocatechin and Epigallocatechin Gallate. *J. Sci. Food Agric.* **2022**, *102*, 6373–6386. [CrossRef]
37. Sidek, N.; Manan, N.S.; Mohamad, S. Efficient Removal of Phenolic Compounds from Model Oil Using Benzyl Imidazolium-Based Ionic Liquids. *J. Mol. Liq.* **2017**, *240*, 794–802. [CrossRef]
38. Su, Y.; Li, L. Structural Characterization and Antioxidant Activity of Polysaccharide from Four Auriculariales. *Carbohydr. Polym.* **2020**, *229*, 115407. [CrossRef]
39. Grzesik, M.; Naparlo, K.; Bartosz, G.; Sadowska-Bartosz, I. Antioxidant Properties of Catechins: Comparison with Other Antioxidants. *Food Chem.* **2018**, *241*, 480–492. [CrossRef]
40. de Morais, F.P.; Pessato, T.B.; Rodrigues, E.; Mallmann, L.P.; Mariutti, L.R.; Netto, F.M. Whey Protein and Phenolic Compound Complexation: Effects on Antioxidant Capacity before and after in Vitro Digestion. *Food Res. Int.* **2020**, *133*, 109104. [CrossRef]
41. Surin, S.; Surayot, U.; Seesuriyachan, P.; You, S.; Phimolsiripol, Y. Antioxidant and Immunomodulatory Activities of Sulphated Polysaccharides from Purple Glutinous Rice Bran (Oryza Sativa L.). *Int. J. Food Sci. Technol.* **2018**, *53*, 994–1004. [CrossRef]
42. PeÑarando, J.; Aranda, E.; RodrÍguez-Ariza, A. Immunomodulatory Roles of Nitric Oxide in Cancer: Tumor Microenvironment Says "NO" to Antitumor Immune Response. *Transl. Res.* **2019**, *210*, 99–108. [CrossRef] [PubMed]

foods

MDPI

Article

Immunomodulatory Effects of Sodium Hyaluronate Health Drink on Immunosuppressed Mice

Xiaozhen Peng [1,2], Xiaoqiao Yao [2], Ya Liu [2] and Bangzhu Peng [2,3,4,*]

1 School of Public Health & Laboratory Medicine, Hunan University of Medicine, Huaihua 418000, China; peng112112@163.com
2 College of Food Science and Technology, Huazhong Agricultural University, Wuhan 430070, China; y1592434697@163.com (X.Y.); liuyaqf@163.com (Y.L.)
3 Shenzhen Institute of Nutrition and Health, Huazhong Agricultural University, Wuhan 430070, China
4 Shenzhen Branch, Guangdong Laboratory for Lingnan Modern Agriculture, Genome Analysis Laboratory of the Ministry of Agriculture, Agricultural Genomics Institute at Shenzhen, Chinese Academy of Agricultural Sciences, Shenzhen 518120, China
* Correspondence: bzpeng@mail.hzau.edu.cn

Abstract: This study aimed to explore the immunomodulatory effects and mechanism of a sodium hyaluronate health drink in immunosuppressed mice. The results showed that the sodium hyaluronate health drink could improve thymus atrophy, repair spleen cell damage, promote the release of IL-2, IL-6 and TNF-α in serum, restore immune deficiency, and enhance immune function. In addition, 16s rRNA sequencing results of intestinal flora showed that different doses of health drink had different effects on the intestinal flora of mice. The low-dose group of mice showed a significant up-regulation of the abundance of *Lactobacillus* and promoted the formation of a new genus *Akkermansia*, while the medium- and high-dose group up-regulated the abundance of *Lactobacillus* and *norank-f-Muribaculaceae*, and stimulated the production of the new genus *Alistipes*. Sodium hyaluronate health drink may enhance the immune function of mice by changing the composition and abundance of intestinal flora, which provided a theoretical basis for the subsequent product development.

Keywords: immunomodulatory; sodium hyaluronate; health drink; intestinal flora; immune deficiency

Citation: Peng, X.; Yao, X.; Liu, Y.; Peng, B. Immunomodulatory Effects of Sodium Hyaluronate Health Drink on Immunosuppressed Mice. *Foods* **2024**, *13*, 842. https://doi.org/10.3390/foods13060842

Academic Editor: Ronaldo Vagner Thomatieli-Santos

Received: 18 January 2024
Revised: 1 March 2024
Accepted: 2 March 2024
Published: 9 March 2024

1. Introduction

The management of immunosuppression utilized in malignancies and severe manifestations of various auto-immune diseases requires highly nuanced care. The prompt identification and treatment of these disorders are crucial in reducing both mortality and morbidity rates. Cyclophosphamide is a common immunosuppressive drug used in clinical practice and trials. However, its therapeutic application is limited by its side effects. The administration of cyclophosphamide has been demonstrated to have deleterious effects on normal tissue DNA, resulting in the destruction of healthy immune cells and the inhibition of the proliferation and differentiation of macrophages, B and T lymphocytes [1]. This can ultimately lead to the suppression of both humoral and cellular immune responses, as well as immune dysfunction and intestinal injury. Specifically, cyclophosphamide has been shown to disrupt the intestinal mucosal barrier, compromise the integrity of the epithelium and neighbouring cell–cell junctions, and impair the absorption of nutrients in the intestine [2–5].

Confronted with these limitations, there is no targeted treatment option, so the intake of health foods that enhance immunity has become the best choice for these people to fight against the decline of immunity. Studies have proved that many polysaccharides have significant immunomodulatory effects, which is the main reason why many foods are currently reported to enhance immune function. Xie et al. demonstrated that *Anoectochilus formosanus* polysaccharide has the potential to mitigate cyclophosphamide-induced immunosuppression, enhance their thymus index and body weight, stimulate cytokine production, and promote intestinal morphology repair [6]. Wang et al., found that Cicadas polysaccharide extracted with ultrasound-assisted extraction could regulate host immune response [7]. Liu found that the polysaccharide of the Opuntia stem could effectively improve the metabolism level of lysine synthesis and decomposition in immunodeficient mice and improve the health level of immunodeficient mice [8].

However, polysaccharides cannot be directly digested and absorbed in the intestine, and are mostly regulated by intestinal flora. Alterations in the polysaccharide sources within the intestinal microbiota frequently play a role in immunomodulatory processes. Studies have found that longan polysaccharides can enhance host immune function under stress conditions by regulating intestinal flora and intestinal metabolites [9]. Ginger polysaccharide has been shown to enhance cytokines IL-2, IL-4, TNF-α, and immunoglobulin Ig-G secretion in the serum of mice, thereby ameliorating immune deficiency. Additionally, ginger polysaccharide has been found to promote the proliferation of beneficial bacterial taxa, such as *Muribaculaceae, Bacteroidaceae*, and *Lactobacillaceae*, while concurrently decreasing the relative abundance of detrimental bacterial populations like *Rikenellaceae* and *Lachnospiraceae* [10]. Cordyceps polysaccharide has been shown to enhance the diversity of the intestinal microbial community, modulate the overall structure of intestinal microflora, induce the secretion of cytokines and the synthesis of transcription factors, and effectively exhibit protective effects against intestinal mucosal immunosuppression and microbial ecological imbalance in mice [11]. Curcuma polysaccharide demonstrated the ability to enhance thymus and spleen indices, elevated whole blood immune cell and lymphoid count indices, and promoted the secretion of serum immunoglobulin IgG. Moreover, it exhibited potential in repairing intestinal immune damage and mitigating intestinal inflammation [12]. Obviously, bioactive polysaccharides supported the growth of beneficial bacterial populations in the gut, bolstered the host immune system, and the gut flora derived from polysaccharides interacts with immune regulation.

Sodium hyaluronate, as a highly physiologically active polysaccharide, was originally isolated from the vitreous humour of bovine eyes [13]. Its known physiological functions include antioxidant [14], cosmetic and moisturizing [15], immune and anti-inflammatory [16,17], anti-tumour [18], and weight loss [19]. Several studies have shown that sodium hyaluronate is involved in regulating the immune response of the body through various channels [20]. This provides new insights into the physiological and pathological mechanisms by which sodium hyaluronate is involved in immune and inflammatory responses. Sodium hyaluronate is often used as a multifunctional antigen carrier for the targeted delivery of nanomedicines, indirectly involved in human immunomodulation [21]. The immunomodulatory effect of sodium hyaluronate is mostly related to the production of stimulated related cytokines, and whether its intake affects the composition of the body's intestinal flora is unclear, nor is it clear whether it participates in the body's immune regulation through the intestinal flora. Therefore, it is of great significance to elucidate the immunomodulatory function of the sodium hyaluronate health drink from the perspective of intestinal flora and explore its immune mechanism.

2. Materials and Methods

2.1. Materials and Reagents

Sodium hyaluronate as a food-grade powder, with a molecular weight of 120 kDa, was purchased from Henan Sanhua Biotechnology Co., Ltd. (Ruzhou, China). Citric acid was purchased from Shandong Weifang Ensign Industrial Co., Ltd. (Weifang, China). White sugar was purchased from Shandong Yahui Sugar Industry Co., Ltd. (HeZhe, China). Vitamin C was purchased from CSPC Group Weisheng Pharmaceutical Co., Ltd. (Shijiazhuang, China). $NaHCO_3$ was purchased from Hebei Zhongnong Bailiang E-Commerce Co., Ltd. (Baoding, China). Paraformaldehyde (4%) was purchased from Wuhan Seville Biotechnology Co., Ltd. (Wuhan, China). Mianyang red blood cells (2%) were purchased from Shanghai Yudore Biotechnology Co., Ltd. (Shanghai, China). Drabkins solution was purchased from Haibiao Technology Co., Ltd. (Xiamen, China). Indian ink was purchased from Soraibao Biotechnology Co., Ltd. (Beijing, China).

2.2. Preparation of Sodium Hyaluronate Health Drink

In this experiment, a health drink based on sodium hyaluronate has been developed as a sample for in vivo experiments in mice. The health drink was prepared as follows: sodium hyaluronate powder was mixed with drinking water and stirred in a magnetic stirrer for 2 h until the sodium hyaluronate was completely dissolved to obtain sodium hyaluronate solution. The formula of the sodium hyaluronate health drink is: 0.03% sodium hyaluronate, 0.03% $NaHCO_3$, 4.5% sugar, 0.1% citric acid, 0.05% vitamin C, and 0.03% edible flavour.

2.3. Immunological Animal Experimental Design

Sixty SPF grade BALB/C male mice (20.0 ± 2.0 g) were purchased from the Animal Centre of Huazhong Agricultural University (Wuhan, China, Certificate number: SCXK (Hubei) 2020-0084). The mice were kept in a clean environment with a stable ambient temperature of 22 ± 2 °C and maintained on a 12 h day/night cycle, and immunological tests were performed after 7 d of acclimatization [6]. The whole process of mouse testing is in strict compliance with the national regulations on test animals. The concentrated health drink was produced according to the manufacturing process of this health drink, in which the sample for the low-dose group was a 2-fold concentrated sodium hyaluronate health drink (sodium hyaluronate concentration of 0.6 g/L); the sample for the medium-dose group was a 4-fold concentrated sodium hyaluronate health drink (sodium hyaluronate concentration of 1.2 g/L); and the sample for the high-dose group was an 8-fold concentrated sodium hyaluronate health drink (sodium hyaluronate concentration of 2.4 g/L). The gavage volume of each mouse was based on its daily body weight (gavage volume (mL) = body weight (g)/100). The mice were randomly and equally divided into 6 groups: normal control (NC), model control (MC), positive control (PC), low dose (LD), medium dose (MD) and high dose (HD). The low-, medium-, and high-dose groups were gavaged with the corresponding concentrations of sodium hyaluronate health drink, the positive control group was gavaged with 100 mg/kg/d of lentinan, and the rest of the groups were gavaged with normal saline. The drug was administered using gavage daily for 35 days (5 weeks). All groups except the normal control group were treated for immune depression using intraperitoneal injection of 80 mg/kg of cyclophosphamide solution according to body weight three days before the first week of gavage administration and two days before the rest of the week, while the normal control group was injected with an equal amount of normal saline. The body weights of the mice were recorded before the start of the test (day 0) and each day after gavage, respectively. The mice were fasted for 12 h after the final dose, and the eyes were removed to obtain blood, and then they were executed. Our study was approved by the Animal Ethics Committee of Huazhong Agricultural University (ID Number: HZAUMO-2021-0163), China.

2.4. Immunological Test Method

After the cervical vertebrae were dislocated and sacrificed, the thymus and spleen were removed as soon as possible, washed with normal saline, dried with filter paper, and weighed, respectively. The thymus index and spleen index of mice were calculated as follows:

$$I = W_1/W_2 \quad (1)$$

where I represents organ index; W_1 represents the mass of the organ; and W_2 is the weight of the mouse.

The spleens of mice were divided into two halves and placed in prenumbered 4% paraformaldehyde fixation solution. After standing and fixing for 24 h, sections were embedded and stained with HE, and then photographed under an inverted microscope.

The humoral immune function of the mice was demonstrated using the measurement of serum haemolysin. Four days before the end of the test, each mouse was immunised using an intraperitoneal injection of 0.2 mL of 2% pressed sheep red blood cells (SRBC) suspension. After 4 days, the eyeballs were removed for blood collection, and the serum was collected using centrifugation at 1000–2000× rpm for 10 min and stored at −80 °C. The serum was diluted 200 times with normal saline in a test tube; 1 mL of the diluted serum was aspirated into a new test tube, and then 0.5 mL of 10% (*v*/*v*) SRBC and 0.5 mL of guinea pig serum complement (diluted 1:8 in saline) were added in turn and mixed well. A control tube without serum was set up (replaced with saline). After mixing, the tubes were placed in a constant temperature water bath at 37 °C for 25 min and then removed to an ice bath for 5 min to terminate the reaction. After the ice bath, the solution in the tubes was transferred to a centrifuge tube and centrifuged at 2000 rpm for 10 min. In total, 1 mL of the supernatant was taken into the tube, and 3 mL of Drabkins solution was added and mixed, while 0.25 mL of 10% (*v*/*v*) SRBC was added to 4 mL of Drabkins solutionand mixed as half of the lysate. The tubes were mixed well and left to stand for 10 min, and the absorbance of each tube was measured at 540 nm using the control tube as a blank. The HC_{50} of the samples was calculated according to the following equation:

$$HC_{50} = As/Ah \quad (2)$$

where HC_{50} represents half hemolysis value; As represents sample tube absorbance value; and Ah represents half lysis tube absorbance value.

The function of monocyte-macrophages was determined by means of a carbon contour test in mice. Mice were injected with Indian ink diluted five times with saline in the tail vein, and the ink was injected and immediately timed. At 2 min and 10 min after ink injection, 20 mL of blood was taken from each of the internal adjacent venous plexus and immediately added to a pre-prepared tube containing 2 mL of 0.1% Na_2CO_3 solution. The absorbance was measured with a spectrophotometer at 600 nm using Na_2CO_3 solution as a blank control. The mice were executed and the liver and spleen were taken, blotted with filter paper to remove blood from the surface of the organs and weighed separately. The phagocytosis index α was calculated according to following equation:

$$K = (\lg OD_1 - \lg OD_2)/(t_2 - t_1) \quad (3)$$

$$\alpha = \text{body weight} \div (\text{liver weight} + \text{spleen weight}) \times 3\sqrt{K} \quad (4)$$

where OD_1 and OD_2 represent the absorbance at 2 min and 10 min, respectively, and t_1 and t_2 represent 2 min and 10 min, respectively.

After eyeball removal, blood was collected and left for 1 h, centrifuged at 3500 rpm for 10 min, and the upper layer of serum was divided and stored in liquid nitrogen. The levels of immune factors IL-2, IL-6, and TNF-α were measured according to the kit.

2.5. Animal Experimental Design for Intestinal Flora

The mice were randomly divided into four groups of 10 mice each, namely the normal control group (NC), low-dose group (LD), medium-dose group (MD), and high-dose group (HD). The low-, medium-, and high-dose groups were gavaged with the corresponding concentrations of sodium hyaluronate health drink, while the rest of the groups were gavaged with normal saline. The drug was administered by gavage for 28 days (4 weeks). Body weight was recorded once a week. The faeces of each mouse were taken in six parallel portions and packed in sterile tubes, snap frozen in liquid nitrogen, and then quickly transferred to −80 °C for storage. The samples of each group were selected as NC (NC1, NC2, NC4, NC5, NC7, NC8), LD (LD1, LD2, LD5, LD6, LD7, LD9), MD (MD1, MD2, MD3, MD4, MD5, MD6), HD (HD3, HD4, HD5, HD6, HD7, HD8). Following the conclusion of the experimental period, all animals underwent a 12 h fasting period prior to euthanasia via an intraperitoneal injection of ketamine hydrochloride (15 mg/kg/i.p.) and xylazine (10 mg/kg/i.p.).

2.6. 16S rRNA Sequencing Method

A total of four groups of six samples each were obtained using 16S rRNA sequencing. The 16S rRNA Illumina sequencing technology was applied to analyse the effect of this health drink on faecal flora. The DNA was first extracted from the faeces, the quality and concentration of the DNA was tested using NanoDrop2000, the integrity of the DNA was tested using a 1% agarose gel, and the DNA obtained was used as a template for PCR amplification.

Primer sequences: forward primer: 5′-ACTCCTACGGGAGGCAGCAG-3′; reverse primer: 5′-GGACTACHVGCCTWTCTAAT-3′. PCR was performed on the V3-V4 region. The amplification procedure was as follows: pre-denaturation at 95 °C for 3 min, 27 cycles (denaturation at 95 °C for 30 s, annealing at 55 °C for 30 s, extension at 72 °C for 45 s), followed by stable extension at 72 °C for 10 min, and finally storage at 10 °C (PCR instrument: ABI GeneAmp® Model 9700, (Thermo Fisher, Waltham, MA, USA). After passing the test, the library was sequenced on an Illumina Hiseq2500 platform (Illumina, San Diego, CA, USA).

2.7. Statistical Analysis

SPSS 17.0 (IBM, Armonk, NY, USA) statistical software was used for the analysis, and one-way ANOVA was chosen to compare statistical differences between groups. All data were expressed as mean ± standard deviation (SD), and $p < 0.05$ was considered as a significant difference.

3. Results

3.1. Body Weight and Organ Index of Mice

The body weights of the mice were recorded in Figure 1A, and the results showed that the body weight of each group of mice decreased significantly after intraperitoneal injection of cyclophosphamide, indicating that cyclophosphamide would cause the body weight of mice to plummet. After administration, the body weight of mice in each group increased slowly, which was not as good as that of the NC group. And, except for the HD group, the other groups did not increase their body weight before modelling, which may be related to the influence of cyclophosphamide on the growth and development of organisms (Figure 1B). In addition, there was no death or abnormal phenomenon in mice during the experiment, indicating that this health drink was harmless to the health of mice.

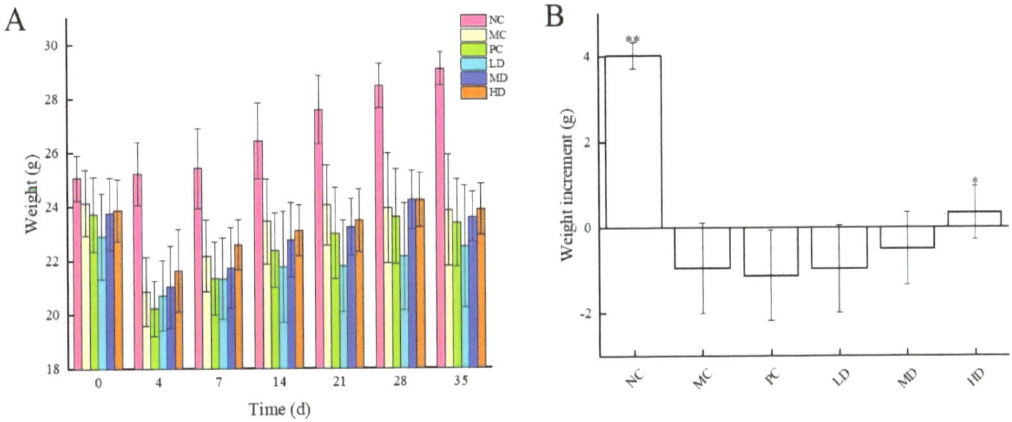

Figure 1. (**A**) Body weight in mice; (**B**) body weight gain of mice. The values are presented as mean ± SD, $n = 10$. Compared to initial weight, * indicates a significant difference ($p < 0.05$), and ** indicates a highly significant difference ($p < 0.01$).

The effect of the sodium hyaluronate health drink on the immune organs of mice is shown in Table 1. There was no significant difference in the spleen index of mice ($p > 0.05$), and the thymus index of mice moulded with cyclophosphamide was significantly lower than that of the NC group ($p < 0.05$), indicating that cyclophosphamide caused atrophy of the thymus. The thymus indexes of mice in the MD and HD groups were significantly higher ($p < 0.05$) compared to those in the MC group, indicating that thymus atrophy in mice in the MD and HD groups improved after 5 weeks of oral gavage of different doses of this health drink.

Table 1. Effect of health drink on spleen index and thymus index in mice.

Groups	Spleen Index (mg/g)	Thymus Index (mg/g)
NC	4.01 ± 0.03	2.54 ± 0.01 *
MC	3.85 ± 0.04	1.67 ± 0.02
PC	4.39 ± 0.06	1.93 ± 0.01
LD	4.08 ± 0.02	1.91 ± 0.01
MD	4.40 ± 0.04	2.07 ± 0.03 *
HD	4.25 ± 0.03	2.06 ± 0.02 *

Note: The values are presented as mean ± SD, $n = 10$. Compared with MC group, * means $p < 0.05$, showing significant difference.

3.2. Spleen Histomorphology of Mice

To further visualise the structural changes in the spleen of immunocompromised mice, the spleen was stained with HE and photographed in sections, and the results are shown in Figure 2A. The two main functional areas of the spleen were called the white and red marrow, and the white and red marrow of the spleen of the NC mice were very well defined, and the splenocytes were well distributed and closely arranged. In contrast, the spleen of the MC group had a blurred border between the red and white marrow, which was almost impossible to distinguish, and there was obvious cellular oedema, similar to ulceration, and a large number of splenocytes were oedematous and distributed in a disorganized manner. Compared with the MC group, the proportion of white marrow in the spleen of mice in the PC group was significantly increased, the cellular edema was relieved, and the arrangement of spleen cells was more compact and orderly. The health drink showed a similar trend to the PC group in all dose groups, but the proportion of white marrow was not as high as that of the PC group.

In order to further observe the distribution and arrangement of splenocytes in the spleen of mice, the morphology of splenocytes in each group of mice was further analysed under high magnification (Figure 2B). The splenocytes of the NC group were arranged in a tight and orderly manner, showing a swirling arrangement. In the MC group, the splenocytes were loosely arranged irregularly, and the number of cells decreased. The arrangements of splenocytes in the various doses of health drink groups were more orderly and compact than those of the model group, and the swirling shape was gradually obvious, approaching the splenocyte state of normal mice, but the arrangement of splenocytes in the PC group was not obvious. The number of splenocytes in the PC group and the health drink group all increased significantly compared with that in the MC group. It showed that this health drink could repair the spleen cell damage caused by cyclophosphamide.

3.3. Immunology Experiments

The haemolytic capacity is judged using colourimetric determination, and the results are usually expressed using the hemolytic value (HC_{50}), with higher HC_{50} values indicating greater haemolytic capacity and immunity. The effect of the sodium hyaluronate health drink on the serum haemolysis of mice is shown in Figure 3A. The HC_{50} of mice in the MC group was significantly lower than that of the NC group ($p < 0.01$), which showed that the moulding was successful and the immune damage caused by cyclophosphamide to the lymphocytes of mice was great. In the MD and HD groups, the serum haemolysin production capacity was significantly enhanced compared to that of the MC group ($p < 0.05$), with the MD group having the most significant effect in enhancing serum haemolysin production capacity. It could be seen that this health drink helped to improve the humoral immune function of mice.

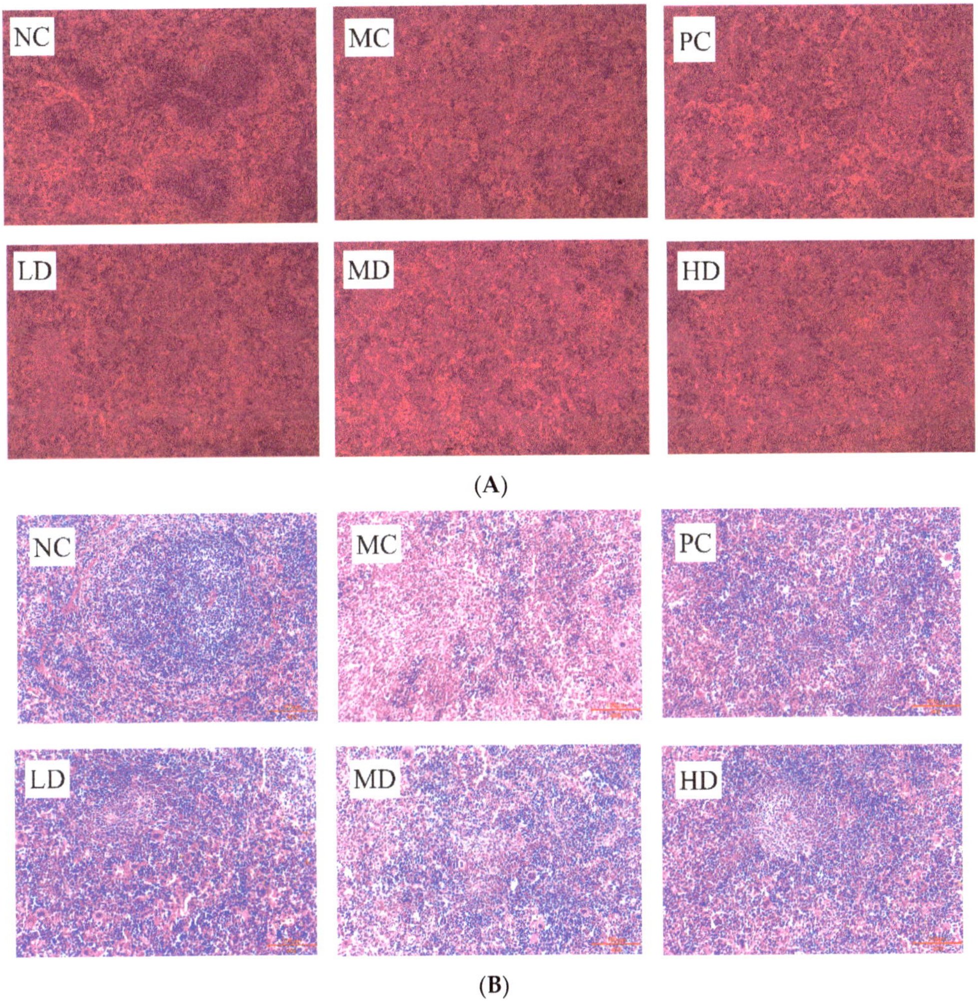

Figure 2. Section of spleen tissue (hematoxylin and eosin staining, $n = 10$): (**A**) (100×); (**B**) (200×).

A delayed allergic reaction is also known as a "delayed hypersensitivity reaction", which occurs 24 h after exposure to an antigen, hence the term "delayed hypersensitivity reaction". In this test, mice were stimulated with sheep red blood cells to induce an allergic reaction. The extent of the reaction was mainly reflected in the swelling of the foot topography, with higher swelling indicating a more severe allergic reaction and stronger immunity. The effect of this health drink on the delayed allergic reaction in mice is shown in Figure 3B. After injection of cyclophosphamide mice stimulated with sheep red blood cells, the degree of foot swelling was significantly lower than that in the NC group, which was due to the cellular immunodeficiency caused by cyclophosphamide. The foot swelling degree of mice after the gastric lavage health drink increased significantly compared with the MC group ($p < 0.01$), indicating that long-term intake of this health drink could repair the immune deficiency caused by cyclophosphamide, among which the tissue swelling level of mice in the LD group was higher than that in the NC group, which showed that oral administration of this health drink not only had the effect of repairing low immunity but had the function of enhancing immunity.

The phagocytosis speed of the phagocytes is directly proportional to the concentration of carbon particles. Considering the different sizes of mouse organs, the phagocytosis index is usually used to express the phagocytic activity per unit of tissue weight, and the level of phagocytosis index reflects the strength of the phagocytic ability of macrophages in mouse organs. The effect of the sodium hyaluronate health drink on the phagocytic index of macrophages in mice is shown in Figure 3C. The phagocytic index of mice in the MC group decreased compared with that in the NC group ($p > 0.05$), which showed that cyclophosphamide had an impairing effect on the phagocytic ability of macrophages in mice. The macrophage phagocytosis of immunocompromised mice was restored after oral administration of this health drink, but there was no significant difference between the dose groups and the MC group ($p > 0.05$), indicating that this health drink did not have a significant effect on enhancing the macrophage phagocytosis of immunocompromised mice.

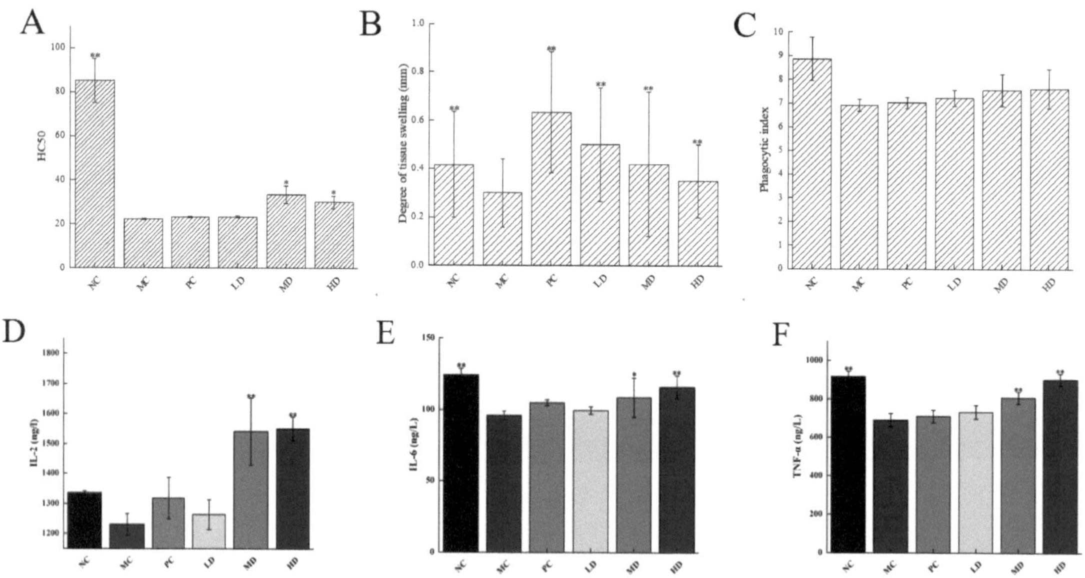

Figure 3. (**A**) Ability to produce serum hemolysin; (**B**) delayed-type hypersensitivity; (**C**) carbon clearance phagocytic index; serum levels of immune factor (**D**) IL-2, (**E**) IL-6, (**F**) TNF-α in mice. The values are presented as mean ± SD, $n = 10$. Compared to NC group, * indicates a significant difference ($p < 0.05$), and ** indicates a highly significant difference ($p < 0.01$).

3.4. Serum Levels of Immune Factors

IL-2, IL-6, and TNF-α are all extremely important immune factors in immune regulation and are involved in the body's immune regulation in a variety of ways [19]. The level of TNF-α immune factor is closely related to tumours, diabetes, and especially autoimmune diseases [22–24]. As can be seen from Figure 3D–F, the content of IL-2 in the serum of mice modelled with cyclophosphamide was significantly lower than that of mice in the NC group ($p > 0.05$); the serum contents of IL-6 and TNF-α in the MC group were significantly lower than those in the NC group ($p < 0.01$). The contents of IL-2, IL-6, and TNF-α in the serum of mice in the MD group and HD group of the sodium hyaluronate health drink after 5 weeks of gavage were significantly enhanced ($p < 0.05$ and $p < 0.01$). It might be seen that the sodium hyaluronate health drink could stimulate the release of immune factors in mice, and among which the IL-6 and TNF-α content in immunocompromised mice gradually returned to normal levels after gavage of the sodium hyaluronate health drink, which indicated that long-term gavage of this sodium hyaluronate health drink could effectively restore the immunodeficiency caused by cyclophosphamide.

3.5. Dilution Curve and Alpha and Beta Diversity

The Rarefaction curve reflects the microbial diversity of each sample at different sequencing volumes. The trend of the curve can reflect the homogeneity, richness, or diversity of a species in a sample, as well as the reasonableness of the amount of sequencing data. The horizontal coordinate of the curve is generally the amount of data randomly selected, while the vertical coordinate corresponds to the Alpha diversity index. The curve flattens out as the sample size increases, indicating that the sample is homogeneous and the amount of data is sufficient. The dilution curve of this experiment was plotted with Sobs index as the vertical coordinate and the amount of randomly selected data as the horizontal coordinate, as shown in Figure 4. The level of Sobs index in the faeces of each group of mice increased rapidly and then levelled off as the number of randomly selected samples increased. The Shannon curve was also plotted with the Shannon index as the vertical coordinate; the Shannon index levelled off and then did not increase significantly as the number of randomly selected samples increased, which indicated that the amount of experimental data was reasonable and large enough to be of analytical significance.

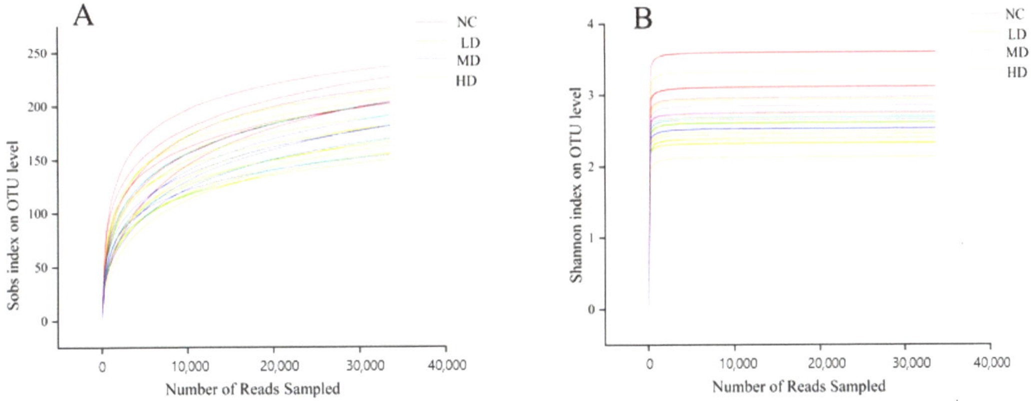

Figure 4. (**A**) Sobs index dilution curve; (**B**) Shannon index dilution curve. $n = 6$.

Alpha diversity analysis reflects the overall richness and diversity of the microbial community, with Chao, Ace, Shannon, and Simpson as the main measures. Table 2 showed the variation of Alpha diversity in the intestinal flora of mice, where the Shannon and Simpson indices were used to reflect the diversity of the community in the sample, while the Chao and Ace indices were used to reflect the abundance of the community. As shown in Table 2, the abundance of intestinal flora in mice after gavage of this health drink was smaller than that of the NC group, but the diversity of flora increased. The Ace and Chao indices in the LD group were significantly lower than those in the NC group ($p < 0.01$ and $p < 0.05$), the Ace and Chao indices in the MD group were not significantly different from those in the NC group ($p > 0.05$), and the Ace and Chao indices in the HD group were significantly lower than those in the NC group ($p < 0.01$). This reflected that gavage of both low and high doses of the health drink resulted in a reduction in the abundance of mouse flora. In contrast, the Simpson index of mice in the LD group was significantly increased ($p < 0.01$) compared with the NC group, the Shannon index and Simpson index of mice in the MD group were significantly decreased ($p < 0.01$) and significantly increased ($p < 0.05$), respectively, and the Shannon index and Simpson index of mice in the HD group were very significantly different from those in the NC group ($p < 0.01$), which showed a decrease in the Shannon index and an increase in the Simpson index, of which the Shannon index decreased less than the increase in Simpson, which indicated that the colony species diversity of mice increased after gastric lavage. Richness and diversity are often regarded as important indicators of the homeostasis of intestinal flora, but not desiring higher richness and diversity, as the more ideal situation is a dynamically balanced homeostatic environment. The coverage of each group in Table 2 was higher than 0.999, indicating that all species in the sample were detected, further verifying that the number for sequencing was sufficient. The difference between groups was tested using a statistical *t* test, as shown in Figure 5A, and there was a significant difference between the index of each dose group and the normal group ($p < 0.05$).

Table 2. Alpha diversity index analysis table.

Sample	Richness		Diversity		Coverage
	Ace	Chao	Shannon	Simpson	
NC	252.59 ± 17.55	255.59 ± 23.59	3.19 ± 0.31	0.09 ± 0.02	0.9992
LD	213.25 ± 20.09 **	217.33 ± 24.75 *	2.52 ± 0.13 **	0.21 ± 0.05 **	0.9993
MD	227.87 ± 25.23	230.46 ± 30.10	2.67 ± 0.11 **	0.14 ± 0.02 *	0.9990
HD	207.86 ± 18.13 **	206.81 ± 17.24 **	2.60 ± 0.41 **	0.16 ± 0.05 **	0.9992

Note: The values are presented as mean ± SD, $n = 6$. Compared to NC group, * indicates a significant difference ($p < 0.05$), and ** indicates a highly significant difference ($p < 0.01$).

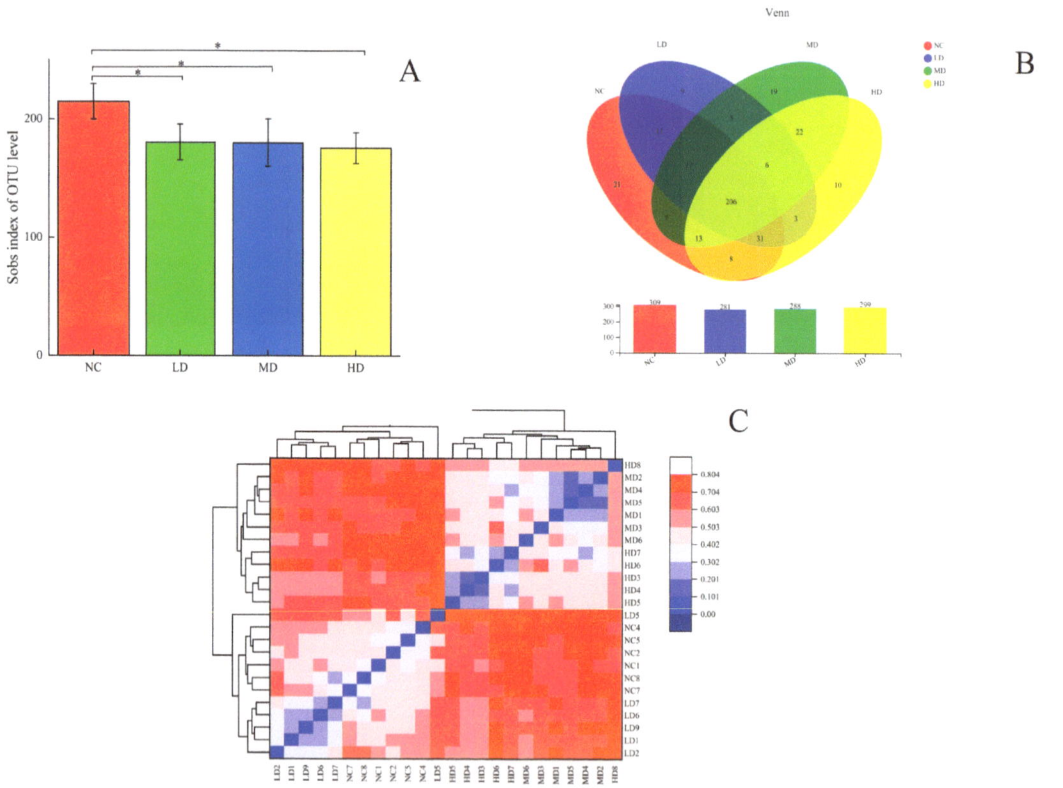

Figure 5. (**A**) Plot of difference between groups at OTU level; (**B**) Venn graph; (**C**) Beta cluster analysis diagram. The values are presented as mean ± SD, $n = 6$. Compared to NC group, * indicates a significant difference ($p < 0.05$). One-way ANOVA was chosen to compare statistical differences between groups.

Similarly, a Beta cluster analysis was performed on the samples to obtain Figure 5C, which showed that the samples in each dose group were significantly different from the NC group, with the samples in the MD group being more similar to those in the HD group.

3.6. Species Composition Analysis

The Venn plot mainly reflects the number of common and unique species (such as OTU) in multiple groups of samples, and it mainly reflects the similarity of species composition (such as OTU) in different environmental samples through overlap. The number of common and unique OTUS between the normal group and each dose group of health drinks was analysed according to the study requirements, and the Venn chart was plotted. The numbers in the overlapping part of the graph represent the number of species common to the overlapping groups and the numbers in the non-overlapping part represent the number of species unique to the corresponding groups. The Venn diagrams for the four groups of samples are shown in Figure 5B, where the number of OTUS common to each group was 206, the number of OUT unique to the NC group was 21, the number of OUT unique to the LD group was 9, the number of OUT unique to the MD group was 19, and the number of OUT unique to the HD group was 10. It could be concluded that some new microbial species would be produced in the intestinal tract of mice after intaking the sodium hyaluronate health drink, and the richness of new bacterial genera was very important for the balance of intestinal flora.

To further investigate the microbial community composition and species richness of the four groups of samples, the species richness of each group of samples was calculated at the phylum, family, and genus levels and plotted as a histogram, as shown in Figure 6. From Figure 6A, it can be seen that at the phylum classification level, the intestinal flora of mice consisted mainly of *Bacteroidetes* (Phylum *Bacteroidetes*) and *Firmicutes* (Phylum Thick-walled Bacteria), which accounted for more than 90% of the total flora. The LD group of health drinks had an increase in *Firmicutes* compared with the NC group, and the number of *Bacteroidetes* decreased compared with the NC group. The abundance of *Firmicutes* in the MD group decreased compared with that in the NC group, corresponding to an increase in *Bacteroidetes*. The HD group showed a similar trend to the MD group. Studies have shown that there is a correlation between the number of *Firmicutes* in the gut and the body's obesity status, with an increase in the body's diet and an increase in body fat content causing a subsequent increase in the number of *Firmicutes*, which is why *Firmicutes* are called "Fat Fungus" [22]. Ley first showed that obesity was associated with intestinal flora and that obese mice had more *Firmicutes* than lean mice [23]. At the same time, Million disputed this view, arguing that it was mostly gut microbes at the species level that correlate more strongly with obesity status in humans [24]. There are several hypotheses as to the specific mechanisms by which gut flora contribute to obesity, but there are no definitive findings. In short, changes in gut flora are involved in the development of obesity. The differences between the remaining phylum groups were small and not discussed.

At the family level, the abundance of *muribaculaceae* in the LD group was lower than that in the NC group, while the abundance of *muribaculaceae* in the MD and HD groups of the health drink was higher than that in the NC group. The abundance of *Lactobacillaceae* in each dose group was increased compared with that in the NC group, and the increase was more obvious in the LD group. The abundance of *Lachnospiraceae* and *Bacteroidaceae* in the MD group was lower than that in the NC group.

At the genus level, compared with the NC group, the abundance of *norank-f-Muribaculaceae* and *Bacteroides* was down-regulated in the LD group, while the abundance of *Lactobacillus* was significantly up-regulated. The MD group up-regulated the abundance of *norank-f-Muribaculaceae* and *Lactobacillus*, and down-regulated the abundance of *Bacteroides*. The HD group up-regulated the abundances of *norank-f-Muribaculaceae* and *Lactobacillus*, and down-regulated the abundances of *Bacteroides* and *Staphylococcus*.

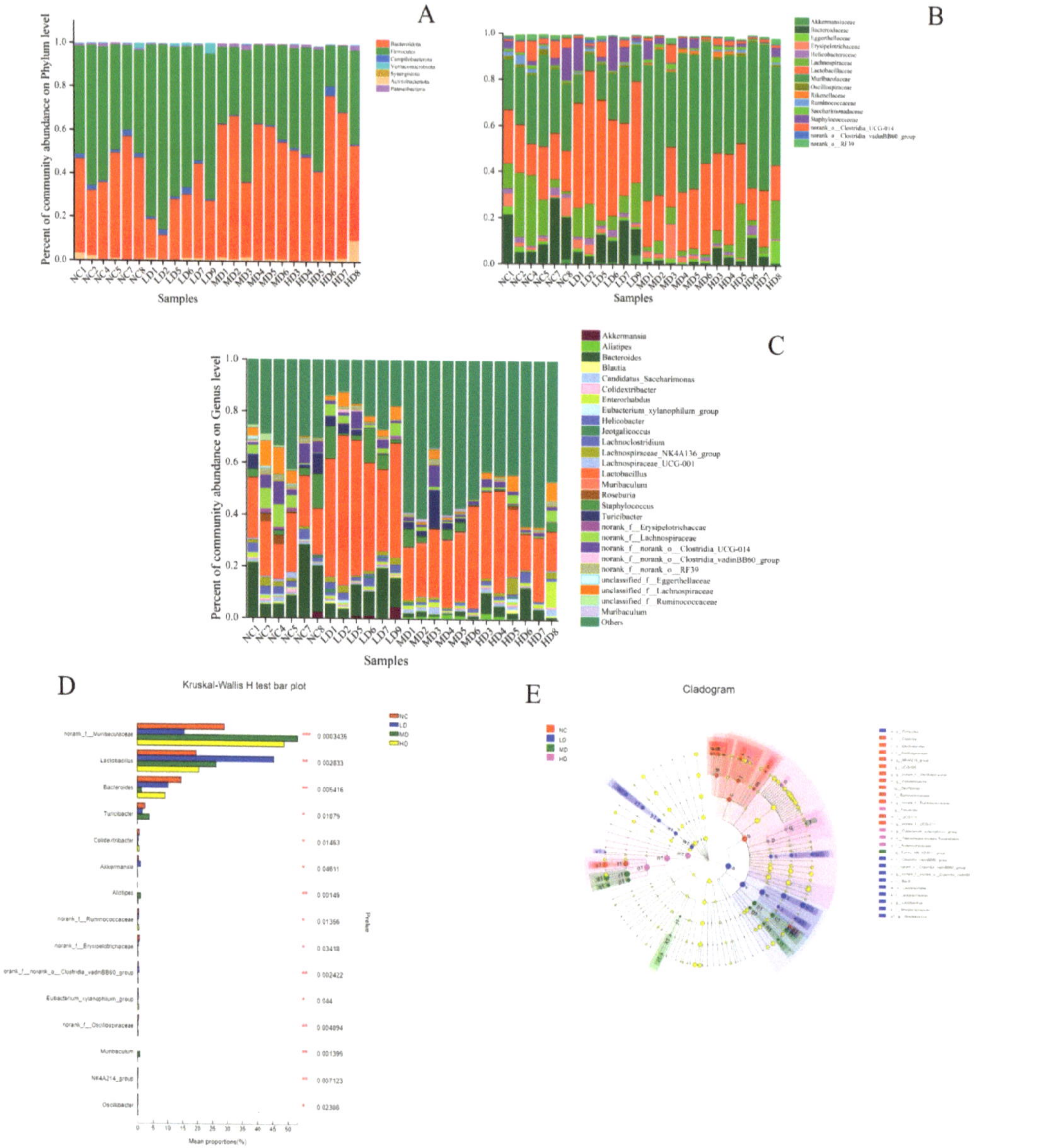

Figure 6. The relative abundances of the major bacteria at (**A**) phylum level, (**B**) family level, (**C**) genus level based 16S rRNA sequences. (**D**) Multispecies difference test bar chart; (**E**) LDA discriminant tree. $n = 6$. One-way ANOVA was chosen to compare statistical differences between groups. Compared to NC group, * indicates a significant difference ($p < 0.05$), and ** indicates a highly significant difference ($p < 0.01$), *** indicates a highly significant difference ($p < 0.001$).

3.7. Species Variation Analysis

The multi-species differences of intestinal microbiota in each group of mice are shown in Figure 6D. As shown in Figure 6D, the abundance of the genus *norank-f-Muribaculaceae* was highly significantly different between the groups ($p < 0.01$). Compared to the NC group, the abundance of this genus was significantly up-regulated in the MD group and HD group, while it was decreased in the LD group. The abundance of *Lactobacillus* in the LD group was significantly up-regulated, and that in the MD and HD group was less up-regulated. The abundance of *Bacteroides* decreased slightly in the LD group compared to the NC group, decreased significantly in the HD group, and decreased very significantly in the MD group; the abundance of *Turicibacter* (involved in fermentation metabolism, and its main metabolites have muscle regulating and anti-fatigue effects) showed significant upward adjustment in the MD group. *Alistipes* is a butyric acid-producing, specialized anaerobic bacterium. The Butler study suggested that its abundance may be inversely associated with obesity [25], and its abundance was significantly up-regulated in the MD group, almost exclusively in mice in the MD and HD groups. The abundance of *Akkermansia* was significantly up-regulated in the LD group, and as a mucin-degradable bacterium, it was inversely associated with obesity, diabetes, inflammation, and metabolic disorders [26], and the up-regulation of the abundance of the genus was beneficial to health. The abundance of the remaining genera is small and will not be discussed here.

LEfSe analysis reflects genetic, metabolic, and taxonomic profiles and is often used to distinguish between multiple groups of organisms. The results of the LDA discriminant plots of the intestinal flora structure of different groups of mice are shown in Figure 6E. It could be seen that the bacteria in the NC group that had a greater influence on the change of flora were *Clostridia, Bacteroidaceae,* and *Bacteroides*. In the LD group, the most influential groups were *Firmicutes, Bacilli, Lactobacillales,* and their species. In the MD group, the groups that had a greater impact on the change in flora were *Muribaculaceae, Bacteroidales, Bacteroidota,* and *Citrobacter*. In the HD group, *Paludicola* had a greater effect on the change in flora. It might be seen that the oral administration of different doses of this health drink could affect the structural changes of the organism's flora, and affect different species of intestinal flora.

4. Discussion

In this study, we found that sodium hyaluronate supplements had a modulating effect on immune function in immunocompromised mice. In addition, we found that sodium hyaluronate drinks had the effect of regulating the intestinal flora of mice, and the effects of different doses on the flora of mice were different.

In the past few decades, a large number of biologically active polysaccharides have been discovered and put into practical use, and immunomodulatory activity is the physiological activity of most polysaccharides. The difference is that different polysaccharides have different structures and are absorbed by the body in different forms, and their mechanism of action in humans are different [27]. Therefore, a large number of scholars have investigated the immunomodulatory mechanism of polysaccharides through animal experiments. The spleen index and thymus index are indicative of the development of immune organs and are commonly utilized as key parameters for assessing overall immune system functionality [28]. In our study, we found that sodium hyaluronate health drinks have immune-enhancing effects on the body, mainly in terms of repairing apparent damage to organs (Table 1, Figure 2). The immune system is a defence system composed of humoral immunity, cell immunity, and non-specific immunity, etc. [29]. The findings of the hemolysin test indicated a significant increase in serum hemolysin production in the MD and HD groups (Figure 3A). Moreover, the impact of the sodium hyaluronate health drinks on the delayed allergic response in mice was demonstrated in Figure 3B. In addition, the phagocytic index of mice in the MC group decreased compared with that in the NC group, which indicated that the health drinks could enhance the phagocytosis of macrophages (Figure 3C). These results indicated that the oral administration of this health drink not only demonstrated efficacy in restoring low immunity levels but also exhibited the capacity to enhance overall immune function. The immune system produces and secretes cytokines that play a role in regulating the immune system to maintain a healthy immune system [10]. To further explore sodium hyaluronate health drinks' immunoregulatory mechanism, the findings indicated that the serum levels of IL-2, IL-6, and TNF-α in mice treated with sodium hyaluronate beverage exhibited alterations compared to those in the control group (Figure 3D–F), suggesting an enhancement in immune response. These results were consistent with the immune-enhancing effects of many polysaccharides.

Numerous studies have shown that most large polysaccharides are not absorbed in the stomach and small intestine; they are often partially or completely hydrolysed and absorbed by the body through the combined action of intestinal microorganisms [30]. It is commonly believed that polysaccharides could be converted into short-chain fatty acids and carbon monoxide via the intestinal flora, which are further involved in the body's energy metabolism and immune response [31], and the absorption mechanism of sodium hyaluronate, as a mucopolysaccharide, in the human body has long been a hot issue for research.

Indeed, the small and large intestines of healthy humans are home to trillions of bacteria, viruses, archaea, and fungi [32]. It is widely accepted that the foetus remains largely sterile during development, the only sterile stage of human growth and development [33]. *Bifidobacterium* and *Lactobacillus* are the main dominant microorganisms colonising the gut in the short term after birth, and these lactic acid metabolising bacteria are mainly of maternal origin, gradually increasing in species with age [34], gradually building on the initial microbial base to form a large gut microbial community [35]. The microbiota significantly contribute to the maintenance of immune homeostasis [10]. This is mainly due to the fact that the gut has a specific immune system and has multiple layers of protection to limit the ability of resident microbes to enter the systemic circulation and thus the systemic immune system [36]. Mariko's study found that after a single oral hyaluronic acid treatment for 8 h in 7–8 week old male SD rats, hyaluronic acid was present in large amounts in the intestinal contents of the rats, implying that sodium hyaluronate is potentially involved in the immunomodulatory processes of the body through the intestinal flora [37]. In this study, the intestinal microbiome of mice was profoundly affected by sodium hyaluronate health drinks. Administration of both low and high doses of the health drink led to a decrease in the abundance of mouse microbiota. Conversely, the diversity of colony species in mice increased following gastric lavage (Table 2). The low-dose group of mice exhibited a noteworthy increase in the prevalence of *Lactobacillus* and facilitated the emergence of a novel taxonomic group, *Akkermansia*, whereas the medium- and high-dose groups demonstrated an increase in the prevalence of *Lactobacillus* and *norank-f-Muribaculaceae*, and induced the production of the new taxonomic group *Alistipes* (Figure 6). A large amount of the literature has documented a positive correlation between the relative abundance of *Lactobacillus* and *Akkermansia* and immunity [10]. *Lactobacillus*, as the most common probiotic, has been used in various beverages for a long time to regulate the balance of the body's intestinal flora, and the proliferation of *Lactobacillus* can promote intestinal absorption, protect the gastric mucosa, and play a positive regulatory role in maintaining human intestinal immunity. In the case of *Alistipes*, a new genus of *Alistipes* emerged in the MD and HD group after long-term gavage of this health drink (Figure 6D), which is considered to be rarely involved in human diseases [38] and may be negatively associated with human obesity [39], mainly due to the positive correlation between *Alistipes* and serum ANGPTL4 and adropin. Serum ANGPTL4, a secreted glycoprotein, has been shown to inhibit lipoprotein lipase activity, while the adropin protein has been shown to play an important role in regulating energy metabolism and maintaining insulin sensitivity, and has been found to be significantly increased in high-fat-fed mice [25]. The rise in both protein levels is closely related to the metabolism of fat, which explains the possible mechanism of the weight loss effect of sodium hyaluronate on the body from the perspective of microorganisms, and the excessive accumulation of fat is related to cardiovascular and cerebrovascular diseases, which in turn affects the human immune system, but how it is possible to affect the immune system from the perspective of protein is not clear. As shown in Figure 6D, a new genus of *Akkermansia* appeared in the mice of the LD group. By colonizing with *Akkermansia muciniphila* in germ-free mice, Derrien confirmed that the genus increased the expression of genes involved in the immune response and determining cell fate, and found that it regulated pathways involved in immune tolerance in the symbiotic microbiota, and participated in the establishment of a homeostatic environment for basal metabolism [40]. In short, the intricate relationship between the sodium hyaluronate health drink, bacterial species, microbial metabolic pathways, and immune indices sheds light on the potential mechanism of host immune response to polysaccharides at the level of intestinal microbiota.

5. Conclusions

In this study, we investigated the immunomodulatory activity of a sodium hyaluronate health drink in immunocompromised mice through immunological tests, and elaborated the effect of the health drink on intestinal microflora of mice from the perspective of intestinal microflora. It was found that long-term intake of the sodium hyaluronate health drink could enhance the immune function of mice. Compared with the model group, the phagocytosis ability of spleen cells of mice was enhanced after drinking the health drink, and the serum hemolysin production ability of mice in the MD and HD groups was significantly higher than that in the model group, increasing by about 10%. In the sequencing of intestinal flora, it was found that the abundance of *Muribaculaceae* and *Lactobacillaceae* in each dose group was significantly increased, and the proportion of *Lactobacillaceae* in intestinal flora was also significantly increased, which confirmed that gavage of the sodium hyaluronate health drink could promote the growth of intestinal probiotics in mice. The intake of different doses of the sodium hyaluronate health drink could stimulate the production of different probiotics in the intestinal tract of mice, thus regulating intestinal balance and participating in the immune regulation of mice.

Author Contributions: X.P. and X.Y. designed and conducted the study; X.Y. and Y.L. analysed the data and wrote the manuscript. B.P. modified the experimental protocol. All authors have read and agreed to the published version of the manuscript.

Funding: This research was funded by HZAU-AGIS Cooperation Fund [SZJY2022021], the Central Government Guides Local Science and Technology Development Fund Project [2023ZYQ138], and the Scientific Research Fund of Hunan Provincial Education Department [21A0614]. And the APC was funded by the Technology Innovation Project of Huaihua [2021R3102].

Institutional Review Board Statement: The animal study protocol was approved by the Institutional Review Board (or Ethics Committee) of Huazhong Agricultural University (ID Number: HZAUMO-2021-0163, approved on 12 September 2021).

Informed Consent Statement: Not applicable.

Data Availability Statement: The original contributions presented in the study are included in the article, further inquiries can be directed to the corresponding author.

Conflicts of Interest: The authors declare no conflicts of interest.

Abbreviations

SPF = service port function; NC = normal control; MC = model control; PC = positive control; LD = low dose; MD = medium dose; HD = high dose; SRBC = sheep red blood cells; SD = standard deviation; OTU = Optical Transform Unit; ANGPTL4 = angiopoietin-like 4.

References

1. Gong, Y.; Wu, J.; Li, S.T. Immuno-enhancement effects of *Lycium ruthenicum* Murr. polysaccharide on cyclophosphamide-induced immunosuppression in mice. *Int. J. Clin. Exp. Med.* **2015**, *8*, 20631–20637. [PubMed]
2. Chen, X.; Nie, W.; Fan, S. A polysaccharide from *Sargassum fusiforme* protects against immunosuppression in cyclophosphamide-treated mice. *Carbohydr. Polym. Sci. Technol. Asp. Ind. Important Polysacch.* **2012**, *90*, 1114–1119. [CrossRef] [PubMed]
3. Li, C.; Duan, S.; Li, Y.; Pan, X.; Han, L. Polysaccharides in natural products that repair the damage to intestinal mucosa caused by cyclophosphamide and their mechanisms: A review. *Carbohydr. Polym.* **2021**, *261*, 117876. [CrossRef] [PubMed]
4. Zhao, Y.; Yan, Y.; Zhou, W.; Chen, D.; Cao, Y. Effects of polysaccharides from bee collected pollen of Chinese wolfberry on immune response and gut microbiota composition in cyclophosphamide-treated mice. *J. Funct. Foods* **2020**, *72*, 104057. [CrossRef]
5. Bai, Y.; Huang, F.; Zhang, R.; Dong, L.; Zhang, M. Longan pulp polysaccharides relieve intestinal injury in vivo and in vitro by promoting tight junction expression. *Carbohydr. Polym.* **2019**, *229*, 115475. [CrossRef]

6. Xie, A.; Wan, H.; Feng, L.; Yang, B.; Wan, Y. Protective Effect of *Anoectochilus formosanus* Polysaccharide against Cyclophosphamide-Induced Immunosuppression in BALB/c Mice. *Foods* **2023**, *12*, 1910. [CrossRef]
7. Wang, W.; Xu, C.; Liu, Z.; Gu, L.; Ma, J.; Hou, J.; Jiang, Z. Physicochemical properties and bioactivity of polysaccharides from *Isaria cicadae* Miquel with different extraction processes: Effects on gut microbiota and immune response in mice. *Food Funct.* **2022**, *13*, 9268–9284. [CrossRef]
8. Liu, Z.; Zhang, J.; Zhao, Q.; Wen, A.; Li, L.; Zhang, Y. The regulating effect of Tibet *Opuntia ficusindica* (Linn.) Mill. polysaccharides on the intestinal flora of cyclophosphamide-induced immunocompromised mice. *Int. J. Biol. Macromol.* **2022**, *207*, 570–579. [CrossRef]
9. Zhang, J.; Yang, G.; Wen, Y.; Liu, S.; Li, W. Intestinal microbiota are involved in the immunomodulatory activities of longan polysaccharide. *Mol. Nutr. Food Res.* **2017**, *61*, 1700466. [CrossRef]
10. Liu, J.P.; Wang, J.; Zhou, S.X.; Huang, D.C.; Qi, G.H.; Chen, G.T. Ginger polysaccharides enhance intestinal immunity by modulating gut microbiota in cyclophosphamide-induced immunosuppressed mice. *Int. J. Biol. Macromol.* **2022**, *223*, 1308–1319. [CrossRef]
11. Ying, M.; Yu, Q.; Zheng, B.; Wang, H.; Wang, J.; Chen, S.; Nie, S.; Xie, M. Cultured Cordyceps sinensis polysaccharides modulate intestinal mucosal immunity and gut microbiota in cyclophosphamide-treated mice. *Carbohydr. Polym.* **2020**, *235*, 115957. [CrossRef]
12. Zhu, Z.; Luo, Y.; Lin, L.; Gao, T.; Yang, Q.; Fan, Y.; Wang, S.; Fu, C.; Liao, W. Modulating Effects of Turmeric Polysaccharides on Immune Response and Gut Microbiota in Cyclophosphamide-Treated Mice. *J. Agric. Food Chem.* **2024**, *72*, 3469–3482. [CrossRef]
13. Meyer, K.; Palmer, J.W. The Polysaccharide of the Vitreous Humor. *J. Biol. Chem.* **1934**, *107*, 629–634. [CrossRef]
14. Tokita, Y.; Sakashita, H.; Okamoto, A.; Kubota, K. Kinetic study of a radical scavenging effect of hyaluronic acid. *Polym. Int.* **2010**, *38*, 161–164. [CrossRef]
15. Sara, S.; Maryam, B. Daily consumption of the collagen supplement Pure Gold Collagen reduces visible signs of aging. *Clin. Interv. Aging* **2014**, *9*, 1747–1758.
16. Brandt, K.D.; Block, J.A.; Michalski, J.P.; Moreland, L.W.; Caldwell, J.R.; Lavin, P.T. Efficacy and safety of intraarticular sodium hyaluronate in knee osteoarthritis. ORTHOVISC Study Group. *Clin. Orthop. Relat. Res.* **2001**, *385*, 130. [CrossRef]
17. Cole, B.J.; Karas, V.; Hussey, K.; Pilz, K.; Fortier, L.A. Hyaluronic Acid Versus Platelet-Rich Plasma: A Prospective, Double-Blind Randomized Controlled Trial Comparing Clinical Outcomes and Effects on Intra-articular Biology for the Treatment of Knee Osteoarthritis. *Am. J. Sports Med.* **2016**, *45*, 339–346. [CrossRef]
18. Huang, G.; Huang, H. Application of hyaluronic acid as carriers in drug delivery. *Drug Deliv.* **2018**, *25*, 766–772. [CrossRef]
19. Park, B.G.; Park, Y.S.; Park, J.W.; Shin, E.; Shin, W.S. Anti-obesity potential of enzymatic fragments of hyaluronan on high-fat diet-induced obesity in C57BL/6 mice. *Biochem. Biophys. Res. Commun.* **2016**, *473*, 290–295. [CrossRef]
20. Petricevic, B.; Wessner, B.; Sachet, M.; Vrbanec, D.; Bergmann, M. CL097, a TLR7/8 ligand, inhibits TLR-4—Dependent activation of IRAK-M and BCL-3 expression. *Shock* **2009**, *32*, 484–490. [CrossRef]
21. Miyazaki, M.; Yuba, E.; Hayashi, H.; Harada, A.; Kono, K. Development of pH-Responsive Hyaluronic Acid-Based Antigen Carriers for Induction of Antigen-Specific Cellular Immune Responses. *ACS Biomater. Sci. Eng.* **2019**, *5*, 5790–5797. [CrossRef] [PubMed]
22. Chadwick, W.; Magnus, T.; Martin, B.; Keselman, A.; Mattson, M.P.; Maudsley, S. Targeting TNF-α receptors for neurotherapeutics. *Trends Neurosci.* **2008**, *31*, 504–511. [CrossRef] [PubMed]
23. Charles, P.; Elliott, M.J.; Davis, D.; Potter, A.; Kalden, J.R.; Antoni, C.; Breedveld, F.C.; Smolen, J.S.; Eberl, G.; de Woody, K.; et al. Regulation of Cytokines, Cytokine Inhibitors, and Acute-Phase Proteins Following Anti-TNF-α Therapy in Rheumatoid Arthritis. *J. Immunol.* **1999**, *163*, 1521–1528. [CrossRef] [PubMed]
24. Ritchlin, C.T.; Haas-Smith, S.A.; Li, P.; Hicks, D.G.; Schwarz, E.M. Mechanisms of TNF-alpha- and RANKL-mediated osteoclastogenesis and bone resorption in psoriatic arthritis. *J. Clin. Investig.* **2003**, *111*, 821–831. [CrossRef] [PubMed]
25. Butler, A.A.; Tam, C.S.; Stanhope, K.L.; Wolfe, B.M.; Ali, M.R.; Majella, O.K.; O'Keeffe, M.; St-Onge, M.-P.; Ravussin, E.; Havel, P.J. Low Circulating Adropin Concentrations with Obesity and Aging Correlate with Risk Factors for Metabolic Disease and Increase after Gastric Bypass Surgery in Humans. *J. Clin. Endocrinol. Metab.* **2012**, *97*, 3783–3791. [CrossRef]
26. Derrien, M.; Vaughan, E.E.; Plugge, C.M.; de Vos, W.M. *Akkermansia muciniphila* gen. Nov., sp. Nov., a human intestinal mucin-degrading bacterium. *Int. J. Syst. Evol. Microbiol.* **2004**, *54*, 1469–1476. [CrossRef]
27. Zi-Feng, S.U.; Zhi-Ming, D.; Jian-Fa, Y. Research Progress on Immune Mechanism of Polysaccharide. *J. Yunnan Agric. Univ.* **2006**, *21*, 205–209.
28. Wang, J.; Tong, X.; Li, P.; Cao, H.; Su, W. Immuno-enhancement effects of Shenqi Fuzheng Injection on cyclophosphamide-induced immunosuppression in Balb/c mice. *J. Ethnopharmacol.* **2012**, *139*, 788–795.
29. Zhao, Y.L.; Wang, J.B.; Shan, L.M.; Jin, C.; Ma, L.; Xiao, X.H. Effect of Radix isatidis polysaccharides on immunological function and expression of immune related cytokines in mice. *Chin. J. Integr. Med.* **2008**, *14*, 207–211. [CrossRef]
30. Yan, H.; Lu, J.; Wang, Y.; Gu, W.; Yang, X.; Yu, J. Intake of total saponins and polysaccharides from *Polygonatum kingianum* affects the gut microbiota in diabetic rats. *Phytomedicine* **2017**, *26*, 45. [CrossRef]
31. Fang, Q.; Hu, J.; Nie, Q.; Nie, S. Effects of polysaccharides on glycometabolism based on gut microbiota alteration. *Trends Food Sci. Technol.* **2019**, *92*, 65–70. [CrossRef]

32. Eckburg, P.B.; Bik, E.M.; Bernstein, C.N.; Purdom, E.; Dethlefsen, L.; Sargent, M.; Gill, S.R.; Nelson, K.E.; Relman, D.A. Diversity of the Human Intestinal Microbial Flora. *Science* **2005**, *308*, 1635. [CrossRef] [PubMed]
33. Jiménez, E.; Marín, M.L.; Martín, R.; Odriozola, J.M.; Olivares, M.; Xaus, J.; Fernández, L.; Rodríguez, J.M. Is meconium from healthy newborns actually sterile? *Res. Microbiol.* **2008**, *159*, 187–193. [CrossRef] [PubMed]
34. Solís, G.; Reyes-Gavilan CG, D.L.; Fernández, N.; Margolles, A.; Gueimonde, M. Establishment and development of lactic acid bacteria and bifidobacteria microbiota in breast-milk and the infant gut. *Anaerobe* **2010**, *16*, 307–310. [CrossRef] [PubMed]
35. Bevins, C.L.; Salzman, N.H. The potter's wheel: The host's role in sculpting its microbiota. *Cell. Mol. Life Sci.* **2011**, *68*, 3675–3685. [CrossRef]
36. Koboziev, I.; Reinoso Webb, C.; Furr, K.L.; Grisham, M.B. Role of the enteric microbiota in intestinal homeostasis and inflammation. *Free Radic. Biol. Med.* **2014**, *68*, 122–133. [CrossRef]
37. Oe, M.; Mitsugi, K.; Odanaka, W.; Yoshida, H.; Matsuoka, R.; Seino, S.; Kanemitsu, T.; Masuda, Y. Dietary hyaluronic acid migrates into the skin of rats. *Sci. World J.* **2014**, *2014*, 378024. [CrossRef]
38. Cobo, F.; Foronda, C.; Pérez-Carrasco, V.; Martin-Hita, L.; Navarro-Marí, J.M. First description of abdominal infection due to *Alistipes onderdonkii*. *Anaerobe* **2020**, *66*, 102283. [CrossRef]
39. Yuan, X.; Chen, R.; Zhang, Y.; Lin, X.; Mccormick, K.L. Gut Microbiota of Chinese Obese Children and Adolescents with and Without Insulin Resistance. *Front. Endocrinol.* **2021**, *12*, 636272. [CrossRef]
40. Derrien, M.; Van Baarlen, P.; Hooiveld, G.; Norin, E.; Müller, M.; De Vos, W.M. Modulation of Mucosal Immune Response, Tolerance, and Proliferation in Mice Colonized by the Mucin-Degrader *Akkermansia muciniphila*. *Front. Microbiol.* **2011**, *2*, 166. [CrossRef]

foods

MDPI

Article

Modulation Effects of *Sargassum pallidum* Extract on Hyperglycemia and Hyperlipidemia in Type 2 Diabetic Mice

Xing Xie [1,2], Chun Chen [1,3,*] and Xiong Fu [1,3]

1 SCUT-Zhuhai Institute of Modern Industrial Innovation, School of Food Science and Engineering, South China University of Technology, Guangzhou 510640, China; lfxfu@scut.edu.cn (X.F.)
2 College of Health, Jiangxi Normal University, Nanchang 330022, China
3 School of Food Science and Engineering, Guangdong Province Key Laboratory for Green Processing of Natural Products and Product Safety, Engineering Research Center of Starch and Vegetable Protein Processing Ministry of Education, South China University of Technology, Guangzhou 510640, China
* Correspondence: chenc@scut.edu.cn

Abstract: The aim of this study was to investigate the antidiabetic effect of the extract from *Sargassum pallidum* (SPPE) on type 2 diabetes mellitus (T2DM) mice. SPPE treatment alleviated hyperglycemia, insulin resistance (IR), liver and pancreatic tissue damage, hyperlipidemia and hepatic oxidative stress resulting from T2DM. SPPE reversed phosphoenolpyruvate carboxylase (PEPCK) and hexokinase (HK) activities to improve gluconeogenesis and glycogen storage in the liver. Furthermore, SPPE modulated glucose metabolism by regulating the levels of mRNA expression involving the PI3K/Akt/FOXO1/G6pase/GLUT2 pathway and could inhibit fatty acid synthesis by reducing the gene expression levels of fatty acid synthase (FAS) and acetyl-CoA carboxylase-1 (ACC-1). A 16 sRNA analysis indicated that SPPE treatment also reversed gut dysbiosis by increasing the abundance of beneficial bacteria (*Bacteroides* and *Lactobacillus*) and suppressing the proliferation of harmful bacteria (*Enterococcus* and *Helicobacter*). Untargeted metabolomics results indicated that histidine metabolism, nicotinate and nicotinamide metabolism and fatty acid biosynthesis were significantly influenced by SPPE. Thus, SPPE may be applied as an effective dietary supplement or drug in the management of T2DM.

Keywords: *Sargassum pallidum*; gut microbiota; serum metabolites; lipid metabolism; glucose metabolism

Citation: Xie, X.; Chen, C.; Fu, X. Modulation Effects of *Sargassum pallidum* Extract on Hyperglycemia and Hyperlipidemia in Type 2 Diabetic Mice. *Foods* **2023**, *12*, 4409. https://doi.org/10.3390/foods12244409

Academic Editors: Jiuliang Zhang, Jingren He and Rui Zhang

Received: 7 November 2023
Revised: 1 December 2023
Accepted: 5 December 2023
Published: 7 December 2023

1. Introduction

Type 2 diabetes mellitus (T2DM) is a chronic disease characterized by hyperglycemia and insulin resistance, and it accounts for over 90% of diabetic cases [1]. As reported by the International Diabetes Federation, the number of diabetic patients was about 537 million in 2021, and underdeveloped and developing countries are experiencing a fast growth rate [2]. Diabetes patients with high blood glucose levels for a long time have an increased risk of diabetes complications, such as kidney disease, retinopathy, cardiopathy and so on [3]. Simultaneously, a large number of diabetes cases show high-blood-lipid profiles, which is the key factor resulting in cardiovascular diseases [4]. Moreover, antidiabetic drugs like insulin, acarbose and metformin exhibit serious side effects, leading to abdominal discomfort, diarrhea and weight gain [5]. Therefore, it is necessary to explore novel antidiabetic medicines that are safe and have high efficiency, and phenolics from natural plants have attracted much attention in this regard.

The IRS1/PI3K/Akt pathway is the critical insulin signaling pathway to maintain glucose homeostasis [6]. Hyperglycemia and hypolipidemia will cause the excessive production of reactive oxygen species and oxidative stress damage [7]. In addition, accumulating studies have confirmed that the gut microbiota is closely related to the development of T2DM and may contribute to the modulation of glucose metabolism pathways [8]. In short, these are important targets for managing T2DM. Phenolics have been reported to exhibit excellent antidiabetic effects through various targets [9]. Kang et al. found that dieckol from *Ecklonia cava* significantly reduced blood glucose and serum insulin levels, increased the activities of antioxidant enzymes, and enhanced the Akt and AMPK phosphorylation of muscle in db/db mice [10]. The phenolic-rich extract from *Hypericum attenuatum* Choisy alleviated T2DM by modulating the AMPK/PI3K/Akt/GSK3β signaling pathway and regulating GLUT4 and PPARγ expression [11]. The phenolic extract of noni fruit could regulate glycolipid metabolism by altering the abundance of gut microbiota in T2DM rats [12].

Sargassum pallidum, which belongs to the genus *Sargassum*, is an edible brown seaweed distributed in Japan and coastal areas of China and has a long history of application as medicines to treat goiter, scrofula and edema [13]. *Sargassum pallidum* is abundant in phenolics, fatty acids, polysaccharides and terpenoids and possesses antioxidant, anti-inflammatory, hypoglycemic and anticancer activities [14]. Cao et al. observed that polysaccharides from *Sargassum pallidum* increased glucose consumption and attenuated insulin resistance by up-regulating PI3K and IRS-1 expression in HepG2 cells [13]. Our previous study also indicated that the phenolic extract of *Sargassum pallidum* displayed strong α-glucosidase inhibition [15]. Nevertheless, to date, the antidiabetic mechanism of the extract from *Sargassum pallidum* is not clear and deserves further investigation.

In this study, high-fat-diet- and streptozocin (HFD/STZ)-induced T2DM mice were used as a model. The effects of the phenolic extract from *Sargassum pallidum* on biochemical indices, histopathology, gut microbiota and serum metabolites in T2DM mice were evaluated. The gene expression involved in glycolipid metabolism in the liver was also investigated. This study presents a logical strategy to promote the application of *Sargassum pallidum* in the treatment of T2DM.

2. Materials and Methods

2.1. Chemicals and Materials

Sargassum pallidum powder was obtained from Qingdao City, Shandong Province. Streptozocin and metformin were purchased from Sigma Aldrich (St Louis, MO, USA). Triglyceride (TG), total cholesterol (TC), low- and high-density lipoprotein cholesterol detection (LDL-C and HDL-C), alanine aminotransferase (ALT), aspartate aminotransferase (AST), phosphoenolpyruvate carboxykinase (PEPCK), hexokinase (HK), catalase (CAT), glutathione peroxidase (GSH-Px), superoxide dismutase (SOD) and malondialdehyde (MDA) kits were provided by Nanjing Jiancheng Bioengineering Institute (Nanjing, China).

2.2. Preparation of Extract from Sargassum Pallidum

Dried powder of *Sargassum pallidum* was mixed with 8.5% aqueous ethanol at a ratio of 1:30 (*m:v*) and extracted by ultrasonication (KQ-300DE, Kunshan ultrasonic instrument Co., Ltd., Kunshan, China) at 42 °C and 490 W for 55 min. The mixtures were filtered by centrifugation, and the residues were re-extracted once under the same conditions. All of the supernatants were collected, concentrated and lyophilized to obtain the extract of *Sargassum pallidum*. The total phenolic content of the extract was determined according to a previous method [15].

2.3. Animals and Experimental Design

Six-week-old male C57BL/6J mice were supplied by Guangdong Sijiajingda Biotechnology Co., Ltd. (Guangdong, China). All mice were acclimated for one week and given free access to water and chow at a specific temperature (22–24 °C) and humidity (50–60%) with a 12 h light/dark cycle. Mice in the control group (n = 8) were fed a low-sugar and low-fat diet, and those in the diabetic group (n = 50) were fed a 60% high-fat diet. After 4 weeks, all mice were fasted for 12 h, and the diabetic and control groups were, respectively, injected with STZ solution (120 mg/kg BW) or the same amount of normal saline. After 1 week, the fasting blood glucose (FBG) levels of mice were determined, and mice with values exceeding 11.1 mmol/L were regarded as T2DM. All T2DM mice were randomly divided into five groups (n = 8): model, SPPE50, SPPE150, SPPE250 and Met groups. The model group consisted of T2DM mice that were untreated except for normal saline. The experimental groups were orally administered 50, 150 or 250 mg/kg/day of SPPE, and the Met group was treated with 250 mg/kg/day of metformin; the treatments lasted for 4 weeks. The body weights and FBG levels of mice were recorded once a week.

At the end of the experimental period, the feces of the mice were collected in cryogenic vials and quickly stored at −80 °C. After fasting overnight, the blood was obtained by abdominal aorta bleeding, and serum was obtained by centrifugation at 3500 r/min for 10 min. Then, mice were sacrificed; one part of the pancreas and liver was collected and kept at −80 °C, and another part was fixed with 4% paraformaldehyde for histopathological examination. All animal procedures were performed in accordance with the guidelines for the care and use of experimental animals in the laboratory animal center of Guangdong Pharmaceutical University and were approved by the Experimental Animal Ethics Committee.

2.4. Oral Glucose Tolerance Test (OGTT)

In the last week, the mice were administered a 1 g/kg BW glucose solution by gavage after fasting for 12 h; then, the blood was collected from the tip of the tail, and the blood glucose levels were evaluated at 0, 30, 60, 90 and 120 min. The areas under the curve (AUCs) were calculated using Origin 2021 software.

Determination of Serum Insulin Levels and HOMA-IR. On the last day, the blood samples were obtained after starvation overnight, and the fasting insulin levels were detected using an insulin ELISA kit (Nanjing Jiancheng Bioengineering Institute). Homeostasis model assessment insulin resistance (HOMA-IR) was calculated according to the formula reported by Wang et al. [16].

2.5. Biochemical Analysis

The concentrations of TG, TC, HDL-C, LDL-C, AST and ALT in the serum were measured using commercial assay kits. The liver tissue of the mice was mixed with suitable physiological saline and crushed into a homogenate, and then the supernatant was obtained by centrifugation. The TG and TC contents in the liver were also determined by the same methods as used for the serum.

2.6. Histopathological Examination

The liver and pancreatic tissues were cut into 3–5 μm thick slices and stained with hematoxylin–eosin staining (H&E). Then, all images were observed and photographed with an orthostatic light microscope (NIKON ECLIPSE CI, Tokyo, Japan).

2.7. Determination of Antioxidant Parameters and Key Glycometabolism Enzymes

The level of MDA and the activities of SOD, CAT and GSH in the supernatant of the liver tissue solution were determined by using commercial assay kits according to instructions.

Briefly, 100 mg of liver tissue was mixed with the extracting solution at a ratio of 1:10 (*m*:*v*) and homogenized in an ice water bath for 5 min. After centrifugation for 15 min, the levels of PEPCK and HK in the supernatant were estimated by using commercial assay kits. The protein concentration of the samples was detected by using a BCA protein assay kit (Nanjing Jiancheng Bioengineering Institute).

2.8. Quantitative Real-Time Polymerase Chain Reaction (RT-PCR) Assay

The total RNA of frozen liver tissues was extracted with Trizol reagent, and its concentration and purity were detected with a NanoDrop 1000 spectrophotometer (Wilmington, DE, USA). The cDNA was synthesized through reverse transcription and amplified by using an assay kit. The levels of cDNA were quantified using a Mini Opticon real-time PCR system (Bio-Rad, Hercules, CA, USA). The sequences of primers are described in Table 1.

The relative expression levels of genes were evaluated by using the $2^{-\Delta\Delta CT}$ method, and β-actin was selected as an internal standard.

Table 1. RT-PCR primer sequences.

Gene	Primer	Sequence (5′ → 3′)
PI3K	sense	ACACCACGGTTTGGACTATGG
	antisense	GGCTACAGT AGTGGGCTTGG
Akt	sense	GCCGGTGACAGACGATACT
	antisense	TGGCATTCACGTTTGTGGAGC
FOXO1	sense	GAGTTAGTGAGCAGGCTACAT
	antisense	TTTGGACTGCTCCTCAGTTCC
G6pase	sense	GGAGTCTTGTCAGGCATTGCT
	antisense	CGGAGGCTGGCATTGTAGAT
GLUT2	sense	GATCGCTCCAACCACACTCA
	antisense	CTGAGGCCAGCAA TC TGACTA
ACC-1	sense	CGCCAACAATGGTATTGCAGC
	antisense	TCG GATTGCACGTTCATTTCG
FAS	sense	GCGGGTTCGTGAAACTGATAA
	antisense	GCAAAATGGGCCTCCTTGATA
β-Actin	sense	GATCGATGCCGGTGCTAAGA
	antisense	TCCTATGGGAGAACGGCAGA

2.9. Cecal Microbiota Analysis

DNA extraction from mice feces was performed by using a QIAamp DNA stool mini kit (Qiagen, Germantown, MD, USA). The V3-V4 hypervariable regions of bacterial 16 rRNA genes were amplified with the following primers: 338F 5′-barcode-ACTCCTACGGGAGGC-AGCAG-3′ and 806R 5′-GGACTACHVG GGTWTCTAAT-3′. The PCR products were sequenced with an Illumina MiSeq platform (Illumina, San Diego, CA, USA). The species information was collected by comparing the results with the Ribosomal Database Project. The data were further analyzed by using the online platform Personalbio Genes Cloud (www.genescloud.cn, accessed on 1 July 2023). After the quantification step, amplicons were pooled in equal amounts and sequenced at Shanghai Personal Biotechnology Co., Ltd. (Shanghai, China), using the Illumina MiSeq platform for paired-end 2 × 250 bp sequencing.

2.10. Untargeted Metabolome Analysis

The non-targeted metabolome analysis of serum was performed using LC-MS (Agilent, Santa Clara, CA, USA). The samples were separated with a HILIC column, and acetonitrile (A) and water containing 25 mM ammonium acetate and ammonia (B) were used as the mobile phase. The flow rate, injection volume and column temperature were 0.5 mL/min, 2 μL and 25 °C, respectively. The elution conditions were as follows: 0–0.5 min, 95% A; 7 min, 65% A; 8–9 min, 40% A; 9.1–12 min, 95% A. TOF MS data were scanned in the m/z 100–1000 in positive and negative ion modes. The other parameters for TOF MS were as follows: curtain gas, 30; source temperature, 600 °C; ion spray voltage floating, ±5500 V; declustering potential, ±60 V; Collision Energy, 35 ± 15 eV. The metabolites were identified by comparing retention times, molecular weights and mass spectrum fragments with the database.

2.11. Statistical Analysis

All data are shown as mean ± standard deviation (SD) and are from experiments repeated at least three times. One-way analysis of variance (ANOVA) and Tukey's test were performed for statistical comparisons between groups, and a p value less than 0.05 was regarded as statistically significant (* $p < 0.05$, ** $p < 0.01$, *** $p < 0.001$).

3. Results and Discussion

3.1. Phenolic Composition of SPPE

The total phenolic content of SPPE was 22.63 ± 0.21 mg GAE/g E. A previous study indicated that the main phenolics in *Sargassum pallidum* are 6-gingerol, quercetin-3-*O*-glucuronide, kuraridine and n-hexacosyl caffeatehexose, which are potential α-glucosidase inhibitors to treat T2DM [15]. In particular, quercetin-3-*O*-glucuronide had an α-glucosidase-inhibitory effect that was 20 times higher than that of acarbose [17].

3.2. Effect of SPPE on Body Weight

As shown in Figure 1A, the body weights of the model group exhibited a declining trend, while those of the SPPE150, SPPE250 and Met groups first decreased and then increased over the whole experimental period. And SPPE50 displayed a higher BW decrease than SPPER250. Compared with the model group, the body weights of the SPPE250 and Met groups were significantly enhanced after 4 weeks of treatment. The results indicated that a high dose of the phenolic extract from *Sargassum pallidum* could improve the weight loss of T2DM mice.

3.3. Effect of SPPE on Glucose Homeostasis

As shown in Figure 1B, the initial FBG levels in diabetic mice showed no significant differences among the groups. After supplementation with SPPE for 4 weeks, the FBG levels in the SPPE50, SPPE150, SPPE250 and Met groups were, respectively, decreased by 8.22%, 15.11%, 42.75% and 34.77%. A high dose of SPPE and Met treatment reduced the FBG level in T2DM mice. Therefore, SPPE250 was the only treatment that displayed significant effects on FBG compared to the model. For the insulin content, the insulin level in T2DM mice was higher than in normal mice, which suggested that diabetic mice had IR. Compared with the model group, the insulin levels in the SPPE and Met groups were reduced by 8.72–22.18% ($p < 0.05$), and that in the SPPE250 group was close to that in normal mice. As shown in Figure 1D, the HOMA-IR index of different groups ranged from 1.79 to 9.24 min mM ($p < 0.05$), and the values in the SPPE250 and Met groups were 2.43 and 2.03 times lower than that in the model group. The results indicated that SPPE could improve the FBG and insulin levels in T2DM mice, and the hypoglycemic effect of SPPE250 was similar to that of metformin.

As given in Figure 1E, the blood glucose concentration in all groups reached the highest value at 30 min and then declined gradually, except for the model group, and the reduction in the SPPE250 group was more significant. The blood glucose levels in all T2DM mice at 120 min were higher than those at 0 min. For AUC profiles, the AUC values in the SPPE group were reduced in a dose-dependent manner. There were no marked differences ($p > 0.05$) between the SPPE50 and model groups. As shown in Figure 1F, the AUC levels in the SPPE250 and Met groups decreased by 860.25 and 727.13 min*mM compared with the model group. These findings suggested that SPPE treatment could increase glucose metabolism and insulin sensitivity in T2DM mice.

3.4. Effect of SPPE on Serum and Hepatic Lipid Profiles

To investigate the influence of SPPE treatment on hyperlipidemia, the levels of serum and hepatic lipids were evaluated. As displayed in Table 2, the serum TG, TC and LDL-c levels in the model group showed an increase compared with the control group and were decreased after SPPE treatment, while the trend of HDL-c levels was the opposite. In particular, for the SPPE250 group, the serum TG and LDL-c levels were 0.95 ± 0.06 and 0.52 ± 0.02 mmol/L, which were consistent with those in normal mice. In addition, compared with the model group, the concentrations of TG and TC in the liver also dropped, and the concentrations in the SPPE250 group were lower than those in the Met group. The results indicated that SPPE treatment had a lipid-lowering impact on T2DM mice.

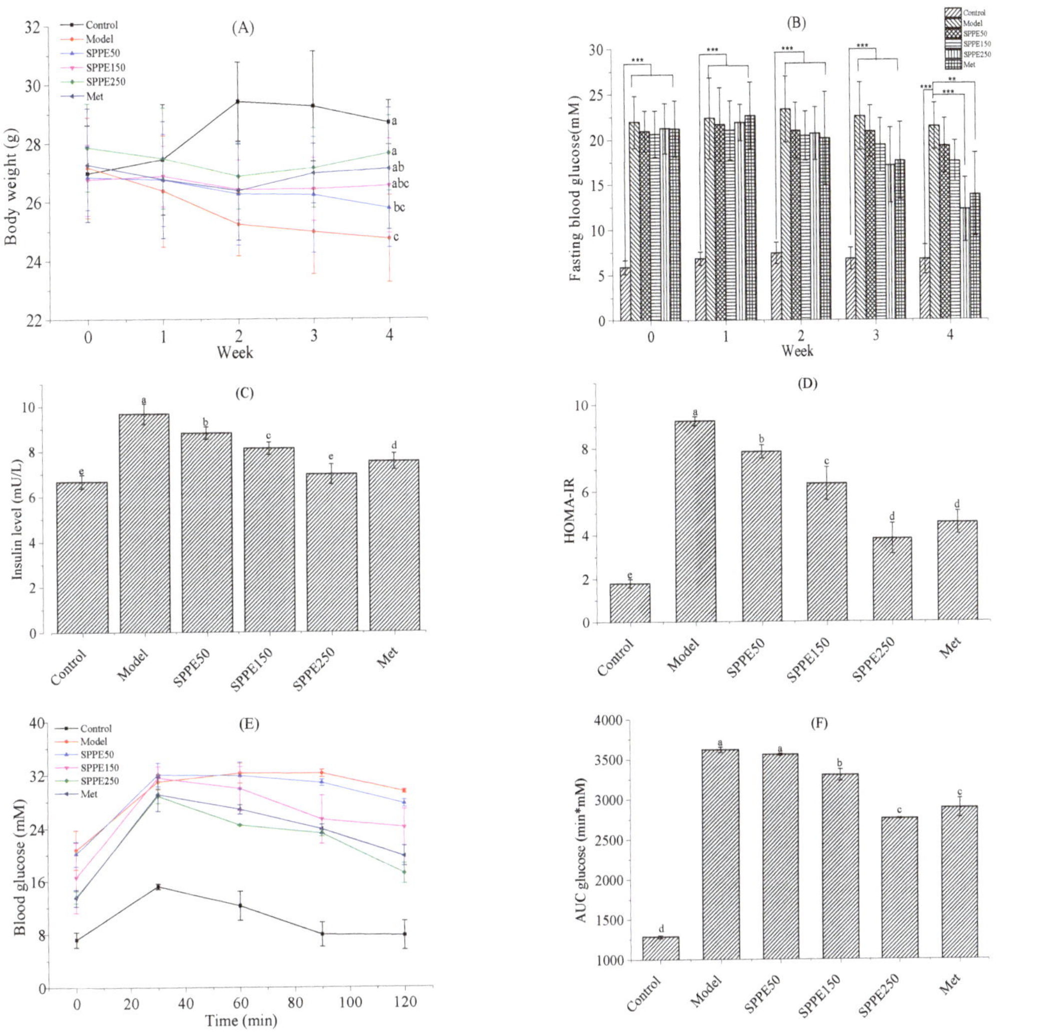

Figure 1. Effect of SPPE on the diabetes-related body indexes of T2DM mice. (**A**) Body weight, (**B**) fasting blood glucose, (**C**) insulin level, (**D**) HOMA-IR, (**E**) oral glucose tolerance test and (**F**) AUC. Different letters in the same pattern represent a significant difference ($p < 0.05$). ** and *** refers to the significance with $p < 0.01$ and $p < 0.001$, respectively.

Table 2. Effect of treatments on lipid profiles of serum TC, serum TC, serum HDL-c, serum LDL-c, hepatic TG and hepatic TC in HFD/STZ-induced T2DM mice.

Group	Serum TG (mmol/L)	Serum TC (mmol/L)	Serum HDL-c (mmol/L)	Serum LDL-c (mmol/L)	Hepatic TG (mmol/L)	Hepatic TC (mmol/L)
Control	0.98 ± 0.05^{d}	2.85 ± 0.22^{d}	2.33 ± 0.10^{a}	0.50 ± 0.03^{cd}	0.26 ± 0.01^{e}	0.13 ± 0.01^{e}
Model	3.36 ± 0.06^{a}	6.37 ± 0.37^{a}	0.82 ± 0.02^{f}	0.78 ± 0.05^{a}	0.50 ± 0.01^{a}	0.40 ± 0.01^{a}
SPPE50	2.71 ± 0.12^{b}	5.68 ± 0.30^{b}	1.15 ± 0.03^{e}	0.70 ± 0.13^{ab}	0.44 ± 0.03^{b}	0.29 ± 0.03^{b}
SPPE150	1.42 ± 0.15^{c}	5.10 ± 0.01^{b}	1.42 ± 0.02^{d}	0.63 ± 0.08^{bc}	0.38 ± 0.02^{c}	0.22 ± 0.02^{c}
SPPE250	0.95 ± 0.06^{d}	4.06 ± 0.07^{c}	1.98 ± 0.07^{b}	0.47 ± 0.08^{d}	0.26 ± 0.03^{e}	0.16 ± 0.01^{d}
Met	1.34 ± 0.11^{c}	4.36 ± 0.19^{c}	1.76 ± 0.14^{c}	0.52 ± 0.02^{cd}	0.32 ± 0.01^{d}	0.18 ± 0.03^{d}

Note: Different letters in the same column represent significant differences ($p < 0.05$).

3.5. Effect of SPPE on the Histomorphology of Liver and Pancreas Tissues

As shown in Figure 2A, the hepatocytes of normal mice were in an orderly arrangement with a clear structure, while those of the model group exhibited an irregular arrangement, fat accumulation and big vacuoles. With the increased intake of SPPE, the structure of liver tissue in T2DM mice was gradually clearer, and the number and size of fat vacuoles also decreased. Moreover, as shown in Figure 2B, the pancreas in the model group exhibited severe atrophy and degeneration with a small area, and the number of insulin cells was small. SPPE treatment repaired the injury to pancreatic tissue, and the recovery in the SPPE250 group was nearly to the level of the control group. The results indicated that the high-dose SPPE treatment could mitigate the liver and pancreas tissue damage caused by T2DM.

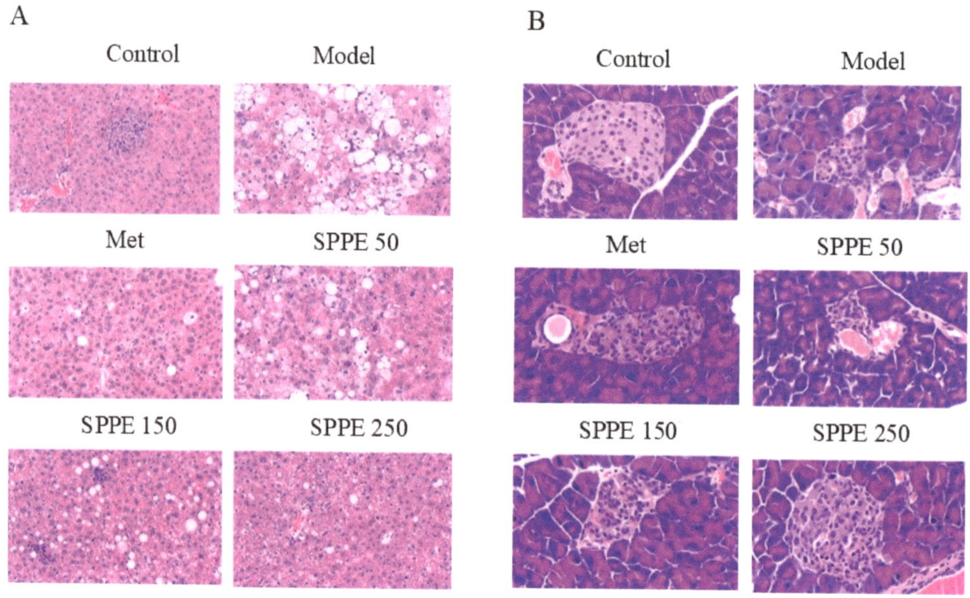

Figure 2. Effect of SPPE on the organs, including liver fat accumulation (**A**) and pancreatic tissue inflammation (**B**), in HFD/STZ-induced T2DM mice.

3.6. Effect of SPPE on AST, ALT and Oxidative Stress

As revealed in Figure 3A,B, in comparison with the model group, the AST and ALT levels in the SPPE groups were, respectively, reduced by 6.93–35.98% ($p < 0.05$) and 5.18–26.86% ($p < 0.05$) and showed a dose effect. The results indicated that SPPE could improve liver function in T2DM mice.

As shown in Figure 3C–E, the SOD, CAT and GSH activities in the model group were weaker than those in other groups, and those in the SPPE250 group were enhanced by 16.06%, 63.90% and 58.90%, respectively. The MDA content in T2DM mice decreased in a dose-dependent manner after SPPE treatment. Specifically, the MDA level in the SPPE250 group was reduced by 26.33% compared with the model group. Moreover, the effect of SPPE and Met treatment at a dose of 250 mg/kg was not significant. The results suggested that SPPE treatment could strengthen the antioxidant enzyme capacity and alleviate oxidative stress in T2DM mice.

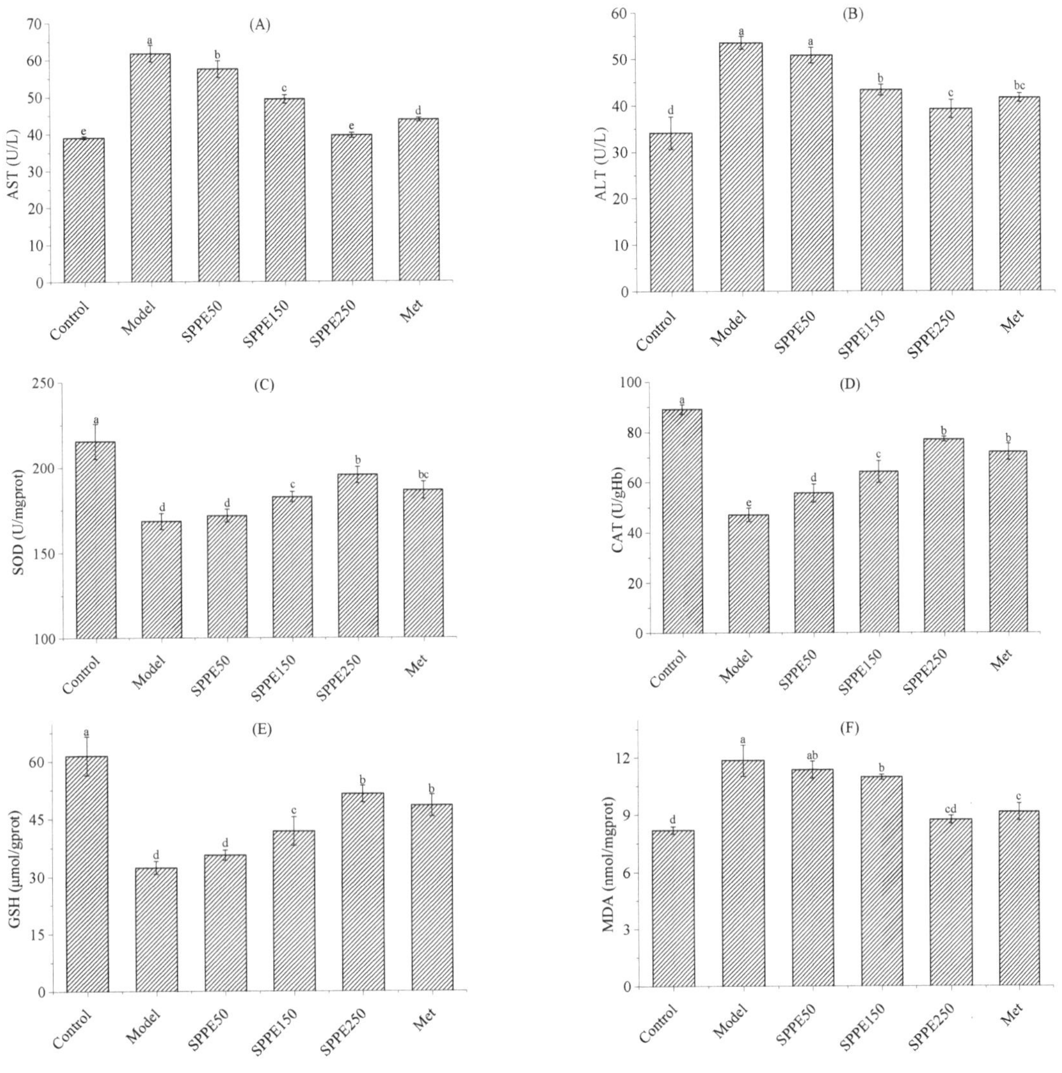

Figure 3. Effect of SPPE on serum AST (**A**), serum ALT (**B**), hepatic SOD (**C**), hepatic CAT (**D**), hepatic GSH (**E**) and hepatic MDA (**F**) levels in HFD/STZ-induced T2DM mice. Different letters in the same pattern represent significant differences ($p < 0.05$).

3.7. Effect of SPPE on Activities of PEPCK and HK

As illustrated in Figure 4A, the activity of PEPCK in the model group was stronger than in the control group, reflecting the occurrence of gluconeogenesis. After SPPE treatment, the activity of PEPCK in diabetic mice was reduced by 6.54–37.32%. Figure 4B shows that the activity of HK in the model group was weaker than in the other groups and increased by 13.23–65.70% after treatment with SPPE. The levels of PEPCK and HK in the SPPE250 and Met groups were high. These findings indicated that SPPE treatment could reverse phosphoenolpyruvate carboxylase (PEPCK) and hexokinase (HK) activities to improve gluconeogenesis and glycogen storage in the livers of T2DM mice.

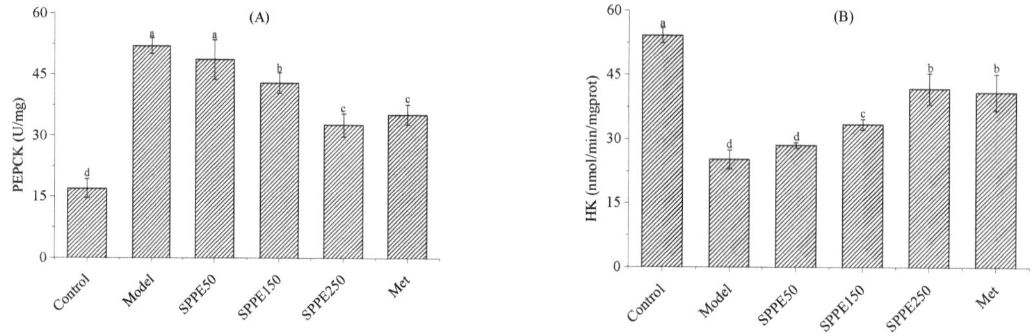

Figure 4. Effect of SPPE on the activity of hepatic PEPCK (**A**) and HK (**B**) in HFD/STZ-induced T2DM mice. Different letters in the same pattern represent significant differences ($p < 0.05$).

3.8. Effect of SPPE on mRNA Expression of Glucose and Lipid Metabolism Genes

As depicted in Figure 5A–E, compared with the control group, lower gene expression levels of PI3K, Akt and GLUT2 were detected in the model group, and the FOXO1 and G6pase expression levels were up-regulated. After 4 weeks of administration, SPPE treatment modulated the gene expression levels of those mRNAs involved in glucose metabolism. The gene expression of FOXO1 and G6pase in the SPPE250 group was reduced by 0.33 and 0.58 times, and that of PI3K, Akt and GLUT2 increased. Moreover, these mRNA expression levels showed no significant differences between the SPPE250 and Met groups, except for FOXO1 ($p > 0.05$). The results indicated that SPPE treatment may regulate glucose metabolism in the livers of T2DM mice by promoting glucose transport and preventing gluconeogenesis through the PI3K/Akt/FOXO1/G6pase/GLUT2 signaling pathway.

As shown in Figure 5F,G, the mRNA expression levels of ACC-1 and FAS in the model group were higher than those in the control group. After treatment with SPPE, the gene expression levels of these two mRNAs declined in T2DM mice. Furthermore, the mRNA expression level of ACC-1 in the SPPE250 group was close to that in the control group. The results suggest that SPPE treatment could inhibit the expression of key lipid synthesis factors in T2DM mice.

3.9. Effect of SPPE on Gut Microbiota

As given in Table 3, the Shannon and Chao indexes differed between the model and control groups, indicating that T2DM changed the composition of the gut microbiota in mice. In contrast, SPPE treatment showed insignificant effects on the richness and diversity of the microbiota in T2DM mice. A difference was found in diversity between the Met and model groups ($p < 0.05$).

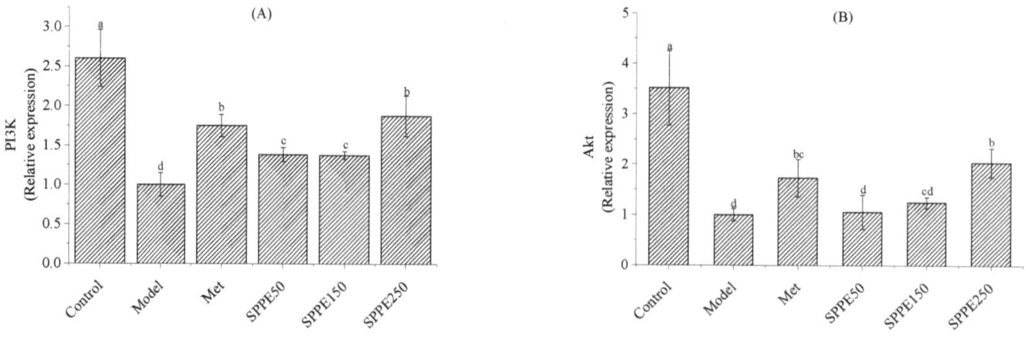

Figure 5. *Cont.*

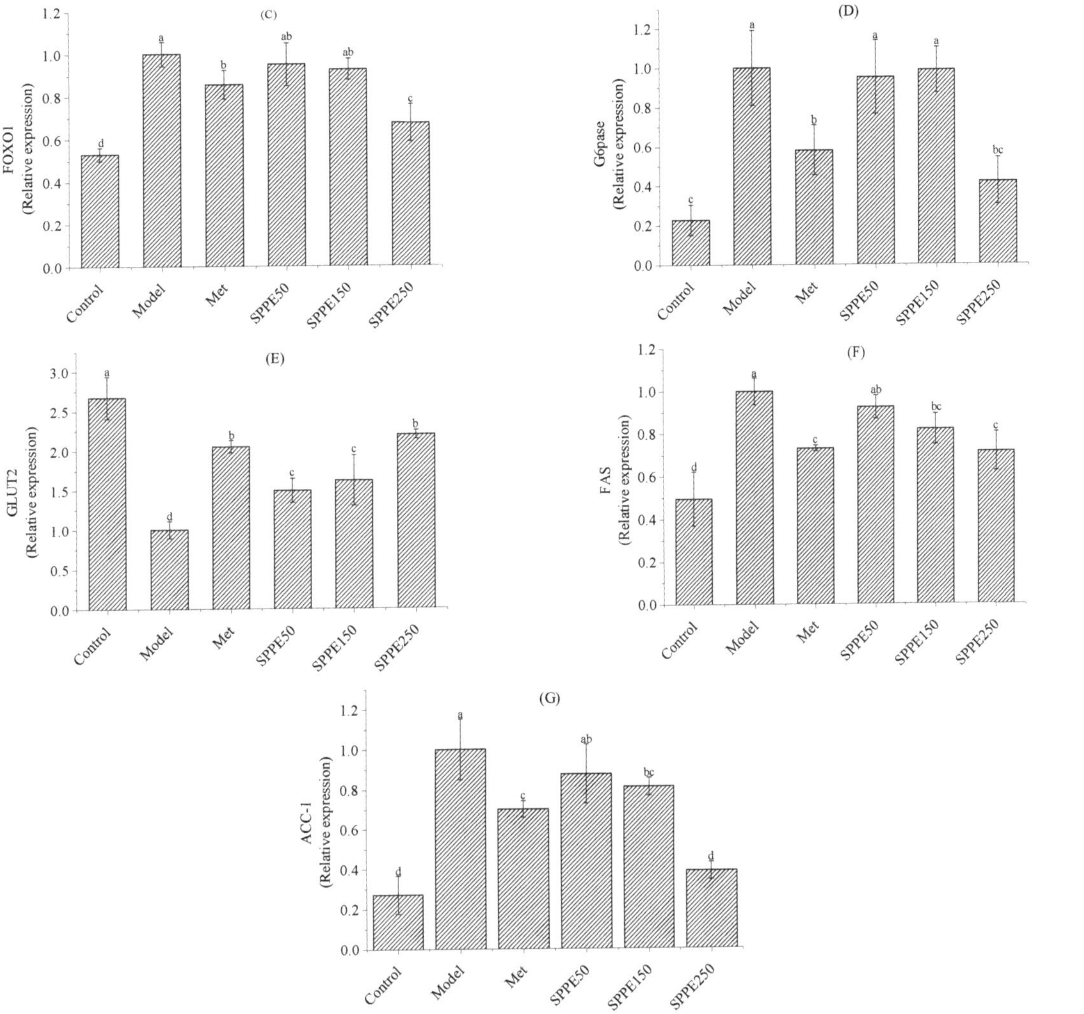

Figure 5. Effect of SPPE treatment on mRNA expression levels of PI3K (**A**), Akt (**B**), FOXO1 (**C**), G6pase (**D**), GLUT2 (**E**), FAS (**F**) and ACC-1 (**G**) in the livers of HFD/STZ-induced T2DM mice. Different letters are significantly different at $p < 0.05$.

The gut microbiota profiles of all groups at different taxonomic levels were further investigated. At the phylum level, *Firmicutes*, *Proteobacteria*, *Actinobacteria* and *Bacteroidetes* were the major bacterial communities in each group (Figure 6A). Compared with the model group, the level of *Proteobacteria* was reduced by 56.21%, and *Bacteroidetes* content was increased by 185.59%. The F/B value of 3.89 in the SPPE250 group was lower than those in the model and Met groups. At the genus level, compared with the model group, the lower contents of *Bacteroides*, *Lactobacillus* and *Parabacteroides* and higher contents of *Enterococcus* and *Helicobacter* were reversed after 250 mg/kg of SPPE treatment (Figure 6B). Furthermore, the relative contents of *Bacteroides* and *Lactobacillus* were augmented by 2.52 and 0.58 times in the SPPE250 group. The results suggest that SPPE treatment could regulate the abundance of beneficial and harmful bacteria in T2DM mice.

Table 3. Effect of SPPE treatment on alpha diversity of gut microbiota in HFD/STZ-induced T2DM mice.

Group	Index	
	Chao	Shannon
Control	767.08 ± 35.34 [c]	5.20 ± 0.25 [b]
Model	916.16 ± 65.12 [b]	6.19 ± 0.23 [a]
SPPE50	952.16 ± 38.82 [b]	5.90 ± 0.15 [a]
SPPE150	966.26 ± 48.21 [b]	5.97 ± 0.21 [a]
SPPE250	1004.20 ± 55.62 [a]	6.72 ± 0.31 [a]
Met	652.75 ± 45.32 [d]	4.60 ± 0.18 [c]

Note: Different letters in the same column represent significant differences ($p < 0.05$).

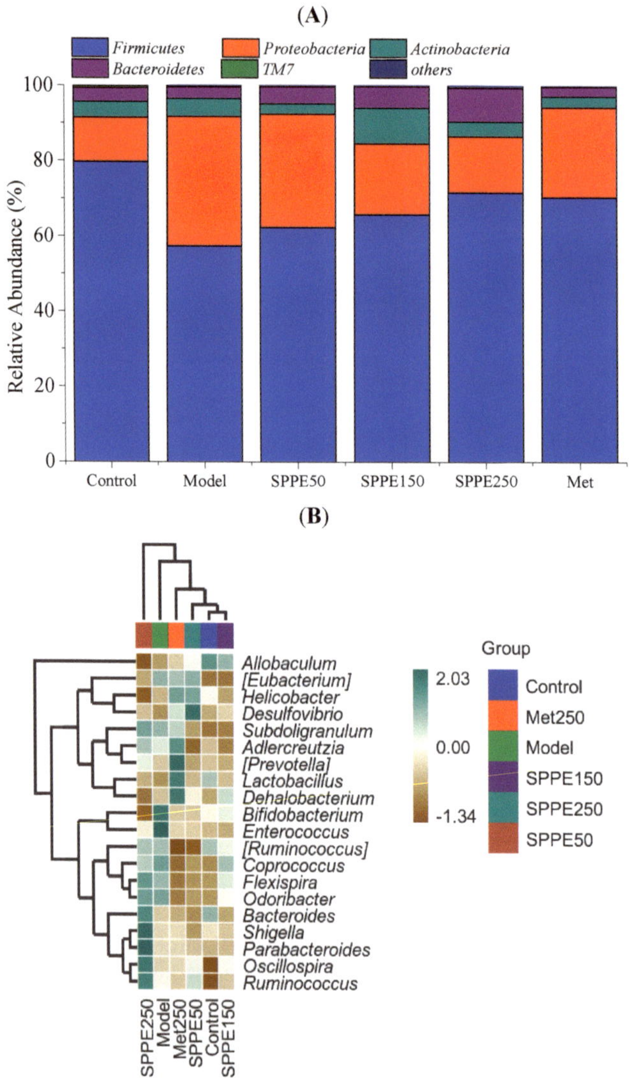

Figure 6. Effect of SPPE treatment on composition of gut microbiota in HFD/STZ-induced T2DM mice. At the phylum level (**A**); heat map of microbial composition at the genus level (**B**).

3.10. Effect of SPPE on Serum Metabolites

The serum metabolites in four selected groups were analyzed by LC-MS. As shown in Figure 7A,B, the PCA scores indicate that the separation was good among the four groups in both negative and positive ion modes, and the quality of the data was acceptable for further study. Based on previous results, the SPPE250 treatment showed the best effect on diabetic mice in all groups, so this study focused on the SPPE250 and model groups. In comparison with the model group, the up- and down-regulation of metabolites in the SPPE250 group are visualized in volcano maps (Figure 7C,D), and the results suggest that 250 mg/kg of SPPE treatment had a significant effect on the metabolism of diabetic mice.

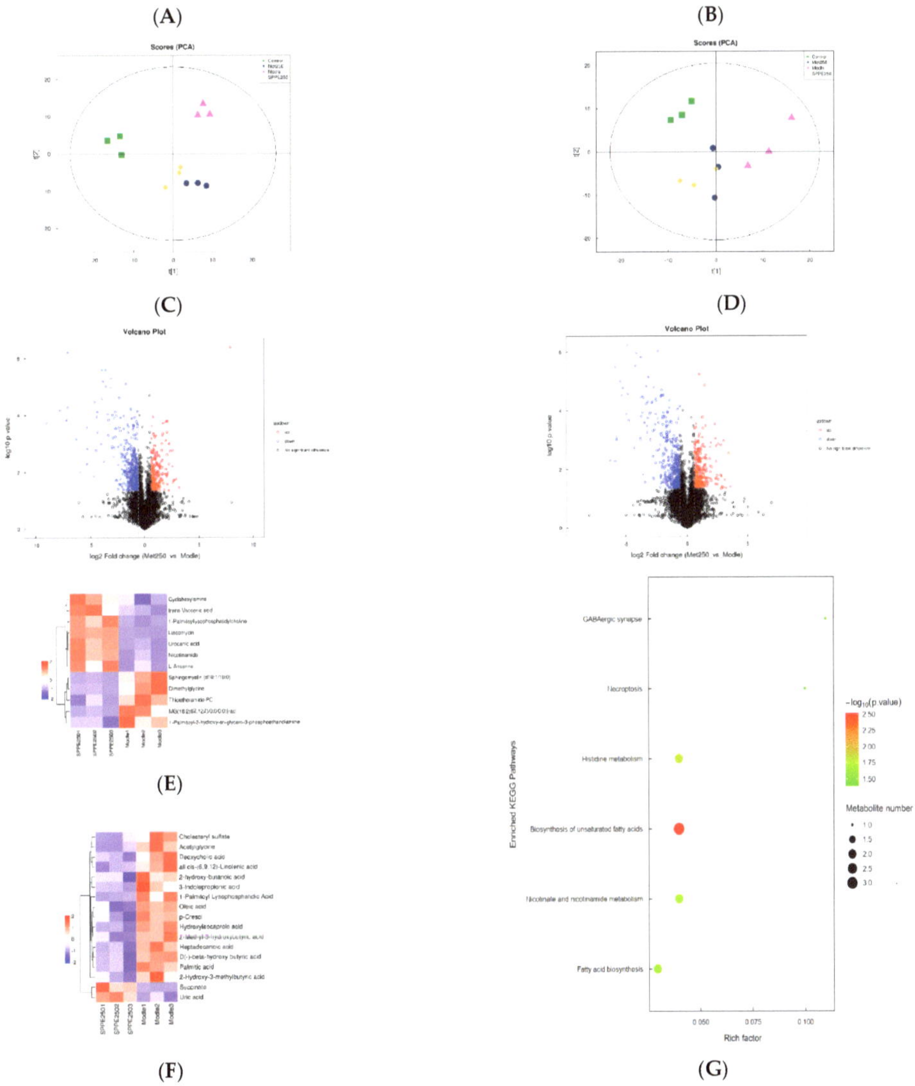

Figure 7. Effect of SPPE treatment on serum metabolites in HFD/STZ-induced T2DM mice. Scatter plot of PCA scores in the negative (**A**) and positive ion modes (**B**). Volcano plot for model vs. SPPE250 groups in the negative (**C**) and positive ion modes (**D**). Hierarchical clustering heat map of significant metabolites for model and SPPE250 groups in the negative (**E**) and positive ion modes (**F**). KEGG enrichment analysis for the model and SPPE250 groups (**G**).

After screening based on VIP scores > 1, the different metabolites in the two groups were displayed in a hierarchical clustering heat map (Figure 7E,F). A total of 17 and 12 metabolites were affected by SPPE treatment in the negative and positive ion modes, respectively. In the negative ion mode, compared with the model group, the levels of Succinate and Uric acid were increased, and the levels of Oleic acid, Heptadecanoic acid, Palmitic acid, 2-Hydroxy-3-methylbutyric acid and 11 other metabolites were significantly reduced. In the positive ion mode, the levels of seven metabolites were increased by SPPE treatment, mainly including Nicotinamide, Urocanic acid and Lincomycin. In addition, the levels of Thioetheramide-PC, MG (18:2(9Z,12Z)/0:0/0:0) [rac], Sphingomyelin (d18:1/18:0) and Dimethylglycine were decreased. Kyoto Encyclopedia of Genes and Genomes (KEGG) pathway analysis was performed based on the differential metabolites, and the enriched pathways in the model and SPPE250 groups are listed in Figure 7G. The results revealed that these metabolites are related to histidine metabolism, nicotinate and nicotinamide metabolism, and biosynthesis of unsaturated fatty acids and fatty acids, which may be important for SPPE to treat T2DM.

3.11. Correlation Analysis between Gut Microbiota and Differential Metabolites

The complex relation between the gut microbiota and differential metabolites in the SPPE250 and model groups was visualized by Spearman's correlation analyses (Figure 8). *Bacteroides* and *Lactobacillus* were positively correlated with Xanthine and 2-Methylbutyroyl-carnitine and negatively correlated with Oleic acid, all cis-(6,9,12)-Linolenic acid and 1-Palmitoyl lysophosphatidic acid. Moreover, Heptadecanoic acid, Palmitic acid and MG (18:2(9Z,12Z)/0:0/0:0) [rac] were highly related to 11 bacterial communities ($p < 0.05$), including *Enterococcus, Oscillospira, Bifidobacterium, Ruminococcus, Odoribacter, Adlercreutzia, Parabacteroides* and so on. *Enterococcus* also exhibited a positive association with 2-Hydroxy-3-methylbutyric acid and Pantothenate.

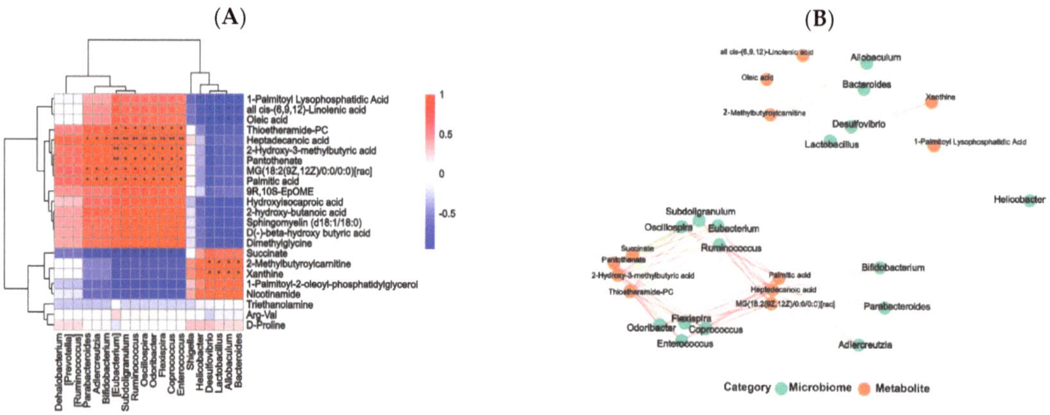

Figure 8. Correlation analysis between serum metabolites and gut microbiota at the genus level. Spearman's analyses (**A**) and visualization of associated network (**B**) for model and SPPE250 groups. * $p \leq 0.05$, ** $p \leq 0.01$.

4. Discussion

T2DM is accompanied by IR, oxidative stress, an imbalance in the gut microbiota, and glucose and lipid metabolism disorders [18]. Synthetic drugs have toxic side effects when administered in the long term, while phenolics from natural plants have been proven as alternative drugs or supplements to prevent and treat T2DM. Our previous study found that SPPE showed excellent hypoglycemic activity in vitro, and 6-gingerol, quercetin-3-*O*-glucuronide, kuraridine and n-hexacosyl caffeatehexose were the main phenolics in

Sargassum pallidum [15]. Therefore, HFD/STZ-induced T2DM mice were used as a model to further assess the hypoglycemic and hypolipidemic abilities of SPPE.

After SPPE treatment for 4 weeks, the weight loss of the model mice was reversed, and a high dose of SPPE was more significant. This result is similar to that reported by Liu et al., who found that Ginsenoside Rk3 treatment could increase the weight of T2DM mice [19]. Furthermore, the levels of FBG, insulin and HOMA-IR declined sharply in the SPPE250 and Met groups compared with the model group. The OGTT results suggested that the consumption of 250 mg/kg of SPPE enhanced the insulin sensitivity of T2DM mice. As previously reported, *Sargassum fusiforme* fucoidan and polyphenol-rich vinegar extract were effective in reducing FBG levels and improving glucose tolerance in T2DM mice, and the same phenomenon was observed in our study [20,21]. Among the phenolics from *Sargassum pallidum*, quercetin-3-O-glucuronide, 6-gingerol and poricoic acid A are the main components with strong inhibitory effects on α-glucosidase and may delay the digestion of carbohydrates, resulting in a decrease in glucose absorption in the small intestine. In addition, the pancreas is an important tissue in the maintenance of glucose homeostasis [22]. SPPE treatment increased the β-cell mass in the pancreas and prevented the degeneration of islets, which could contribute to improving pancreatic damage and FBG levels in T2DM mice.

The development of T2DM can disturb lipid metabolism and result in dyslipidemia [23]. The serum lipid profiles showed that SPPE treatment could modulate dyslipidemia by significantly reducing the levels of TG, TC and LDL-c while increasing the level of HDL-c in T2DM mice. Wu et al. [24] confirmed that the ethanol extract of *Sargarsum fusiforme* normalized the lipid levels in T2DM mice, which is in line with our study. Quercetin derivatives, as the main phenolics in *Sargassum pallidum*, have been demonstrated to improve lipid metabolism by accelerating β-oxidation in mice [25]. Based on the combined data, SPPE showed strong hypoglycemic and hypolipidemic effects on T2DM mice.

As is well known, the liver is a vital organ involved in glycogen synthesis, gluconeogenesis, IR, lipid synthesis and oxidation [26]. Liver damage can reflect the glucose and lipid metabolism abilities of T2DM mice. AST and ALT are markers of liver function and are often used to evaluate it [27]. In this research, SPPE treatment decreased the serum AST and ALT contents in T2DM mice. Moreover, histomorphological observations of liver tissue showed that SPPE could reduce the number of fat vacuoles and fat accumulation. These findings suggested that SPPE could repair liver injury caused by T2DM in mice. In addition, excessive ROS generation in DM patients leads to oxidative stress and is highly related to IR and liver damage [28]. Our results showed that higher SOD, CAT and GSH levels and lower MDA content were detected in the SPPE group, and the effects in the SPPE250 and Met250 groups were comparable, indicating that SPPE may alleviate oxidative stress by enhancing antioxidant enzyme activities. As reported by Gheda et al., phenolic extracts from *Cystoseira compressa* also improved oxidative stress injury in the livers of diabetic rats by increasing MDA and GSH levels [29]. In addition, PEPCK and HK are important enzymes in gluconeogenesis, glycogen synthesis and glycolysis and are associated with glucose homeostasis [6]. SPPE could significantly decrease the PEPCK and HK concentrations to reduce gluconeogenesis and increase glycogen synthesis, acting to maintain hepatic glucose homeostasis and blood glucose levels. These results indicate that SPPE can regulate glycolipid metabolism in the livers of T2DM mice in various ways.

In order to explore the underlying mechanism, the expression of key mRNAs was evaluated to clarify the effect of SPPE treatment on hepatic glucose and lipid metabolism in T2DM mice. The PI3K-Akt pathway plays an important role in regulating glucose metabolism in T2DM; G6pase and FOXO are key enzymes for glucose synthesis and are implicated in hepatic gluconeogenesis [30,31]. GLUT2 is responsible for the transport of glucose and can alleviate IR in the body [32]. The consumption of SPPE up-regulated the expression levels of PI3K, Akt and GLUT2 and down-regulated the expression levels of FOXO1 and G6pase, implying that the PI3K/Akt/FOXO1/G6pase/GLUT2 pathway may contribute to the hypoglycemic effect of SPPE. A similar result was obtained using the phenolic extract from brown rice, which activated the PI3K/Akt signaling pathway to improve hepatic glucose metabolism [33]. In addition, FAS and ACC-1 are important enzymes involved in lipid metabolism, which, respectively, participate in catalyzing the synthesis of long-chain fatty acids into adipose tissue and fatty acid synthesis [34]. Lower expression levels of ACC-1 and FAS were observed in T2DM mice after treatment with SPPE, and these mRNA levels in the SPPE250 group were close to those in the normal group, indicating that SPPE could modulate liver lipid metabolism by suppressing fatty acid synthesis and fat accumulation. These results reveal that the antidiabetic mechanism of SPPE occurs through the modulation of the expression of key genes related to the PI3K/Akt pathway and lipogenesis in the liver. SPPE is rich in phenolics like 6-gingerol and quercetin derivatives. Some studies have reported that 6-gingerol and quercetin derivatives could enhance the mRNA expression of the PI3k/Akt pathway to alleviate insulin resistance in T2DM mice [35,36] and decrease lipogenesis by down-regulating the mRNA levels of ACC and FAS in mice [37,38]. Therefore, the restored expression of these glycolipid-metabolism-related mRNAs in T2DM mice may be due to the combined synergetic effect of the bioactive compounds, including phenolics, in *Sargassum pallidum*.

A growing list of studies have shown that the composition and diversity of the gut microtia play critical roles in the development of T2DM [39]. *Proteobacteria* are abundant in DM patients and can result in an inflammatory response in the body [40]. *Bacteroidetes* and the F/B value were positively related to the plasma glucose concentration and lipid metabolism order [41]. In this study, the effect of SPPE on diversity and richness was not significant. Increased levels of *Bacteroidetes* and decreased levels of *Proteobacteria* and F/B were observed in T2DM mice after a high dose of SPPE treatment. Xia et al. [21] also found that a polyphenol-rich vinegar extract showed an antidiabetic effect by regulating the levels of *Proteobacteria* and F/B. *Lactobacillus*, as beneficial bacteria, can modulate glycolipid metabolism by mitigating intestinal inflammation to suppress metabolic diseases like DM [42]. *Parabacteroides* contributed to improvements in IR, inflammation and intestinal integrity [43]. *Enterococcus* and *Helicobacter* were highly associated with the occurrence of intestinal diseases [44,45]. After SPPE treatment, *Lactobacillus* and *Parabacteroides* were enriched, while the abundance of *Enterococcus* and *Helicobacter* declined in T2DM mice. Zhang et al. [46] reported that RBDF phenolics could reduce the level of *Enterococcu* in db/db diabetic mice. Qi et al. confirmed that Fu brick tea aqueous extract treatment increased the levels of *Lactobacillus* and *Parabacteroides*, exhibiting anti-inflammatory and hypoglycemic effects [47]. These results imply that SPPE could change the abundance of beneficial and harmful bacteria to exert an antidiabetic effect.

Serum metabolites are the end products of gene expression involved in the regulation of metabolism and can reflect the effect of SPPE on the metabolism of T2DM mice. Niacinamide is the product of niacin amidation and is closely associated with glucose glycolysis and fat metabolism [48]. Supplementation with SPPE reversed the levels of Niacinamide and Succinate, thus regulating nicotinate and nicotinamide metabolism. Histidine metabolism could improve oxidative stress and inflammatory responses in the body [49]. After SPPE treatment, the levels of histidine metabolites like L-Anserine and Urocanic acid were increased, which may enhance the antioxidant enzyme activities and inhibit inflammatory reactions in T2DM mice. Sphingomyelin (d18:1/18:0), as a sphingolipid, is positively related to insulin resistance, β-cell dysfunction and inflammation [50]. The consumption of SPPE significantly reduced the level of Sphingomyelin (d18:1/18:0) in T2DM mice and may contribute to the balance of glucose homeostasis. Moreover, after SPPE treatment, lower levels of saturated fatty acids, like Palmitic and Heptadecanoic acid, were observed in T2DM mice, and the content of Oleic acid slightly declined. SPPE may inhibit the biosynthesis of saturated fatty acids and thereby reduce the risk of T2DM, with the same phenomenon also found by Liu et al. [51], but could not promote the production of unsaturated fatty acids like Oleic acid. Dimethylglycine is positively related to HOMA-IR and incident T2D [52], and SPPE treatment decreased the concentration of dimethylglycine in T2DM mice. In addition, Spearman's correlation analyses showed that *Lactobacillus* was positively correlated with 2-Methylbutyroyl carnitine, suggesting that it may play a vital role in the improvement of fatty acid metabolism disorders [53]. *Enterococcus* was positively related to Heptadecanoic and Palmitic acids and may enhance the synthesis of saturated fatty acids. These findings indicate that SPPE could modulate serum metabolite disorders caused by T2DM, and the gut microbiota had a significant effect on the composition of serum metabolites. This suggests that SPPE may be used as a functional food for managing diabetes. However, which metabolites play the most important role still needs to be explored.

5. Conclusions

In conclusion, the present study indicated that SPPE could modulate glycolipid metabolism in T2DM mice. In addition, SPPE could improve gut dysbiosis by increasing the abundance of *Bacteroides*, *Lactobacillus* and *Parabacteroides* and reducing the abundance of *Enterococcus* and *Helicobacter*. The untargeted metabolome analysis indicated that a high dose of SPPE significantly changed the concentrations of 29 metabolites in diabetic mice and displayed a moderating effect on histidine metabolism, nicotinate and nicotinamide metabolism and fatty acid biosynthesis. The gut microbiota exhibited a high correlation with serum metabolites. These findings could provide a scientific basis for the clinical analysis of SPPE in T2DM.

Author Contributions: Writing—original draft, X.X.; Project administration, C.C. and X.F. All authors have read and agreed to the published version of the manuscript.

Funding: Financial and moral assistance to conduct the project was provided by the Guangzhou Science and Technology Planning Project (2023A04J0763), Guangdong Science and Technology Planning Project (2023A0505050132), Guangdong Basic and Applied Basic Research Foundation (2022A1515010730), Guangdong Basic and Applied Basic Research Foundation (2023B1515040014), National Natural Science Foundation of China (32001647), Natural Science Foundation of Jiangxi Province (20224BAB215051), National Natural Science Foundation of China (31972022) and 111 Project (B17018).

Institutional Review Board Statement: The study was conducted in accordance with Guangdong Pharmaceutical University Experimental Animal Ethics Committee. NO. 00297564.

Data Availability Statement: Data are contained within the article.

Conflicts of Interest: The authors have no conflict of interest to declare.

References

1. Chen, C.; Fu, X. Spheroidization on Fructus Mori polysaccharides to enhance bioavailability and bioactivity by anti-solvent precipitation method. *Food Chem.* **2019**, *300*, 125245. [CrossRef] [PubMed]
2. Chen, C.; You, L.-J.; Abbasi, A.M.; Fu, X.; Liu, R.H. Optimization for ultrasound extraction of polysaccharides from mulberry fruits with antioxidant and hyperglycemic activity in vitro. *Carbohydr. Polym.* **2015**, *130*, 122–132. [CrossRef] [PubMed]
3. Wang, P.-P.; Wang, W.-D.; Chen, C.; Fu, X.; Liu, R.-H. Effect of *Fructus Mori.* bioactive polysaccharide conjugation on improving functional and antioxidant activity of whey protein. *Int. J. Biol. Macromol.* **2020**, *148*, 761–767.
4. Tran, H.G.; Vu, M.T.; Nguyen, V.; Hoang, T.M.H.; Ngo, T.H.T.; Phan, T.H.; Nguyen, X.N.; Nguyen, T.M.H. Two new neolignans from the roots of *Platycodon grandiflorus* and their hypolipidemic effects. *Phytochem. Lett.* **2023**, *57*, 167–171. [CrossRef]
5. Dou, Z.-M.; Chen, C.; Huang, Q.; Fu, X. Comparative study on the effect of extraction solvent on the physicochemical properties and bioactivity of blackberry fruit polysaccharides. *Int. J. Biol. Macromol.* **2021**, *183*, 1548–1559. [CrossRef] [PubMed]
6. Deng, N.; Guo, R.; Zheng, B.; Li, T.; Liu, R.H. IRS-1/PI3K/Akt pathway and miRNAs are involved in whole grain highland barley (*Hordeum vulgare* L.) ameliorating hyperglycemia of db/db mice. *Food Funct.* **2020**, *11*, 9535–9546. [CrossRef] [PubMed]
7. Elsayed, R.H.; Kamel, E.M.; Mahmoud, A.M.; El-Bassuony, A.A.; Bin-Jumah, M.; Lamsabhi, A.M.; Ahmed, S.A. *Rumex dentatus* L. phenolics ameliorate hyperglycemia by modulating hepatic key enzymes of carbohydrate metabolism, oxidative stress and PPARγ in diabetic rats. *Food Chem. Toxicol.* **2020**, *138*, 111202. [CrossRef]
8. Chen, C.; Zhang, B.; Huang, Q.; Fu, X.; Liu, R.H. Microwave-assisted extraction of polysaccharides from Moringa oleifera Lam. leaves: Characterization and hypoglycemic activity. *Ind. Crops Prod.* **2017**, *100*, 1–11. [CrossRef]

9. Zhao, C.; Wan, X.; Zhou, S.; Cao, H. Natural Polyphenols: A Potential Therapeutic Approach to Hypoglycemia. *eFood* **2020**, *1*, 107–118. [CrossRef]
10. Kang, M.-C.; Wijesinghe, W.A.J.P.; Lee, S.-H.; Kang, S.-M.; Ko, S.-C.; Yang, X.; Kang, N.; Jeon, B.-T.; Kim, J.; Lee, D.-H.; et al. Dieckol isolated from brown seaweed *Ecklonia cava* attenuates type II diabetes in db/db mouse model. *Food Chem. Toxicol.* **2013**, *53*, 294–298. [CrossRef]
11. Lv, Y.; Hao, J.; Liu, C.; Huang, H.; Ma, Y.; Yang, X.; Tang, L. Anti-diabetic effects of a phenolic-rich extract from *Hypericum attenuatum* Choisy in KK-Ay mice mediated through AMPK /PI3K/Akt/GSK3β signaling and GLUT4, PPARγ, and PPARα expression. *J. Funct. Foods* **2019**, *61*, 103506. [CrossRef]
12. Wang, R.; Zhang, L.; Zhang, Q.; Zhang, J.; Liu, S.; Li, C.; Wang, L. Glycolipid Metabolism and Metagenomic Analysis of the Therapeutic Effect of a Phenolics-Rich Extract from Noni Fruit on Type 2 Diabetic Mice. *J. Agric. Food Chem.* **2022**, *70*, 2876–2888. [CrossRef]
13. Cao, C.; Zhang, B.; Li, C.; Huang, Q.; Fu, X.; Liu, R.H. Structure and in vitro hypoglycemic activity of a homogenous polysaccharide purified from *Sargassum pallidum*. *Food Funct.* **2019**, *10*, 2828–2838. [CrossRef] [PubMed]
14. Yuan, D.; Li, C.; You, L.; Dong, H.; Fu, X. Changes of digestive and fermentation properties of Sargassum pallidum polysaccharide after ultrasonic degradation and its impacts on gut microbiota. *Int. J. Biol. Macromol.* **2020**, *164*, 1443–1450. [CrossRef] [PubMed]
15. Xie, X.; Chen, C.; Fu, X. Screening α-glucosidase inhibitors from four edible brown seaweed extracts by ultra-filtration and molecular docking. *LWT* **2021**, *138*, 110654. [CrossRef]
16. Wang, L.; Li, C.; Huang, Q.; Fu, X. Polysaccharide from *Rosa roxburghii* Tratt Fruit Attenuates Hyperglycemia and Hyperlipidemia and Regulates Colon Microbiota in Diabetic db/db Mice. *J. Agric. Food Chem.* **2020**, *68*, 147–159. [CrossRef] [PubMed]
17. Xing, X.; Chun, C.; Qiang, H.; Xiong, F.; Rui-Hai, L. Investigation into the mechanisms of quercetin-3-O-glucuronide inhibiting α-glucosidase activity and non-enzymatic glycation by spectroscopy and molecular docking. *Food Funct.* **2021**, *12*, 7825–7835. [CrossRef] [PubMed]
18. Chen, C.; You, L.-J.; Huang, Q.; Fu, X.; Zhang, B.; Liu, R.-H.; Li, C. Modulation of gut microbiota by mulberry fruit polysaccharide treatment of obese diabetic db/db mice. *Food Funct.* **2018**, *9*, 3732–3742. [CrossRef]
19. Liu, Y.; Deng, J.; Fan, D. Ginsenoside Rk3 ameliorates high-fat-diet/streptozocin induced type 2 diabetes mellitus in mice via the AMPK/Akt signaling pathway. *Food Funct.* **2019**, *10*, 2538–2551. [CrossRef]
20. Wu, Q.; Wu, S.; Cheng, Y.; Zhang, Z.; Mao, G.; Li, S.; Yang, Y.; Zhang, X.; Wu, M.; Tong, H. Sargassum fusiforme fucoidan modifies gut microbiota and intestinal metabolites during alleviation of hyperglycemia in type 2 diabetic mice. *Food Funct.* **2021**, *12*, 3572–3585. [CrossRef]
21. Xia, T.; Zhang, Z.; Zhao, Y.; Kang, C.; Zhang, X.; Tian, Y.; Yu, J.; Cao, H.; Wang, M. The anti-diabetic activity of polyphenols-rich vinegar extract in mice via regulating gut microbiota and liver inflammation. *Food Chem.* **2022**, *393*, 133443. [CrossRef] [PubMed]
22. Du, Y.; Li, D.; Chen, J.; Li, Y.-H.; Zhang, Z.; Hidayat, K.; Wan, Z.; Xu, J.-Y.; Qin, L.-Q. Lactoferrin improves hepatic insulin resistance and pancreatic dysfunction in high-fat diet and streptozotocin-induced diabetic mice. *Nutr. Res.* **2022**, *103*, 47–58. [CrossRef]
23. Luo, D.; Mu, T.; Sun, H. Sweet potato (*Ipomoea batatas* L.) leaf polyphenols ameliorate hyperglycemia in type 2 diabetes mellitus mice. *Food Funct.* **2021**, *12*, 4117–4131. [CrossRef] [PubMed]
24. Wu, S.; Zuo, J.; Cheng, Y.; Zhang, Y.; Zhang, Z.; Wu, M.; Yang, Y.; Tong, H. Ethanol extract of *Sargarsum fusiforme* alleviates HFD/STZ-induced hyperglycemia in association with modulation of gut microbiota and intestinal metabolites in type 2 diabetic mice. *Food Res. Int.* **2021**, *147*, 110550. [CrossRef] [PubMed]
25. Sun, X.; Yamasaki, M.; Katsube, T.; Shiwaku, K. Effects of quercetin derivatives from mulberry leaves: Improved gene expression related hepatic lipid and glucose metabolism in short-term high-fat fed mice. *Nutr. Res. Pract.* **2014**, *9*, 137–143. [CrossRef] [PubMed]
26. Zhou, Y.-J.; Xu, N.; Zhang, X.-C.; Zhu, Y.-Y.; Liu, S.-W.; Chang, Y.-N. Chrysin Improves Glucose and Lipid Metabolism Disorders by Regulating the AMPK/PI3K/AKT Signaling Pathway in Insulin-Resistant HepG2 Cells and HFD/STZ-Induced C57BL/6J Mice. *J. Agric. Food Chem.* **2021**, *69*, 5618–5627. [CrossRef] [PubMed]
27. Zhang, S.; Zheng, L.; Dong, D.; Xu, L.; Yin, L.; Qi, Y.; Han, X.; Lin, Y.; Liu, K.; Peng, J. Effects of flavonoids from *Rosa laevigata* Michx fruit against high-fat diet-induced non-alcoholic fatty liver disease in rats. *Food Chem.* **2013**, *141*, 2108–2116. [CrossRef]
28. Singh, A.; Kukreti, R.; Saso, L.; Kukreti, S. Mechanistic Insight into Oxidative Stress-Triggered Signaling Pathways and Type 2 Diabetes. *Molecules* **2022**, *27*, 950. [CrossRef]
29. Gheda, S.; Naby, M.A.; Mohamed, T.; Pereira, L.; Khamis, A. Antidiabetic and antioxidant activity of phlorotannins extracted from the brown seaweed *Cystoseira compressa* in streptozotocin-induced diabetic rats. *Environ. Sci. Pollut. Res.* **2021**, *28*, 22886–22901. [CrossRef]
30. Liu, Q.; Zhang, F.G.; Zhang, W.S.; Pan, A.; Yang, Y.L.; Liu, J.F.; Li, P.; Liu, B.L.; Qi, L.W. Ginsenoside Rg1 Inhibits Glucagon-Induced Hepatic Gluconeogenesis through Akt-FoxO1 Interaction. *Theranostics* **2017**, *7*, 4001–4012. [CrossRef]

31. Zhang, Y.; Cao, Y.; Chen, J.; Qin, H.; Yang, L. A New Possible Mechanism by Which Punicalagin Protects against Liver Injury Induced by Type 2 Diabetes Mellitus: Upregulation of Autophagy via the Akt/FoxO3a Signaling Pathway. *J. Agric. Food Chem.* **2019**, *67*, 13948–13959. [CrossRef] [PubMed]
32. Kang, O.-H.; Shon, M.-Y.; Kong, R.; Seo, Y.-S.; Zhou, T.; Kim, D.-Y.; Kim, Y.-S.; Kwon, D.-Y. Anti-diabetic effect of black ginseng extract by augmentation of AMPK protein activity and upregulation of GLUT2 and GLUT4 expression in db/db mice. *BMC Complement. Altern. Med.* **2017**, *17*, 341. [CrossRef] [PubMed]
33. Gao, Y.; Zhang, M.; Zhang, R.; You, L.; Li, T.; Liu, R.H. Whole Grain Brown Rice Extrudate Ameliorates the Symptoms of Diabetes by Activating the IRS1/PI3K/AKT Insulin Pathway in db/db Mice. *J. Agric. Food Chem.* **2019**, *67*, 11657–11664. [CrossRef] [PubMed]
34. Sahin, K.; Orhan, C.; Kucuk, O.; Tuzcu, M.; Sahin, N.; Ojalvo, S.P.; Komorowski, J. Effects of Magnesium Biotinate Supplementation on Serum Insulin, Glucose, and Lipid Parameters Along with Gene Expressions of Intermediary Metabolism in Rats. *Res. Sq.* **2020**, *34*, 9–19. [CrossRef]
35. Chakraborty, D.; Mukherjee, A.; Sikdar, S.; Paul, A.; Ghosh, S.; Khuda-Bukhsh, A.R. [6]-Gingerol isolated from ginger attenuates sodium arsenite induced oxidative stress and plays a corrective role in improving insulin signaling in mice. *Toxicol. Lett.* **2012**, *210*, 34–43. [CrossRef] [PubMed]
36. Sharma, A.; Kashyap, D.; Sak, K.; Tuli, H.S.; Sharma, A.K. Therapeutic charm of quercetin and its derivatives: A review of research and patents. *Pharm. Pat. Anal.* **2018**, *7*, 15–32. [CrossRef] [PubMed]
37. Okamoto, M.; Irii, H.; Tahara, Y.; Ishii, H.; Hirao, A.; Udagawa, H.; Hiramoto, M.; Yasuda, K.; Takanishi, A.; Shibata, S.; et al. Synthesis of a New [6]-Gingerol Analogue and Its Protective Effect with Respect to the Development of Metabolic Syndrome in Mice Fed a High-Fat Diet. *J. Med. Chem.* **2011**, *54*, 6295–6304. [CrossRef] [PubMed]
38. Jiang, H.; Horiuchi, Y.; Hironao, K.-Y.; Kitakaze, T.; Yamashita, Y.; Ashida, H. Prevention effect of quercetin and its glycosides on obesity and hyperglycemia through activating AMPKα in high-fat diet-fed ICR mice. *J. Clin. Biochem. Nutr.* **2020**, *67*, 75–83. [CrossRef]
39. Baothman, O.A.; Zamzami, M.A.; Taher, I.; Abubaker, J.; Abu-Farha, M. The role of Gut Microbiota in the development of obesity and Diabetes. *Lipids Health Dis.* **2016**, *15*, 108. [CrossRef]
40. Jeong, M.-Y.; Jang, H.-M.; Kim, D.-H. High-fat diet causes psychiatric disorders in mice by increasing Proteobacteria population. *Neurosci. Lett.* **2019**, *698*, 51–57. [CrossRef]
41. Yan, F.; Li, N.; Shi, J.; Li, H.; Yue, Y.; Jiao, W.; Wang, N.; Song, Y.; Huo, G.; Li, B. Lactobacillus acidophilus alleviates type 2 diabetes by regulating hepatic glucose, lipid metabolism and gut microbiota in mice. *Food Funct.* **2019**, *10*, 5804–5815. [CrossRef] [PubMed]
42. Dang, F.; Jiang, Y.; Pan, R.; Zhou, Y.; Wu, S.; Wang, R.; Zhuang, K.; Zhang, W.; Li, T.; Man, C. Administration of Lactobacillus paracasei ameliorates type 2 diabetes in mice. *Food Funct.* **2018**, *9*, 3630–3639. [CrossRef] [PubMed]
43. Wu, T.-R.; Lin, C.-S.; Chang, C.-J.; Lin, T.-L.; Martel, J.; Ko, Y.-F.; Ojcius, D.M.; Lu, C.-C.; Young, J.D.; Lai, H.-C. Gut commensal *Parabacteroides goldsteinii* plays a predominant role in the anti-obesity effects of polysaccharides isolated from *Hirsutella sinensis*. *Gut* **2019**, *68*, 248–262. [CrossRef] [PubMed]
44. Baldelli, V.; Scaldaferri, F.; Putignani, L.; Del Chierico, F. The Role of Enterobacteriaceae in Gut Microbiota Dysbiosis in Inflammatory Bowel Diseases. *Microorganisms* **2021**, *9*, 697. [CrossRef] [PubMed]
45. Germani, Y.; Dauga, C.; Duval, P.; Huerre, M.; Levy, M.; Pialoux, G.; Sansonetti, P.; Grimont, P.A.D. Strategy for the detection of Helicobacter species by amplification of 16S rRNA genes and identification of *H. felis* in a human gastric biopsy. *Res. Microbiol.* **1997**, *148*, 315–326. [CrossRef] [PubMed]
46. Zhang, X.; Dong, L.; Jia, X.; Liu, L.; Chi, J.; Huang, F.; Ma, Q.; Zhang, M.; Zhang, R. Bound Phenolics Ensure the Antihyperglycemic Effect of Rice Bran Dietary Fiber in db/db Mice via Activating the Insulin Signaling Pathway in Skeletal Muscle and Altering Gut Microbiota. *J. Agric. Food Chem.* **2020**, *68*, 4387–4398. [CrossRef] [PubMed]
47. Qi, B.; Ren, D.; Li, T.; Niu, P.; Zhang, X.; Yang, X.; Xiao, J. Fu Brick Tea Manages HFD/STZ-Induced Type 2 Diabetes by Regulating the Gut Microbiota and Activating the IRS1/PI3K/Akt Signaling Pathway. *J. Agric. Food Chem.* **2022**, *70*, 8274–8287. [CrossRef] [PubMed]
48. Zhao, X.-Q.; Guo, S.; Lu, Y.-Y.; Hua, Y.; Zhang, F.; Yan, H.; Shang, E.-X.; Wang, H.-Q.; Zhang, W.-H.; Duan, J.-A. *Lycium barbarum* L. leaves ameliorate type 2 diabetes in rats by modulating metabolic profiles and gut microbiota composition. *Biomed. Pharmacother.* **2020**, *121*, 109559. [CrossRef]
49. Yu, B.; Li, A.H.; Muzny, D.; Veeraraghavan, N.; Vries, P.S.D.; Bis, J.C.; Musani, S.K.; Alexander, D.; Morrison, A.C.; Franco, O.H.; et al. Association of Rare Loss-Of-Function Alleles in *HAL*, Serum Histidine. *Circ. Cardiovasc. Genet.* **2015**, *8*, 351–355. [CrossRef]
50. Yun, H.; Sun, L.; Wu, Q.; Zong, G.; Qi, Q.; Li, H.; Zheng, H.; Zeng, R.; Liang, L.; Lin, X. Associations among circulating sphingolipids, β-cell function, and risk of developing type 2 diabetes: A population-based cohort study in China. *PLoS Med.* **2020**, *17*, e1003451. [CrossRef]
51. Liu, H.; Zhang, Z.; Li, J.; Liu, W.; Warda, M.; Cui, B.; El-Aty, A.M.A. Oligosaccharides derived from *Lycium barbarum* ameliorate glycolipid metabolism and modulate the gut microbiota community and the faecal metabolites in a type 2 diabetes mouse model: Metabolomic bioinformatic analysis. *Food Funct.* **2022**, *13*, 5416–5429. [CrossRef]

52. Svingen, G.F.T.; Schartum-Hansen, H.; Pedersen, E.R.; Ueland, P.M.; Tell, G.S.; Mellgren, G.; Njølstad, P.R.; Seifert, R.; Strand, E.; Karlsson, T.; et al. Prospective Associations of Systemic and Urinary Choline Metabolites with Incident Type 2 Diabetes. *Clin. Chem.* **2016**, *62*, 755–765. [CrossRef]
53. Yang, Z.; Dan, W.; Li, Y.; Zhou, X.; Liu, T.; Shi, C.; Li, R.; Zhang, Y.; Zhang, J.; Yan, J.; et al. Untargeted metabolomics analysis of the anti-diabetic effect of Red ginseng extract in Type 2 diabetes Mellitus rats based on UHPLC-MS/MS. *Biomed. Pharmacother.* **2022**, *146*, 112495. [CrossRef]

foods

MDPI

Article

Protective Effects of Different Selenium Green Tea Polysaccharides on the Development of Type 2 Diabetes in Mice

Weilan Gao [1,2], Zhan Zheng [1], Xuehua Wang [1,*,†], Li Wang [1], Na Zhang [1], Haiyuan Liu [3], Xin Cong [1,3], Shuyi Li [1] and Zhenzhou Zhu [1,†]

1 National R&D Center for Se-Rich Agricultural Products Processing, Hubei Engineering Research Center for Deep Processing of Green Se-Rich Agricultural Products, School of Modern Industry for Selenium Science and Engineering, Wuhan Polytechnic University, Wuhan 430048, China; weilangao@126.com (W.G.); 17671055616@163.com (Z.Z.); lwang@whpu.edu.cn (L.W.); 15038327156@163.com (N.Z.); congxinwhpu@whpu.edu.cn (X.C.); shuyi.li198708@gmail.com (S.L.); zhenzhouzhu@126.com (Z.Z.)
2 College of Food Science and Engineering, Wuhan Polytechnic University, Wuhan 430023, China
3 Enshi Se-Run Material Engineering Technology Co., Ltd., Enshi 445000, China; liuhaiyuan51126@163.com
* Correspondence: wangxuehua@whpu.edu.cn
† These authors contributed equally to this work.

Abstract: Selenium polysaccharides have attracted significant interest due to their superior function to that of individual polysaccharides. However, limited research has compared the protective effects of different selenium polysaccharides from different selenization methods on diabetes. This work aims to compare the preventive effects of natural selenium-enriched green tea polysaccharides (NSe-TPS), synthetic selenized green tea polysaccharides (PCSe-TPS), and a mixture of sodium selenite and green tea polysaccharides (ordinary tea polysaccharides (Ord-TPS)+Se) on the development of diabetes. While establishing a diabetes model induced by a high-sugar, high-fat diet combined with streptozotocin, different selenium polysaccharides were administered daily by gavage for nine weeks. Our findings indicate that PCSe-TPS exhibited superior preventive effects on developing type 2 diabetes compared to NSe-TPS and Ord-TPS+Se. PCSe-TPS effectively regulated glucose metabolism and insulin resistance by activating the PI3K/Akt pathway, thereby preventing elevated blood glucose levels. Additionally, PCSe-TPS mitigated oxidative damage and inflammatory responses in liver tissues. Notably, PCSe-TPS intervention reversed the decline in bacterial species richness and the abundance of *unclassified_Oscillospiraceae* during the development of diabetes in mice. These results provide valuable insights into the protective effects of PCSe-TPS against diabetes development, highlighting its advantages over NSe-TPS and Ord-TPS+Se.

Keywords: tea polysaccharides; selenium-enriched; synthetic selenized; selenium form; type 2 diabetes prevention

Citation: Gao, W.; Zheng, Z.; Wang, X.; Wang, L.; Zhang, N.; Liu, H.; Cong, X.; Li, S.; Zhu, Z. Protective Effects of Different Selenium Green Tea Polysaccharides on the Development of Type 2 Diabetes in Mice. *Foods* **2023**, *12*, 4190. https://doi.org/10.3390/foods12234190

Academic Editor: Souhail Besbes

Received: 16 October 2023
Revised: 6 November 2023
Accepted: 17 November 2023
Published: 21 November 2023

1. Introduction

According to the International Diabetes Federation, by 2022, 537 million adults worldwide, or one in ten adults, will be living with diabetes [1]. Diabetes mellitus has emerged as a significant chronic disease, ranking third in terms of threatening human health, following cancer and cardiovascular diseases. Approximately 95% of diabetes is type 2 diabetes (T2D) [2]. Insulin resistance (IR), mainly due to dysregulation of insulin signaling pathways, is generally considered to be the crucial pathogenesis of T2D. Impaired glucose and lipid metabolism, pancreatic islet cell damage, oxidative stress, inflammation, and weight loss are the typical features of T2D [3]. The main modifiable risk factors for T2D are poor diet, sedentary lifestyle, and obesity [4]. Previous studies have found a negative association between T2D and consuming fish, low-fat dairy products, nuts, and fresh fruits and vegetables [5]. This suggests that dietary intervention may slow or even halt the progression of T2D; therefore, using functional foods is a recommended strategy for preventing the onset of diabetes. Furthermore, adopting proactive measures to prevent the development of diabetes is more straightforward and more effective than treating the disease after it has already manifested.

Selenium (Se), an essential trace element, is known to have implications for T2D risk, with both inadequate and excessive dietary selenium intake found to increase this risk [6]. The World Health Organization recommends 60–200 mg of Se per day for healthy adults, with a maximum tolerable intake of 400 mg [7]. There are two natural forms of Se: inorganic and organic. Organic Se has been reported to have a higher toxicity tolerance dose and bioavailability than inorganic Se [8]. Selenium polysaccharides, consisting of Se and polysaccharides, are an essential form of organic Se. These selenium polysaccharides can be classified into two types: natural selenium polysaccharides derived from selenium-containing plants or microorganisms and synthetic selenium polysaccharides synthesized by the selenization reaction of ordinary polysaccharides and selenium [9]. The selenium content in ordinary selenium-enriched tea polysaccharides is around 0.1–10 mg/kg, while the selenium content in synthetic selenized tea polysaccharides can be up to 10–100 times higher [9]. A recent study found that ordinary and selenium-enriched black tea showed similar alleviation effects on IR and hyperglycemia [10]. Meanwhile, another study found that the selenized polysaccharides were more effective in inhibiting α-amylase and α-glucosidase than individual polysaccharides from *Ribes nigrum* L. [11]. In addition, a recent study showed that the structural characteristics and α-glucosidase inhibition activity of tea polysaccharides (TPS) were significantly affected by different methods of selenization [12]. These findings suggest that the dose of selenium or different selenization methods of polysaccharides are essential for their hypoglycemic effect. However, few studies have compared the hypoglycemic effect of different Se polysaccharides from different artificial selenization methods in vivo, primarily a comprehensive comparison between the hypoglycemic effects of the ordinary selenium-enriched, synthetic selenized and mixed-selenium polysaccharides.

Therefore, this study aimed to compare the protective effects of selenium-enriched, synthetic selenized, and mixed-selenium green tea polysaccharides on the development of T2D in mice. Our study explores the mechanisms of different selenium green tea polysaccharides in preventing the onset of diabetes and their relationships and differences and provides a theoretical basis for utilizing synthetic selenized green tea polysaccharides for special populations.

2. Materials and Methods

2.1. Preparation of Different Selenium Green Tea Polysaccharides

Ordinary green tea and naturally selenium-enriched green tea were purchased from Blue Baked Tea Co., Ltd. (Enshi, Hubei, China). The selenium content of the selenium-enriched tea was 2.16 ± 0.08 μg/g, while that of the ordinary tea was 0.06 ± 0.01 μg/g. Ordinary tea polysaccharides (Ord-TPS) and natural selenium-rich tea polysaccharides (NSe-TPS) were prepared according to our previous study [13]. The chemically synthetic selenized tea polysaccharide with pulsed electric fields (PCSe-TPS) was prepared based on the last study with minor modifications [14]. In brief, Ord-TPS (0.5 g) was dissolved in 50 mL HNO3 (0.5%, v/v), then Na_2SeO_3 (0.5 g) was added. Pulsed electric fields (5 kV/cm/30 °C/4 times) were introduced to promote selenylation. The mixture was cooled (room temperature), and the pH value was adjusted to 6–7 ($NaHCO_3$, 1 mol/L). The solution was collected and dialyzed (Mw cutoff: 3500 Da, Beijing Solarbio Technology Co. Ltd., Beijing, China) with ultrapure water for 48 h to remove the unreacted reagents until the solution was colorless when ascorbic acid was added. Afterward, four times the volume of anhydrous ethanol was added to precipitate the tea polysaccharides. The precipitates were then concentrated and lyophilized to obtain synthetic selenized polysaccharides (PCSe-TPS). The selenium content of PCSe-TPS was 246.93 ± 3.95 μg/g. The inorganic selenium-mixed polysaccharide (Ord-TPS+Se) was prepared by mixing Ord-TPS and Na2SeO3 according to the ratio of polysaccharide and selenium in the PCSe-TPS.

2.2. Animal Experiment

C57BL/6 male mice (6-week-old) were obtained from the Hubei Province Laboratory Animal Research Center and housed under standard pathogen-free conditions with a 12 h dark/light cycle (23 °C, 55 ± 5% humidity). After one week of adaptation, the mice were divided into seven groups (n = 12): normal control (NC), diabetes model control (DC), positive control (MET), Ord-TPS, NSe-TPS, PCSe-TPS, and Ord-TPS+Se (Figure S1). The NC group was fed a basic diet, while the other groups were fed a high-sucrose and high-fat diet (formula: basic feed + 10% lard + 20% sucrose + 2.5% cholesterol + 1% sodium cholate, the diet provided 3.941 kcal/g, of which 15.6% was from protein, 31.1% from fat, and 53.3% from carbohydrate.). The MET was gavaged with metformin, and the intervention groups were gavaged with different selenium tea polysaccharides (200 mg/kg, the dose chosen through a preliminary experiment (Figure S2)) for six weeks. The daily selenium intake of mice in the SSe-TPS and Ord-TPS+Se groups was calculated via adults (70 kg, 400 μg/kg) by body surface area [15]. After intervention for seven weeks, the mice received intraperitoneal injections of streptozotocin (STZ) for two consecutive days at 100 mg/kg according to the previous study with minor modification [16], and the NC group was administered the exact dosage of sodium citrate buffer. Following two weeks of intraperitoneal injection, fasting blood glucose (FBG) was measured in mice after 12 h of fasting. The FBG values of all mice in the DC group were above 11.1 mmol/L, suggesting that the diabetic mice model was successfully established. At the end of the experiment, the FBG and oral glucose tolerance test (OGTT) were measured; samples of cecum content, plasma, liver, pancreas, and other internal organs were collected. Fresh organs were weighed promptly to calculate the organ index, as described in (1).

$$\text{Organ index} = \text{organ wet weight/body weight of mice} \tag{1}$$

Animal experiments were conducted at the Hubei Province Laboratory Animal Research Center (Hubei, China). This project passed the ethical review of laboratory animal welfare by the Hubei Center for Disease Control and Prevention and the Laboratory Animal Management and Use Committee, and the approval number is Safety Evaluation Center dynamic (Fu) No. 202210207. Experiments were conducted in compliance with national regulations and local guidelines.

2.3. Biochemical Analysis

All mice were fasted for 12 h the night before the OGTT test. Mice were given glucose (1.0 g/kg) according to body weight as described in a previous study [17], and the blood glucose values were measured at 30, 60, and 120 min. The changes in blood glucose concentration and the area under the curve (AUC) of the blood glucose at each time point were observed. The formula is shown in Equation (2):

$$\text{AUC (mmol/L)} = (0.5(\text{A} + \text{B}) + 1.5(\text{B} + \text{C}))/2 \tag{2}$$

where A, B, and C represent the blood glucose values at 0, 30, and 120 min, respectively.

The serum insulin content (INS) was measured using an enzyme-linked immunosorbent assay (ELISA). INS ELISA kit was purchased from Jiangsu Enzyme Immunoassay Industry Co., LTD (Yancheng, China), and the insulin resistance index (HOMA-IR) was evaluated using the provided Equation (3) below.

$$\text{HOMA} - \text{IR} = (\text{FBG} \times \text{INS})/22.5 \tag{3}$$

where FBG represents the fasting blood glucose (mmol/L), and INS stands for insulin content (mIU/L).

Serum triglyceride (TG), total cholesterol (TC), high-density lipoprotein cholesterol (HDL-C), and low-density lipoprotein cholesterol (LDL-C) were measured using a serum biochemistry analyzer. The enzyme activities of aspartate aminotransferase (AST), alanine aminotransferase (ALT), and alkaline phosphatase (ALP) were measured using ELISA. Kits were purchased from Radu Life Science Co., Ltd. (Shenzhen, China).

2.4. Histopathological Analysis of the Pancreas and Liver

The pancreas and a portion of the liver were fixed in paraformaldehyde (4%), dehydrated in gradient ethanol, washed in xylene, embedded in paraffin, sectioned (4 μm), and baked. Afterward, the tissue sections were stained with hematoxylin and eosin and observed under a microscope.

2.5. Measurement of Oxidative Stress and Inflammatory Parameters in the Liver

The superoxide dismutase (SOD), glutathione peroxidase (GSH-Px), malondialdehyde (MDA), and total antioxidant capacity (T-AOC) in the liver homogenate were detected by commercial kits and following the manufacturers' instructions. Determination kits were purchased from Nanjing Jiancheng Bioengineering Institute (Nanjing, China). The protein expression levels of interleukin-1β (IL-1β), interleukin-6 (IL-6), and tumor necrosis factor-α (TNF-α) in the liver homogenate were determined using ELISA; ELISA kits were purchased from Jiangsu Enzyme Immunoassay Industry Co., Ltd. (Yancheng, China).

2.6. Real-Time Quantitative PCR

Liver tissue RNA was extracted using the kit (Novica Zhan Biotechnology Co., LTD., Nanjing, China). The quality of the RNA was estimated using the OD260/OD280 ratio. The experimental requirements were met if the ratio was between 1.8 and 2.0. The total RNA was stored in a refrigerator at −80 °C until use. Reverse transcribed into cDNA using a two-step process. In the first step, the gDNA was removed. The reaction system contained 4 μL of 4× gDNA wiper rMix and 4 μg of total RNA added to 16 μL of nuclease-free water, which was gently blown and mixed with a pipetting gun. The reaction was performed (42 °C, 2 min). RNA was reverse transcribed in the second step, and the reaction system consisted of 4 μL of 5× HiScrip II qRT SuperMix II and the first step reaction solution. The cDNA was obtained using the following conditions: 50 °C: 15 min→85 °C: 5 s. The reaction system, reaction procedure, and primers are shown in supplementary materials. (Tables S1–S3). The relative quantification was calculated by the comparative $2^{-\Delta\Delta CT}$ method.

2.7. Gut Microbiota Analysis

Total DNA was extracted from the cecal contents, and the DNA extraction quality was assessed using electrophoresis on a 1% agarose gel. Subsequently, the v3-v4 region of the 16S rRNA gene was amplified using PCR with the forward primer 338F (5′-ACTCCTACGGGAGGCAGCA-3′) and the reverse primer 806R (5′-GGACTACHVGGGT WTCTAAT-3′). The Illumina NovaSeq6000 platform was employed for constructing a small fragment library for sequencing. The obtained sequences underwent quality filtering using Trimmomatic v0.33 software. Primer sequences were identified and removed with Cutadapt 1.9.1 software, resulting in the acquisition of high-quality sequences. QIIME2 2020.6 was used for sequence denoising, followed by eliminating chimeric sequences. Subsequently, operational taxonomic units (OTUs) were identified at a 97% similarity threshold. Sequencing data were deposited in the NCBI Sequence Read Archive (SRA) (http://www.ncbi.nlm.nih.gov/sra/ (accessed on 16 October 2023)) under the Bioproject ID PRJNA1028394. The α diversity index of the samples, encompassing the Chao1, Ace, Simpson Index, Shannon Index, and Rarefaction Curves, was evaluated using QIIME software. To compare the β diversity of species among different samples, principal coordinate analysis (PCoA) was performed. Lastly, linear discriminant analysis effect size (LEfSe) and linear discriminant analysis (LDA) were conducted to identify biomarkers with statistically significant differences between various groups.

2.8. Statistical Analysis

Results are expressed as mean ± standard error of the mean (SEM). The Kruskal–Wallis test is performed to compare medians between nonnormally distributed groups. Comparisons among groups were performed by one-way analysis of variance (ANOVA) followed by the Turkey post hoc test and graphed by GraphPad Prism 8.

3. Results

3.1. Protective Effects of Different Selenium Tea Polysaccharides on Basic Physiological Indices in the Development of Diabetic Mice

After STZ injection at week 7, a significant decrease in body weight was observed in mice, followed by significantly higher food and water intake in the DC group compared to the NC group at week 9, indicating typical symptoms of diabetes [4] (Figure 1). The positive control group and different selenium tea polysaccharides groups showed promising inhibition of body weight loss, polydipsia, and polyphagia (Figure 1). The DC group exhibited significantly increased indices of the liver, kidney, and pancreas (Table 1), suggesting swelling and dysfunction of these organs in diabetic mice. The supplementation of different selenium tea polysaccharides demonstrated improvements in the mentioned symptoms (Table 1), corroborating the liver and pancreas histopathological analysis. As shown in Figure 1D, histological examination revealed hepatocyte nuclei swelling, disordered arrangement, cytoplasmic vacuolization (black arrow), and hepatic steatosis in the DC group. However, the selenium tea polysaccharide groups showed less degeneration of nuclei and inflammatory infiltration, with the best protective effect observed in the PCSe-TPS group. Furthermore, the intervention of selenium tea polysaccharides demonstrated a notable mitigation of the reduction in β-cell count and the degeneration and cavitation of islets. These beneficial effects were particularly pronounced in the PCSe-TPS and NSe-TPS groups (Figure 1E).

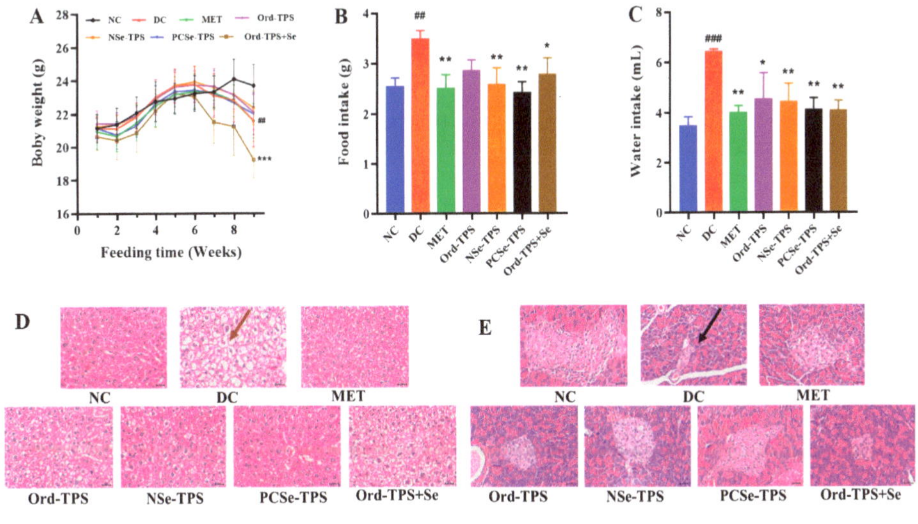

Figure 1. Effects of different selenium green tea polysaccharides on (**A**) body weight (n = 12), (**B**) food intake (n = 12), (**C**) water intake (n = 12), (**D**) liver, and (**E**) pancreas histopathology in mice (n = 3). Cytoplasmic vacuolization (red arrow) and β-cell count (black arrow). The magnification was 400×, and the ruler was 100 μm. NC: normal control; DC: diabetes model control; MET: metformin positive control; Ord-TPS: ordinary tea polysaccharides; NSe-TPS: natural selenium-rich tea polysaccharides; PCSe-TPS: the chemically synthetic selenized tea polysaccharide with pulsed electric fields; Ord-TPS+Se: Ord-TPS and Na_2SeO_3 mixture. ## $p < 0.01$, ### $p < 0.001$, compared to normal control (NC) group, * $p < 0.05$, ** $p < 0.01$, *** $p < 0.001$ compared to diabetes model (DC) group.

Table 1. Effects of different selenium green tea polysaccharides on organ index of mice.

Groups	Heart (%)	Liver (%)	Spleen (%)	Lungs (%)	Kidney (%)	Thymus (%)	Pancreas (%)	Stomach (%)
NC	0.52 ± 0.06	3.36 ± 0.29	0.21 ± 0.02	0.56 ± 0.03	1.03 ± 0.10	0.19 ± 0.05	0.61 ± 0.15	0.82 ± 0.13
DC	0.53 ± 0.09	4.99 ± 0.42 ###	0.24 ± 0.02	0.62 ± 0.05	1.31 ± 0.27 ###	0.18 ± 0.05	0.93 ± 0.05 ###	0.70 ± 0.19
MET	0.53 ± 0.05	4.25 ± 0.34 ***	0.26 ± 0.05	0.62 ± 0.04	1.12 ± 0.03 *	0.18 ± 0.06	0.72 ± 0.16 **	0.79 ± 0.18
Ord-TPS	0.56 ± 0.07	4.43 ± 0.51 *	0.27 ± 0.04	0.62 ± 0.04	1.15 ± 0.10 *	0.17 ± 0.03	0.82 ± 0.20	0.78 ± 0.09
NSe-TPS	0.55 ± 0.10	4.33 ± 0.31 **	0.23 ± 0.03	0.59 ± 0.03	1.12 ± 0.07 *	0.14 ± 0.03	0.81 ± 0.06	0.69 ± 0.09
PCSe-TPS	0.54 ± 0.07	4.13 ± 0.40 ***	0.26 ± 0.05	0.61 ± 0.06	1.11 ± 0.06 **	0.15 ± 0.04	0.79 ± 0.09	0.76 ± 0.06
Ord-TPS+Se	0.65 ± 0.11 ##	4.46 ± 0.57 *	0.49 ± 0.06 ###	0.77 ± 0.12 ###	1.17 ± 0.13	0.15 ± 0.06	0.82 ± 0.16	0.77 ± 0.10

Mean values ± standard deviation (n = 12). NC: normal control; DC: diabetes model control; MET: metformin positive control; Ord-TPS: ordinary tea polysaccharides; NSe-TPS: natural selenium-rich tea polysaccharides; PCSe-TPS: the chemically synthetic selenized tea polysaccharide with pulsed electric fields; Ord-TPS+Se: Ord-TPS and Na_2SeO_3 mixture. ## $p < 0.01$, ### $p < 0.001$, compared to normal control (NC) group, * $p < 0.05$, ** $p < 0.01$, *** $p < 0.001$ compared to diabetes model (DC) group.

It was observed that the Ord-TPS+Se group exhibited a more pronounced weight loss compared to the other groups at week 9 ($p < 0.001$) (Figure 1A). Moreover, the heart, spleen, and lung indices were significantly elevated in the Ord-TPS+Se group compared to the NC and DC groups (Table 1). This phenomenon could be attributed to the prolonged consumption of inorganic selenium, which has a higher toxicity level than organic selenium [8].

3.2. Protective Effects of Different Selenium Tea Polysaccharides on Glucose and Lipid Metabolism in the Development of Diabetic Mice

The FBG levels of the model group were consistently higher than those of the other groups from the 5th week, indicating high-sucrose and high-fat diet-induced hyperglycemia (Figure 2A). By the 9th week, the FBG level of the DC group had reached 15.06 mmol/L, confirming the successful establishment of the diabetic model (Figure 2B). Different selenium tea polysaccharide interventions inhibited FBG elevation, and PCSe-TPS showed a better protective effect with lower FBG levels. In addition, the results of the OGTT demonstrated that the PCSe-TPS and Ord-TPS+Se interventions effectively improved glucose intolerance (Figure 2C). This was evidenced by the significantly lower AUC values compared to the control group (Figure 2D). Higher hepatic glycogen was also found in the PCSe-TPS groups than in the DC group (Figure 2K). This suggests that glucose in the blood is converted to synthesize liver glycogen. The HOMA-IR index was utilized to assess insulin resistance. The HOMA-IR level in the diabetic group was 110.00% higher than that in the NC group, indicating the occurrence of insulin resistance. Compared to the DC group, interventions with different selenium tea polysaccharides reduced the HOMA-IR levels, especially the PCSe-TPS and Ord-TPS+Se interventions, which showed significance (Figure 2F). It suggested that PCSe-TPS and Ord-TPS+Se could be beneficial for alleviating IR in the development of diabetic mice.

In addition, different selenium tea polysaccharides tended to counteract the increase in TG and LDL-C and the decrease in HDL-C; however, there was no statistical significance elicited by these groups (Figure 2G–J). Only PCSe-TPS significantly inhibited the TC content elevation in the development of diabetic mice (Figure 2G). It suggests that PCSe-TPS exhibits a significant effect in alleviating hyperlipidemia in the development of diabetic mice.

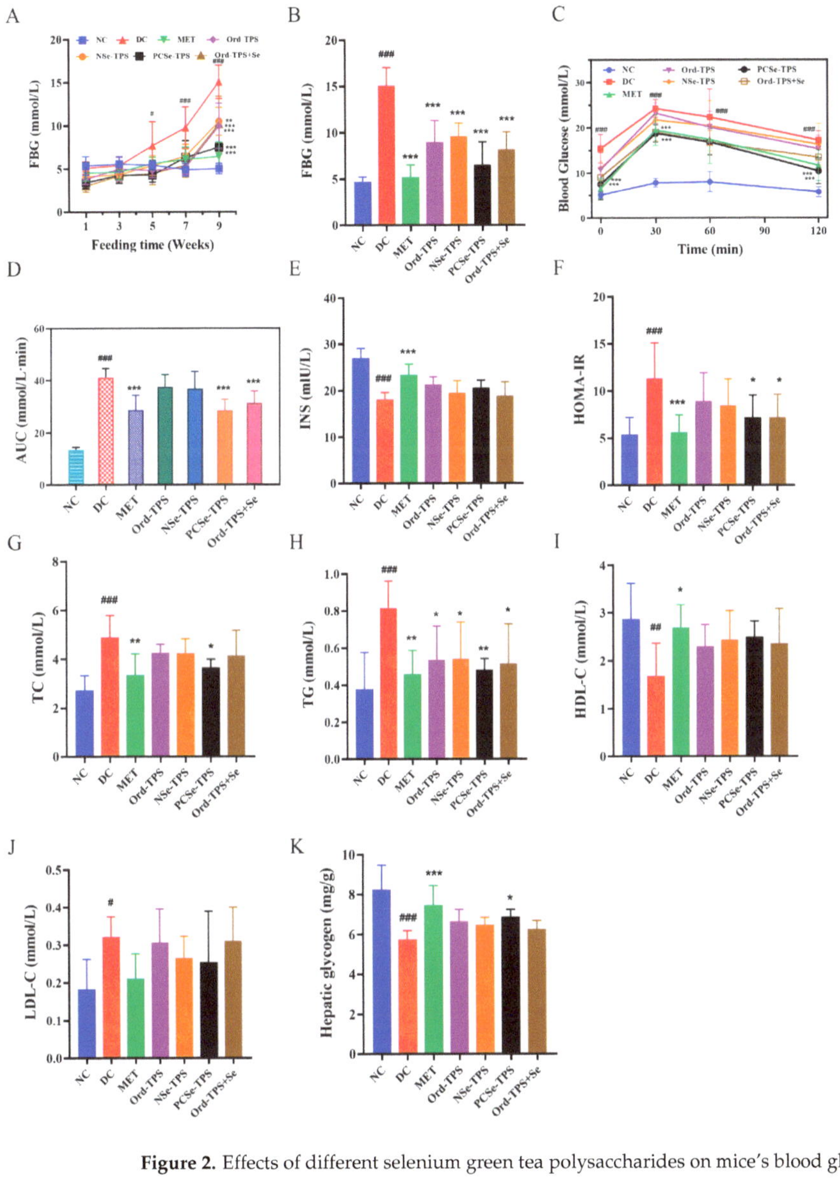

Figure 2. Effects of different selenium green tea polysaccharides on mice's blood glucose and blood lipids (n = 8): (**A**) blood glucose changes, (**B**) FBG, (**C**) INS, (**D**) HOMA-IR, (**E**) oral glucose tolerance test (OGTT), (**F**) area under curve (AUC), (**G**) TC, (**H**) TG, (**I**) HDL-C, (**J**) LDL-C, and (**K**) hepatic glycogen. NC: normal control; DC: diabetes model control; MET: metformin positive control; Ord-TPS: ordinary tea polysaccharides; NSe-TPS: natural selenium-rich tea polysaccharides; PCSe-TPS: the chemically synthetic selenized tea polysaccharide with pulsed electric fields; Ord-TPS+Se: Ord-TPS and Na_2SeO_3 mixture. # $p < 0.05$, ## $p < 0.01$, ### $p < 0.001$, compared to normal control (NC) group, * $p < 0.05$, ** $p < 0.01$, *** $p < 0.001$ compared to diabetes model (DC) group.

3.3. Different Selenium Tea Polysaccharides Enhanced Insulin Signaling Pathway PI3K/Akt

The PI3K/AKT signaling pathway mediated by insulin is essential in the pathogenesis of diabetes mellitus [18]. To explore the possible mechanism of different selenium tea polysaccharides in alleviating blood glucose elevation and insulin resistance, we quantified the mRNA expression levels of key targets involved in insulin sensitivity, which included the PI3K/AKT pathway and its upstream components (phosphatase and tensin homolog (PTEN) and insulin receptor substrate 1 (IRS-1)) as well as downstream elements (glycogen synthase kinase-3β (GSK-3β) and glucose transporter-2 (GLUT-2)) (Figure 3G).

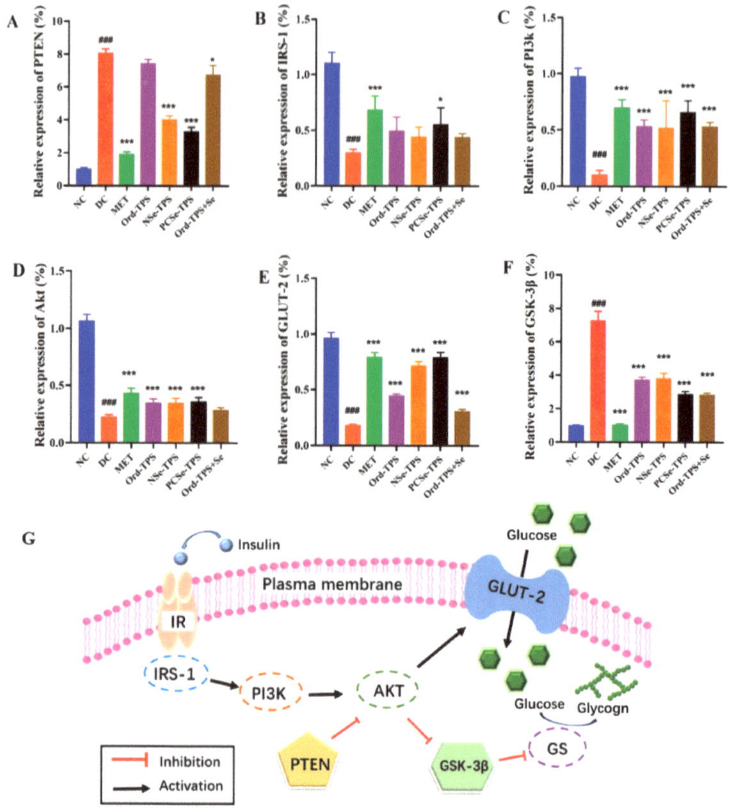

Figure 3. Effects of different selenium green tea polysaccharides on the expression of insulin resistance-related genes in mice (n = 4). (**A**) PTEN, (**B**) IRS-1, (**C**) PI3K, (**D**) Akt, (**E**) GLUT-2, (**F**) GSK-3β, (**G**) mechanisms of regulation. NC: normal control; DC: diabetes model control; MET: metformin positive control; Ord-TPS: ordinary tea polysaccharides; NSe-TPS: natural selenium-rich tea polysaccharides; PCSe-TPS: the chemically synthetic selenized tea polysaccharide with pulsed electric fields; Ord-TPS+Se: Ord-TPS and Na_2SeO_3 mixture. ### $p < 0.001$, compared to normal control (NC) group, * $p < 0.05$, *** $p < 0.001$ compared to diabetes model (DC) group.

The mRNA expression levels of PTEN and GSK-3β were significantly increased, while IRS-1, PI3k, Akt, and GLUT-2 were decreased considerably in the DC group compared to the NC group ($p < 0.001$), demonstrating the dysregulation of the insulin-mediated PI3K/AKT pathway in diabetes. Selenium tea polysaccharide interventions, especially NSe-TPS and PCSe-TPS, prevented the down-regulation of these genes (Figure 3). Notably, only PCSe-TPS was found to significantly inhibit all of the above gene expression changes in the development of diabetic mice. These results suggested that PCSe-TPS could significantly improve blood glucose levels by regulating glucose uptake and metabolism through the PI3K/Akt pathway.

3.4. Different Selenium Tea Polysaccharides Reduced Liver Dysfunction, Oxidative Stress, and Inflammation

Persistent hyperglycemia and hyperlipidemia can cause liver dysfunction, oxidative stress, and inflammation [4]. Liver dysfunction, reflected by increased ALT, AST, and ALP activities, was observed in the DC group compared to the normal group (Figure 4A–C). However, treatment with MET and different selenium tea polysaccharides alleviated these elevated enzyme activities, with PCSe-TPS significantly inhibiting their increase. Oxidative stress was also observed in the DC group, indicated by decreased SOD, T-AOC, and GSH-Px activities and increased MDA levels. As expected, different selenium tea polysaccharides inhibited T-AOC decline and MDA elevation (Figure 4D,G). Notably, only PCSe-TPS showed a significant protective effect on alleviating the activities of SOD and GSH-Px (Figure 4E,F). Elevated pro-inflammatory factors in liver tissue are an essential factor leading to inflammation and liver dysfunction. Selenium tea polysaccharide interventions attenuated the high levels of IL-6, IL-1β, and TNF-α induced by a high-sucrose and high-fat diet combined with STZ. However, only PCSe-TPS showed significantly down-regulated effects on all these pro-inflammatory factors in the development of diabetic mice (Figure 4H–J). In summary, PCSe-TPS can potentially prevent the onset of diabetes by alleviating hepatic damage, oxidative stress, and inflammation.

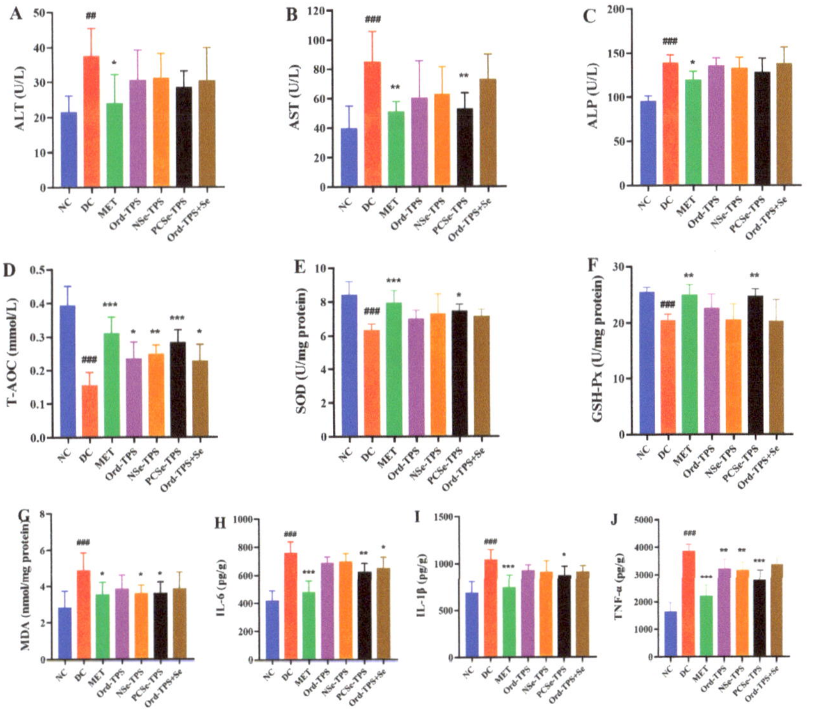

Figure 4. Effects of different selenium green tea polysaccharides on (**A**) ALT, (**B**) AST, (**C**) ALP, (**D**) T-AOC, (**E**) SOD, (**F**) GSH-PX, (**G**) MDA, (**H**) IL-6, (**I**) IL-1β and (**J**)TNF-α in the liver of mice (n = 8). NC: normal control; DC: diabetes model control; MET: metformin positive control; Ord-TPS: ordinary tea polysaccharides; NSe-TPS: natural selenium-rich tea polysaccharides; PCSe-TPS: the chemically synthetic selenized tea polysaccharide with pulsed electric fields; Ord-TPS+Se: Ord-TPS and Na_2SeO_3 mixture. ## $p < 0.01$, ### $p < 0.001$, compared to normal control (NC) group, * $p < 0.05$, ** $p < 0.01$, *** $p < 0.001$ compared to diabetes model (DC) group.

3.5. Different Selenium Tea Polysaccharides Alter the Gut Microbiota in Diabetic Mice

Since T2D leads to dysbiosis of the gut microbiota and tea polysaccharides can regulate the structure of the gut microbiota [19], the effects of different selenium tea polysaccharides on the gut microbiota structure during the development of diabetic mice were examined.

The bacterial alpha diversity refers to species richness and evenness within a given ecosystem. As shown in Figure 5, the Ace, Chao1, Shannon, and Simpson indices decreased in the DC group compared to the NC group. Although no significant alterations were observed in the Ace, Chao1, Shannon, and Simpson indices between the intervention and DC groups, PCSe-TPS significantly prevented the reduction in the Chao1 index (Figure 5). This suggests that PCSe-TPS can potentially ameliorate the decline in bacterial species richness.

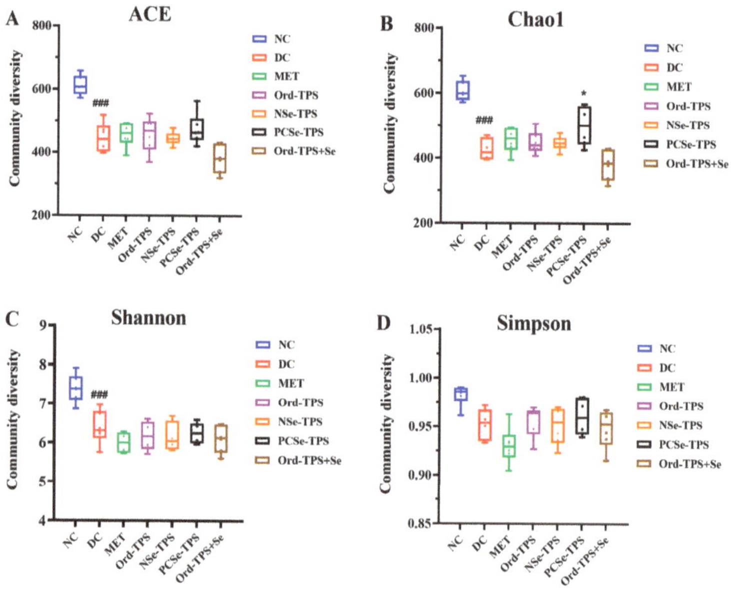

Figure 5. Effects of different selenium green tea polysaccharides on α-diversity of gut microbiota in mice (n = 6): (**A**) ACE, (**B**) Chao1, (**C**) Shannon, and (**D**) Simpson. NC: normal control; DC: diabetes model control; MET: metformin positive control; Ord-TPS: ordinary tea polysaccharides; NSe-TPS: natural selenium-rich tea polysaccharides; PCSe-TPS: the chemically synthetic selenized tea polysaccharide with pulsed electric fields; Ord-TPS+Se: Ord-TPS and Na_2SeO_3 mixture. ### $p < 0.001$, compared to normal control (NC) group, * $p < 0.05$ compared to diabetes model (DC) group.

PCoA based on OTUs revealed a distinct clustering of microbiota composition among the DC, MET, and NC groups (Figure 6A). This clustering indicated that the combination of a high-sucrose and high-fat diet with STZ injection had a significant impact on the structure of the gut microbiota community in mice, leading to gut microbiota dysbiosis. As anticipated, the dysbiosis of the gut microbiota was modulated by MET and different selenium tea polysaccharide interventions, although a slight overlap between the group ellipses of the DC and Ord-TPS, as well as the DC and NSe-TPS groups, was observed in the PCoA plot. The PCSe-TPS and Ord-TPS+Se intervention groups also displayed a similar clustering of the gut bacterial community relative to the MET group, indicating sound protective effects of PCSe-TPS and Ord-TPS+Se interventions.

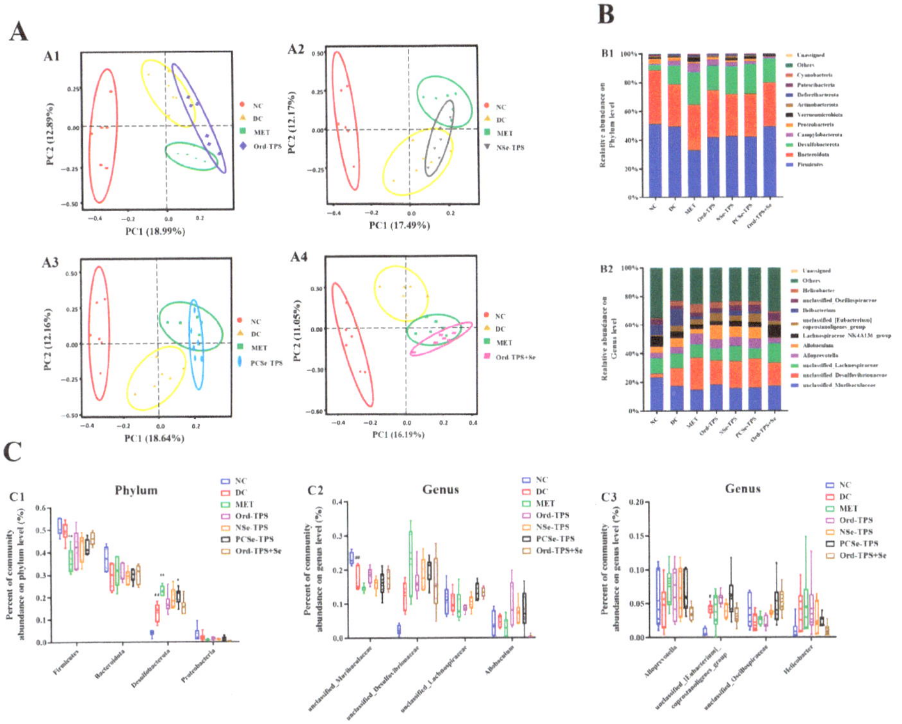

Figure 6. Effect of different selenium green tea polysaccharides on β-diversity and relative abundance of gut microbiota in mice (n = 6): (**A**) PCoA analysis based on binary Jaccard: (**A1**) NC, DC, MET and Ord-TPS groups; (**A2**) NC, DC, MET and NSe-TPS groups; (**A3**) NC, DC, MET and PCSe-TPS groups; (**A4**) NC, DC, MET and Ord-TPS+Se groups; (**B**) the gut bacterial composition at the (**B1**) phylum and (**B2**) genus level; (**C**) the relative abundance of some bacterial flora at (**C1**) phylum level and genus level of (**C2**) *unclassified_Muribaculaceae, unclassified_Desulfovibrionaceae, unclassified_Lachnospiraceae, and Allobaculum,* (**C3**) *Alloprevotella, unclassified_[Eubacterium]_coprostanoligenes_group, unclassified_Oscillospiraceae, and Helicobacter*. NC: normal control; DC: diabetes model control; MET: metformin positive control; Ord-TPS: ordinary tea polysaccharides; NSe-TPS: natural selenium-rich tea polysaccharides; PCSe-TPS: the chemically synthetic selenized tea polysaccharide with pulsed electric fields; Ord-TPS+Se: Ord-TPS and Na_2SeO_3 mixture. # $p < 0.05$, ## $p < 0.01$, compared to normal control (NC) group, * $p < 0.05$, ** $p < 0.01$, compared to diabetes model (DC) group.

The bacterial communities and their relative abundances were further investigated. At the phylum level, the DC group exhibited a significant increase in the relative abundance of *Desulfobacterota*, along with a decrease in the relative abundance of *Bacteroidota* and *Proteobacteria* compared to the NC group (Figure 6(B1)). Both the MET and selenium tea polysaccharide groups demonstrated a significant decrease in the relative abundance of *Firmicutes* and a considerable increase in the relative abundance of *Desulfobacterota* compared to the DC group. At the genus level, the relative abundances of *unclassified_Desulfovibrionaceae* and *Helicobacter* were higher, and the relative abundances of *unclassified_Muribaculaceae* and *unclassified_Oscillospiraceae* were lower in the DC group than those in the NC group (Figure 6(B2,C3)). The relative abundances of *unclassified_[Eubacterium]_coprostanoligenes_group*, *unclassified_Lachnospiraceae*, and *unclassified_Oscillospiraceae increased* in the Ord-TPS and PCSe-TPS intervention groups compared with those in the DC group. Furthermore, the relative abundance of *Allobaculum* increased in the Ord-TPS, NSe-TPS, and PCSe-TPS intervention groups but decreased in the Ord-TPS+Se intervention group compared with that of the DC group.

The LEfSe algorithm was applied to identify the specific microbial taxa that were enriched in mice with different selenium tea polysaccharides (Figure 7). Our results showed that, compared with Ord-TPS, NSe-TPS was associated with the enrichment of *Ileibacterium* (Figure 7A), PCSe-TPS was related to the enrichment of *Oscillospiraceae* (Figure 7B), and Ord-TPS+Se was associated with the enrichment of *Lachnospiraceae_NK4A136_group, unclassified-Lachnospiraceae,* and *unclassified_Oscillospiraceae* (Figure 7C). We also compared the specific microbial taxa that were enriched in mice with organic selenium tea polysaccharides (PCSe-TPS) and the mixture of inorganic selenium and tea polysaccharides (Ord-TPS+Se). This showed that PCSe-TPS was associated with the enrichment of *Erysipelotrichaceae, Allobaculum,* and *Alloprevotella,* while Ord-TPS+Se was related to the enrichment of *Firmicutes, Roseburia, unclassified-Lachnospiraceae,* and *Lachnospiraceae_NK4A136_group* (Figure 7D).

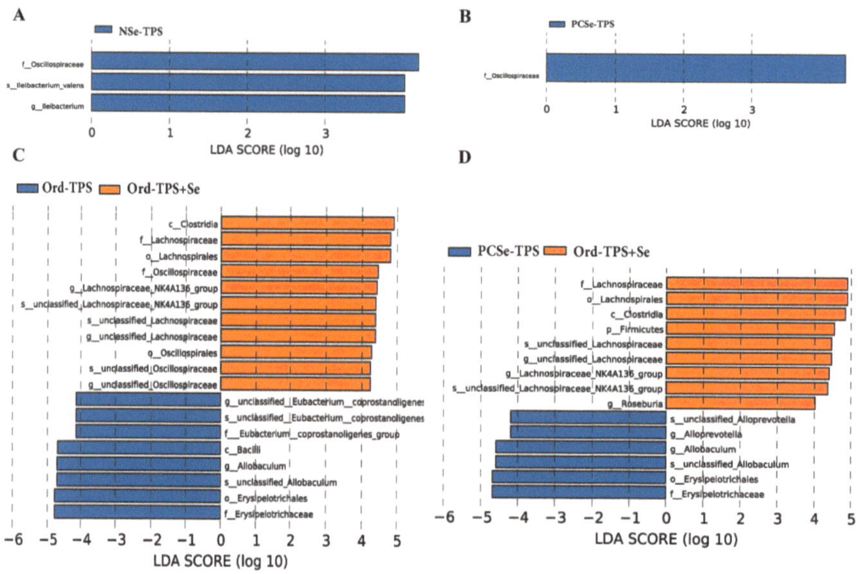

Figure 7. LEfSe analysis to identify taxonomic differences in the gut microbiota of the mice administrated with different selenium green tea polysaccharides (n = 6): (**A**) TPS and NSe-TPS, (**B**) TPS and PCSe-TPS, (**C**) TPS and Ord-TPS+Se, and (**D**) PCSe-TPS and Ord-TPS+Se (linear discriminant analysis score (log10) $\geq$ 4.0). NOrd-TPS: ordinary tea polysaccharides; NSe-TPS: natural selenium-rich tea polysaccharides; PCSe-TPS: the chemically synthetic selenized tea polysaccharide with pulsed electric fields; Ord-TPS+Se: Ord-TPS and Na_2SeO_3 mixture.

4. Discussion

In recent years, selenium polysaccharides have garnered interest due to their role as organic selenium sources and superior function to that of individual polysaccharides, such as hypoglycemic effects [11,12]. However, the protective effects of different selenium polysaccharides obtained through various selenization methods in the development of diabetes have not been comprehensively compared. Therefore, studies were carried out to compare the effects of ordinary selenium-enriched, synthetic selenized, and selenium-mixed polysaccharides from green tea on preventing the onset of diabetes.

The essential features of T2D include higher FBG levels, lower glucose tolerance and IR, and disturbances in glucolipid metabolism [4]. In our study, different selenium tea polysaccharides significantly prevented the upregulation of FBG. However, only PCSe-TPS and Ord-TPS+Se significantly prevented the upregulation of OGTT and HOMA-IR in the development of diabetic mice. This is related to the higher selenium content in PCSe-TPS and Ord-TPS+Se. In our work, the daily selenium intake of mice in the PCSe-TPS and Ord-TPS+Se groups was around 100 times higher than that of the NSe-TPS group. A previous study found that the selenized mycelial polysaccharide from *C. ventricosum* with a medium selenium content had the best anti-diabetic efficacy in mice [20]. Another study found that ordinary and selenium-enriched black tea showed similar glucose lowering effects in diabetic mice [10]. Insulin resistance can result in the dysfunction of the pancreas [21]. PCSe-TPS intervention significantly attenuated pancreatic lesions, and the intervention was best compared to the effects of the other polysaccharide groups. TC and TG levels in the blood reflect lipid metabolism [22]. Only PCSe-TPS significantly inhibited both TC and TG content elevation in the development of diabetic mice. The liver and kidney are vital metabolic organs, and the pancreas is the tissue that controls insulin secretion. As a form of inorganic selenium, Ord-TPS+Se showed its side effects by reducing body weight and increasing the organ index. Thus, the protective effects of PCSe-TPS were better in the essential characteristics of the development of diabetic mice compared to those of the other groups. It is related to the appropriate dose and form of selenium, as the toxicity tolerance dose and bioavailability of organic selenium are higher than that of inorganic selenium [8].

It is known that the abnormality of the PI3K-Akt signaling pathway is one of the main reasons for the onset of insulin resistance [4,9,23]. PTEN and IRS-1 are upstream molecules within the PI3K-Akt signaling pathway [24,25]. GLUT-2 and GSK-3β, which play crucial roles in glucose transport, absorption, and glycogen synthase activity, are regarded as downstream molecules within the PI3K-Akt signaling pathway [26]. In our study, the up-regulation of PTEN and GSK-3β and the down-regulation of IRS-1, PI3k, Akt, and GLUT-2 were effectively inhibited by different selenium tea polysaccharides, promoting liver glycogen synthesis and lowering blood glucose levels. Thus, different selenium tea polysaccharides protect against the development of diabetes in mice via the PI3K/Akt signaling pathway. However, only PCSe-TPS showed significance in regulating all the abovementioned genes. This suggests that the regulation effects of PCSe-TPS on the PI3K/Akt signaling pathway are better than those of the other selenium tea polysaccharides, which are consistent with the fundamental indices mentioned above (i.e., TC, hepatic glycogen, and pancreatic lesions (Figures 2 and 3)). This was related to the different regulatory mechanisms of organic and inorganic selenium on the development of diabetic mice because of the different levels of remission exhibited.

Diabetes is associated with the disruption of antioxidant defense mechanisms due to low levels of antioxidant enzymes and excess free radicals under a hyperglycemic environment [27]. GSH-Px is an important peroxidative catabolic enzyme widely found in the organism. The active center of GSH-Px is selenocysteine, and the magnitude of its activity reflects the selenium level of the organism [28]. As expected, the administration of PCSe-TPS resulted in a significant increase in GSH-Px and SOD levels and a decrease in MDA levels. However, the effects of Ord-TPS+Se on these antioxidant enzymes' levels were not statistically significant. It was also found that organic selenium supplementation exhibited higher erythrocyte GSH-Px activity than inorganic selenium supplementation [29]. Additionally, the activity of GSH-Px and the selenium content in the whole blood of lambs that were fed organic selenium (selenium-enriched yeast and selenium-enriched probiotics) were found to be higher compared to lambs that were fed sodium selenite [28]. This suggests that the antioxidant effect of organic selenium is more effective than inorganic selenium in vivo, which may be related to the bioavailability of selenium in different forms [8]. Our study revealed that the onset of oxidative stress is accompanied by inflammation, leading to increased levels of pro-inflammatory factors such as IL-6, IL-1β, and TNF-α in diabetic mice. Elevated levels of pro-inflammatory factors in liver tissue contribute to the inflammatory response that leads to liver dysfunction [30]. Serum ALT, AST, and ALP levels were considered as liver function damage markers [31]. As expected, PCSe-TPS intervention significantly decreased the levels of IL-6, IL-1β, and TNF-α, inhibited the elevation of ALT, AST, and ALP, and alleviated the liver tissue lesions in the development of diabetic mice.

Analysis of the fecal microbiota demonstrated that selenium tea polysaccharides significantly alleviated gut microbiota dysbiosis in the development of diabetic mice. Our findings demonstrated that although only PCSe-TPS exhibited an inhibitory effect on the decreased richness of gut microbiota in diabetic mice, all interventions of selenium tea polysaccharides caused alterations in the β-diversity of gut bacteria during the development of diabetes in mice. The abundance of *Oscillospira* was significantly reduced in the gut microbiota of diabetes mellitus patients [32]. In our study, the declining abundance of *unclassified_Oscillospiraceae* in the development of diabetic mice was reversed by Ord-TPS and PCSe-TPS intervention. *Unclassified_Oscillospiraceae* was found to possess the ability to synthesize short-chain fatty acids (SCFAs), specifically butyrate. SCFAs are essential in regulating energy metabolism and maintaining the intestinal environment's homeostasis [33]. In addition, acetate and butyrate have been found to show a protective effect on β-cell survival in vitro [34]. Sodium-butyrate treatment indirectly increased insulin secretion, mainly by participating in glucagon-like peptide-1 (GLP-1) secretion [35]. GLP-1 promotes the secretion of insulin and has good hypoglycemic function. Thus, Ord-TPS and PCSe-TPS intervention could up-regulate the relative abundance of *unclassified Oscillospiraceae* to promote butyrate production and then enhance the secretion of GLP-1 to inhibit FBG elevation in the development of diabetic mice. The abundance of unclassified *Oscillospiraceae* and the level of GLP-1 were both found to be up-regulated in diabetic rats with pectic polysaccharide intervention [36]. Furthermore, compared with the DC group, the Ord-TPS and PCSe-TPS intervention up-regulated the relative abundance of *unclassified_[Eubacterium]_coprostanoligenes_group*, *unclassified_Lachnospiraceae*, and *Allobaculum*, while Ord-TPS+Se intervention decreased the relative abundance of *Allobaculum*. *Eubacterium_coprostanoligenes_group* and *Allobaculum* were identified as producers of SCFAs. It was observed that *Eubacterium_coprostanoligenes_group* demonstrated anti-dyslipidemia effects in response to a high-fat diet [37], while *Allobaculum* mitigated intestinal dysbiosis in diabetic mice [38]. *Unclassified Lachnospiraceae*, a potentially beneficial bacterium, was found to reduce the risk of T2D by reducing ketone body levels [39]. This indicates that different gut microbiota structures exist between different selenium tea polysaccharide interventions, although both alleviated gut microbiota dysbiosis in the development of diabetic mice. A previous study found that the relative abundance of *Ileibacterium* was significantly higher in the deacetylated konjac glucomannan group than in the konjac glucomannan group in mice fed a high-fat diet, although both exhibited hypolipidemic effects [40]. This suggests that modification of polysaccharides can change the gut microbiota structure regulation. Thus, we further identified the specific microbial taxa enriched in mice with different selenium green tea polysaccharides. Compared with the TPS group, the NSe-TPS group was characterized by *Ileibacterium*, the PCSe-TPS group was characterized by *Oscillospiraceae*, and the Ord-TPS+Se group was characterized by *Lachnospiraceae_NK4A136_group*, *unclassified-Lachnospiraceae*, and *unclassified_Oscillospiraceae*. These bacteria were found to be important in alleviating hyperglycemia and complications caused by diabetes [32,37,40]. Lastly, the gut microbiota structure in mice with organic and inorganic selenium was compared due to the different mitigating effects shown on the development of diabetes mellitus. *Erysipelotrichaceae*, *Allobaculum*, and *Alloprevotella* characterized PCSe-TPS, while Ord-TPS+Se was characterized by *Firmicutes*, *Roseburia*, *unclassified-Lachnospiraceae*, and *Lachnospiraceae_NK4A136_group*. *Erysipelotrichaceae* was found to be strongly associated with diabetic nephropathy [41]. *Erysipelotrichaceae UCG-003* increased in abundance in the healthy aging cohorts compared to the non-healthy aging cohorts [42]. *Alloprevotella*, *Allobaculum*, and *Roseburia* were reported to be SCFA producers [43,44]. Tea polysaccharides were found to restore the relative abundance of *Roseburia* and *Lachnospira* reduced by diabetes [19]. It suggests that synthetic selenylation alters the bioavailability of tea polysaccharides by the gut microbiota, which is related to the change in the structural properties of tea polysaccharides after selenylation [12,40].

5. Conclusions

Our study investigated the protective effects of different selenium green tea polysaccharides on the development of T2D in mice. The results showed that PCSe-TPS exhibited better protection against the development of T2D than NSe-TPS and Ord-TPS+Se. PCSe-TPS inhibited elevated blood glucose by regulating the PI3K/Akt signaling pathway and alleviated liver tissue oxidative damage and inflammatory responses. In addition, PCSe-TPS intervention reversed the decline in bacterial species richness and the abundance of *unclassified Oscillospiraceae* in the development of diabetes in mice. Our work provides a comprehensive understanding of the protective effects of different selenium tea polysaccharides on diabetes onset and lays the theoretical foundation for applying PCSe-TPS-related products.

Supplementary Materials: The following supporting information can be downloaded at: https://www.mdpi.com/article/10.3390/foods12234190/s1, Figure S1: Experimental design; Figure S2: Protective effects of different doses of synthetic selenized green tea polysaccharides (PCSe-TPS) on the development of diabetic mice; Table S1: Reaction system of RT-qPCR; Table S2: Reaction procedure of RT-qPCR; Table S3: Primer sequences.

Author Contributions: Methodology, W.G., Z.Z. (Zhan Zheng), N.Z., L.W., H.L. and X.C.; software, W.G.; validation, W.G. and Z.Z. (Zhan Zheng); formal analysis, W.G., Z.Z. (Zhan Zheng) and N.Z.; investigation, W.G.; resources, H.L., X.C., S.L. and Z.Z. (Zhenzhou Zhu); data curation, W.G.; writing—original draft preparation, W.G., X.W. and L.W.; writing—review and editing, W.G., X.W. and H.L.; visualization, W.G., L.W., X.C., X.W. and Z.Z. (Zhenzhou Zhu); supervision, N.Z., S.L., X.W. and Z.Z. (Zhenzhou Zhu); project administration, S.L. and Z.Z. (Zhenzhou Zhu); funding acquisition, X.W. and Z.Z. (Zhenzhou Zhu). All authors have read and agreed to the published version of the manuscript.

Funding: This work was supported by Outstanding Young and Middle-aged Science and Technology Innovation Team in Hubei Province (T2020012), Key Research and Development Program of Hubei Province (2020BBA043), the Scientific and Technology Project of Enshi Tujia and Miao Autonomous Prefecture (D20220088), and Research and Innovation Initiatives of Wuhan Polytechnic University (No. 2023RZ027).

Institutional Review Board Statement: Animal experiments were conducted at the Hubei Province Laboratory Animal Research Center (Hubei, China) under the Safety Assessment Centre (Animal Welfare) approval No. 202210207, following national legislation and local guidelines.

Data Availability Statement: The data used to support the findings of this study can be made available by the corresponding author upon request.

Conflicts of Interest: The authors declare no conflict of interest. Author Haiyuan Liu and Xin Cong are employed by Enshi Se-Run Material Engineering Technology Co., Ltd. The remaining authors declare that the research was conducted in the absence of any commercial or financial relationships that could be construed as a potential conflict of interest.

References

1. International Diabetes Federation. *IDF Diabetes Atlas*, 10th ed.; International Diabetes Federation: Brussels, Belgium, 2021.
2. World Health Organization. Diabetes. Available online: https://www.who.int/zh/news-room/fact-sheets/detail/diabetes (accessed on 5 April 2023).
3. Yang, Q.; Vijayakumar, A.; Kahn, B.B. Metabolites as Regulators of Insulin Sensitivity and Metabolism. *Nat. Rev. Mol. Cell Biol.* **2018**, *19*, 654–672. [CrossRef]
4. American Diabetes Association. 2. Classification and Diagnosis of Diabetes: Standards of Medical Care in Diabetes-2020. In *Diabetes Care*; American Diabetes Association: Arlington, VA, USA, 2020; Volume 43, pp. S14–S31.
5. Basiak-Rasała, A.; Różańska, D.; Zatońska, K. Food Groups in Dietary Prevention of Type 2 Diabetes. *Rocz. Panstw. Zakl. Hig.* **2019**, *70*, 347–357. [CrossRef] [PubMed]
6. Huang, Y.C.; Combs, G.F., Jr.; Wu, T.L.; Zeng, H.; Cheng, W.H. Selenium Status and Type 2 Diabetes Risk. *Arch. Biochem. Biophys.* **2022**, *730*, 109400. [CrossRef]
7. World Health Organization. *Trace Elements in Human Nutrition and Health*; World Health Organization: Geneva, Switzerland, 1996.
8. Hadrup, N.; Ravn-Haren, G. Absorption, Distribution, Metabolism and Excretion (ADME) of Oral Selenium from Organic and Inorganic Sources: A Review. *J. Trace Elem. Med. Biol.* **2021**, *67*, 126801. [CrossRef]

9. Duan, W.-X.; Yang, X.-H.; Zhang, H.-F.; Feng, J.; Zhang, M.-Y. Chemical Structure, Hypoglycemic Activity, and Mechanism of Action of Selenium Polysaccharides. *Biol. Trace Elem. Res.* **2021**, *200*, 4404–4418. [CrossRef]
10. Shang, L.; Li, F.; Zhu, J.; Sun, C.; Wang, Y. Selenium-Enriched and Ordinary Black Teas Regulate the Metabolism of Glucose and Lipid and Intestinal Flora of Hyperglycemic Mice. *Plant Foods Hum. Nutr.* **2022**, *78*, 61–67. [CrossRef]
11. Zhao, M.; Bai, J.; Bu, X.; Yin, Y.; Wang, L.; Yang, Y.; Xu, Y. Characterization of Selenized Polysaccharides from *Ribes nigrum* L. and its Inhibitory Effects on α-amylase and α-glucosidase. *Carbohydr. Polym.* **2021**, *259*, 117729. [CrossRef]
12. Zhu, J.; Yu, C.; Han, Z.; Chen, Z.; Wei, X.; Wang, Y. Comparative Analysis of Existence form for Selenium and Structural Characteristics in Artificial Selenium-Enriched and Synthetic Selenized Green Tea Polysaccharides. *Int. J. Biol. Macromol.* **2020**, *154*, 1408–1418. [CrossRef]
13. Gao, W.; Zhang, N.; Li, S.; Li, S.; Zhu, S.; Cong, X.; Cheng, S.; Barba, F.J.; Zhu, Z. Polysaccharides in Selenium-Enriched Tea: Extraction Performance under Innovative Technologies and Antioxidant Activities. *Foods* **2022**, *11*, 2545. [CrossRef]
14. Górska, S.; Maksymiuk, A.; Turło, J. Selenium-Containing Polysaccharides—Structural Diversity, Biosynthesis, Chemical Modifications and Biological Activity. *Appl. Sci.* **2021**, *11*, 3717. [CrossRef]
15. Li, S.; Chen, H.; Wang, J.; Wang, X.; Hu, B.; Lv, F. Involvement of the PI3K/Akt signal pathway in the hypoglycemic effects of tea polysaccharides on diabetic mice. *Int. J. Biol. Macromol.* **2015**, *81*, 967–974. [CrossRef]
16. Nair, A.; Jacob, S. A simple practice guide for dose conversion between animals and human. *Br. J. Clin. Pharmacol.* **2016**, *7*, 27. [CrossRef]
17. Zhao, Z.; Chen, Y.; Li, X.; Zhu, L.; Wang, X.; Li, L.; Sun, H.; Han, X.; Li, J. Myricetin relieves the symptoms of type 2 diabetes mice and regulates intestinal microflora. *Biomed. Pharmacother.* **2022**, *153*, 113530. [CrossRef]
18. Andrikopoulos, S.; Blair, A.R.; Deluca, N.; Fam, B.C.; Proietto, J.; Du Toit, E.F.; Tai, F.; Cox, A.; O'Connor, D.; Griffith, T.A.; et al. Evaluating the glucose tolerance test in mice. *Am. J. Physiol. Endocrinol. Metab.* **2008**, *295*, 1323–1332. [CrossRef]
19. Li, H.; Fang, Q.; Nie, Q.; Hu, J.; Yang, C.; Huang, T.; Li, H.; Nie, S. Hypoglycemic and Hypolipidemic Mechanism of Tea Polysaccharides on Type 2 Diabetic Rats via Gut Microbiota and Metabolism Alteration. *J. Agric. Food Chem.* **2020**, *68*, 10015–10028. [CrossRef]
20. Liu, Y.; You, Y.; Li, Y.; Zhang, L.; Yin, L.; Shen, Y.; Li, C.; Chen, H.; Chen, S.; Hu, B.; et al. The Characterization, Selenylation and Anti-diabetic Activity of Mycelial Polysaccharides from *Catathelasma ventricosum*. *Carbohydr. Polym.* **2017**, *174*, 72–81. [CrossRef]
21. Lempesis, I.G.; Georgakopoulou, V.E. Physiopathological Mechanisms Related to Inflammation in Obesity and Type 2 Diabetes Mellitus. *World J. Exp. Med.* **2023**, *13*, 7–16. [CrossRef]
22. Fu, Y.; Yin, R.; Guo, E.; Cheng, R.; Diao, X.; Xue, Y.; Shen, Q. Protein Isolates from Raw and Cooked Foxtail Millet Attenuate Development of Type 2 Diabetes in Streptozotocin-Induced Diabetic Mice. *Mol. Nutr. Food Res.* **2021**, *65*, e2000365. [CrossRef]
23. Wu, J.; Shi, S.; Wang, H.; Wang, S. Mechanisms Underlying the Effect of Polysaccharides in the Treatment of Type 2 Diabetes: A Review. *Carbohydr. Polym.* **2016**, *144*, 474–494. [CrossRef]
24. Deng, Q.; Ma, D.; Sun, G.; Yuan, X.; Wang, Z.; Liu, G. PTEN Influences Insulin and Lipid Metabolism in Bovine Hepatocytes in Vitro. *J. Dairy Res.* **2019**, *86*, 73–76. [CrossRef]
25. Hu, X.; Wang, S.; Xu, J.; Wang, D.B.; Chen, Y.; Yang, G.Z. Triterpenoid Saponins from *Stauntonia chinensis* Ameliorate Insulin Resistance via the AMP-Activated Protein Kinase and IR/IRS-1/PI3K/Akt Pathways in Insulin-Resistant Hepg2 Cells. *Int. J. Mol. Sci.* **2014**, *15*, 10446–10458. [CrossRef]
26. Huang, X.; Liu, G.; Guo, J.; Su, Z. The PI3K/AKT Pathway in Obesity and Type 2 Diabetes. *Int. J. Biol. Sci.* **2018**, *14*, 1483–1496. [CrossRef]
27. Luc, K.; Schramm-Luc, A.; Guzik, T.J.; Mikolajczyk, T.P. Oxidative Stress and Inflammatory Markers in Prediabetes and Diabetes. *J. Physiol. Pharmacol.* **2019**, *70*, 809–824. [CrossRef]
28. Qin, S.; Gao, J.; Huang, K. Effects of Different Selenium Sources on Tissue Selenium Concentrations, Blood GSH-Px Activities and Plasma Interleukin Levels in Finishing Lambs. *Biol. Trace Elem. Res.* **2007**, *116*, 91–102. [CrossRef]
29. Clausen, J.; Nielsen, S.A. Comparison of Whole Blood Selenium Values and Erythrocyte Glutathione Peroxidase Activities of Normal Individuals on Supplementation with Selenate, Selenite, L-Selenomethionine, and High Selenium Yeast. *Biol. Trace Elem. Res.* **1988**, *15*, 125–138. [CrossRef]
30. Gong, P.; Wang, X.; Liu, M.; Wang, M.; Wang, S.; Guo, Y.; Chang, X.; Yang, W.; Chen, X.; Chen, F. Hypoglycemic Effect of a Novel Polysaccharide from *Lentinus Edodes* on STZ-Induced Diabetic Mice via Metabolomics Study and Nrf2/HO-1 Pathway. *Food Funct.* **2022**, *13*, 3036–3049. [CrossRef]
31. Ke, W.; Wang, P.; Wang, X.; Zhou, X.; Hu, X.; Chen, F. Dietary Platycodon grandiflorus Attenuates Hepatic Insulin Resistance and Oxidative Stress in High-Fat-Diet Induced Non-Alcoholic Fatty Liver Disease. *Nutrients* **2020**, *12*, 480. [CrossRef]
32. Feng, W.; Liu, J.; Ao, H.; Yue, S.; Peng, C. Targeting Gut Microbiota for Precision Medicine: Focusing on the Efficacy and Toxicity of Drugs. *Theranostics* **2020**, *10*, 11278–11301. [CrossRef]
33. Morrison, D.J.; Preston, T. Formation of Short Chain Fatty Acids by the Gut Microbiota and Their Impact on Human Metabolism. *Gut Microbes* **2016**, *7*, 189–200. [CrossRef]
34. Hu, S.; Kuwabara, R.; de Haan, B.J.; Smink, A.M.; de Vos, P. Acetate and Butyrate Improve beta-cell Metabolism and Mitochondrial Respiration under Oxidative Stress. *Int. J. Mol. Sci.* **2020**, *21*, 1542. [CrossRef]
35. Mayorga-Ramos, A.; Barba-Ostria, C.; Simancas-Racines, D.; Guaman, L.P. Protective Role of Butyrate in Obesity and Diabetes: New Insights. *Front. Nutr.* **2022**, *9*, 1067647. [CrossRef]

36. Rang, Y.; Liu, H.; Cheng, X.; Li, W.; Shi, J.; Ou, G.; Huang, H.; Chen, C.; Xiao, X.; Liu, C. Structural Characterization of Pectic Polysaccharides from *Amaranth caudatus* Leaves and the Promotion Effect on Hippocampal Glucagon-Like Peptide-1 Level. *Int. J. Biol. Macromol.* **2023**, *242*, 124967. [CrossRef]
37. Wei, W.; Jiang, W.; Tian, Z.; Wu, H.; Ning, H.; Yan, G.; Zhang, Z.; Li, Z.; Dong, F.; Sun, Y.; et al. Fecal *g. Streptococcus* and *g. Eubacterium_coprostanoligenes_group* Combined with Sphingosine to Modulate the Serum Dyslipidemia in High-Fat Diet Mice. *Clin. Nutr.* **2021**, *40*, 4234–4245. [CrossRef]
38. Zhou, W.; Yang, T.; Xu, W.; Huang, Y.; Ran, L.; Yan, Y.; Mi, J.; Lu, L.; Sun, Y.; Zeng, X.; et al. The Polysaccharides from the Fruits of *Lycium Barbarum L.* Confer Anti-Diabetic Effect by Regulating Gut Microbiota and Intestinal Barrier. *Carbohydr. Polym.* **2022**, *291*, 119626. [CrossRef]
39. Kim, K.; Lee, S.; Park, S.-C.; Kim, N.-E.; Shin, C.; Lee, S.K.; Jung, Y.; Yoon, D.; Kim, H.; Kim, S.; et al. Role of an *Unclassified lachnospiraceae* in the Pathogenesis of Type 2 Diabetes: A Longitudinal Study of the Urine Microbiome and Metabolites. *Exp. Mol. Med.* **2022**, *54*, 1125–1132. [CrossRef]
40. Zou, X.; Deng, J.; Wang, Z.; Zhang, M.; Sun, Y.; Li, M. Gut Microbiota Plays a Predominant Role in Affecting Hypolipidemic Effect of Deacetylated Konjac Glucomannan (Da-KGM). *Int. J. Biol. Macromol.* **2022**, *208*, 858–868. [CrossRef]
41. Kikuchi, K.; Saigusa, D.; Kanemitsu, Y.; Matsumoto, Y.; Thanai, P.; Suzuki, N.; Mise, K.; Yamaguchi, H.; Nakamura, T.; Asaji, K.; et al. Gut Microbiome-Derived Phenyl Sulfate Contributes to Albuminuria in Diabetic Kidney Disease. *Nat. Commun.* **2019**, *10*, 1835. [CrossRef]
42. Singh, H.; Torralba, M.G.; Moncera, K.J.; DiLello, L.; Petrini, J.; Nelson, K.E.; Pieper, R. Gastro-Intestinal and Oral Microbiome Signatures Associated with Healthy Aging. *GeroScience* **2019**, *41*, 907–921. [CrossRef]
43. Kong, C.; Gao, R.; Yan, X.; Huang, L.; Qin, H. Probiotics Improve Gut Microbiota Dysbiosis in Obese Mice Fed a High-Fat or High-Sucrose Diet. *Nutrition* **2019**, *60*, 175–184. [CrossRef]
44. Gurung, M.; Li, Z.; You, H.; Rodrigues, R.; Jump, D.B.; Morgun, A.; Shulzhenko, N. Role of Gut Microbiota in Type 2 Diabetes Pathophysiology. *EBioMedicine* **2020**, *51*, 102590. [CrossRef]

foods

Article

Protective Effects of Naringenin and Apigenin in Ameliorating Skin Damage via Mediating the Nrf2 and NF-κB Pathways in Mice

Jie Li [1,2], Bingyong Mao [1,2], Xin Tang [1,2], Qiuxiang Zhang [1,2], Jianxin Zhao [1,2], Hao Zhang [1,2,3,*] and Shumao Cui [1,2]

1 State Key Laboratory of Food Science and Resources, Jiangnan University, Wuxi 214122, China; 7210112021@stu.jiangnan.edu.cn (J.L.); maobingyong@jiangnan.edu.cn (B.M.); xintang@jiangnan.edu.cn (X.T.); zhangqx@jiangnan.edu.cn (Q.Z.); zhaojianxin@jiangnan.edu.cn (J.Z.); cuishumao@jiangnan.edu.cn (S.C.)
2 School of Food Science and Technology, Jiangnan University, Wuxi 214122, China
3 National Engineering Research Center for Functional Food, Jiangnan University, Wuxi 214122, China
* Correspondence: zhanghao61@jiangnan.edu.cn

Abstract: Naringenin and apigenin are common flavonoids derived from edible plants with the potential to alleviate inflammation and improve skin antioxidation. This study aimed to evaluate the effects of naringenin and apigenin on oleic acid-induced skin damage in mice and compare their underlying mechanisms of action. Triglycerides and non-esterified fatty acids were significantly decreased by naringenin and apigenin, while apigenin intervention resulted in a better recovery of skin lesions. Naringenin and apigenin improved the antioxidative abilities of the skin by increasing catalase and total antioxidant capacity levels and decreasing malondialdehyde and lipid peroxide levels. The release of skin proinflammatory cytokines, such as interleukin (IL)-6, IL-1β, and tumor necrosis factor α, was inhibited after naringenin and apigenin pretreatments, but naringenin only promoted the excretion of IL-10. Additionally, naringenin and apigenin regulated antioxidant defense and inflammatory response by activating nuclear factor erythroid-2 related factor 2-dependent mechanisms and suppressing the expression of nuclear factor-kappa B. In summary, naringenin and apigenin are prospective ingredients that contribute to the amelioration of skin damage by activating anti-inflammatory and antioxidative responses.

Keywords: naringenin; apigenin; antioxidative system; inflammation; signaling pathway

Citation: Li, J.; Mao, B.; Tang, X.; Zhang, Q.; Zhao, J.; Zhang, H.; Cui, S. Protective Effects of Naringenin and Apigenin in Ameliorating Skin Damage via Mediating the Nrf2 and NF-κB Pathways in Mice. *Foods* **2023**, *12*, 2120. https://doi.org/10.3390/foods12112120

Academic Editor: Carla Gentile

Received: 23 April 2023
Revised: 16 May 2023
Accepted: 23 May 2023
Published: 24 May 2023

1. Introduction

Citurs sinensis L.Osbeck and *Oenanthe javanica* (Blume) DC are accepted almost around the world as the food sources of humans to provide nutritional and functional ingredients [1,2]. They are reported to contain an abundance of antioxidant nutraceuticals or phytochemicals that potentially regulate human metabolism with constant ingestion in a favorable manner [2,3]. Their flavonoid components have been extracted and reported to have a variety of pharmacological activities that prevent chronic and degenerative diseases and even improve skin conditions [3,4]. Naringenin derived from *C. sinensis* L.Osbeck has been reported to result in significant improvement in skin inflammation, skin damage, and skin oxidative reaction in a UVB induced-skin model [5–7]. Recent studies have also pointed out that naringenin has potential effects on the suppression of skin cancer in mouse skin tumor models [8]. Additionally, apigenin is the most common flavonoid found in the plant ingredients of *O. javanica* (Blume) DC that has been used in a wide variety of medicines, foods, and other fields [9,10]. In vitro studies have shown that apigenin has anti-inflammatory and skin-protective activities via regulating the phosphorylation of mitogen-activated protein kinase (MAPK) signaling molecules [11]. Apigenin has benefits on flap tissue survival and is useful in plastic surgery [12]. Previous in vitro studies have demonstrated the usefulness of naringenin and apigenin in protecting the skin from the invasion of UV, tumors, and aging, indicating the potential protective effects of naringenin and apigenin against skin damage. Given the complex metabolic processes in body, in vitro evidence cannot be directly equated to obtaining an in vivo conclusion. Thus, more studies are required to confirm whether these two compounds could exhibit similar or even superior efficacy when administered as dietary supplements.

An increasing number of reports have demonstrated the close relationship between dietary habits and skin metabolism [13,14]. The adverse effect of milk and foods with a high glycemic and/or fat index on skin condition has been verified, showing that these foods promote the development of acne vulgaris [15]. The effect of lifestyle factors on carotenoid levels in the skin demonstrates that a positive relationship exists between increased antioxidant and inflammatory levels and intake of carotenoid-rich foods [16]. Dietary supplementation of plant-derived composition can defend against skin aging after exposure to UV radiation [6,17].

Oleic acid (OA) is regarded as the main trigger for inflammation and oxidation, and sensitivity to this free fatty acid plays a vital role in the pathogenesis of skin disease; the effect of OA has been demonstrated in a mouse model of skin damage caused by over-secretion of skin lipid [18]. The activation of the transcription factor nuclear factor (erythroid-derived 2)-like 2 (Nrf2) is an important mechanism for controlling stress responses and antioxidant defense in organs, including the skin. Nuclear factor-kappa B (NF-κB) is the most important signaling factor for regulating inflammatory responses in mammals and causes the transcription of pro-inflammatory mediators, such as interleukin (IL)-6, tumor necrosis factor-α (TNF-α), IL-1, and intracellular adhesion cyclooxygenase-2 (COX2). Previous studies have shown that naringenin and apigenin regulate the Nrf2 and NF—κB pathway to defend against oxidative and inflammatory processes [19,20].

Thus, the objective of this study is to explore the protective effects of naringenin and apigenin in improving skin damage and to compare their underlying mechanisms of action. We analyzed the contents of the main lipid components to evaluate sebum production in sebaceous glands. The skin levels of inflammatory factors and antioxidative stress were determined. Histopathological examination was used to evaluate skin injury and inflammatory responses in the back of mice. mRNA expression related to inflammation and antioxidation was further determined to compare the mechanisms of naringenin and apigenin supplementation.

2. Materials and Methods

2.1. Reagents and Materials

Naringenin (NAR, purity $\geq$ 97%) derived from C. L.Osbeck was obtained from Chengdu Herbpurify Co., Ltd. (Chengdu, China). Apigenin (API, purity $\geq$ 98%) was derived from *O. javanica* (Blume) DC and purchased from Aladdin® Co., Ltd. (Shanghai, China). Cryptotanshinone (CRY, extracted from Salvia miltiorrhiza, purity $\geq$ 98%) was obtained from Chengdu Herbpurity Co., Ltd. (Chengdu, China). Naringin and apigenin were dissolved in methyl alcohol and determined using UHPLC-MS/MS (Q Exactive, Thermo Fisher Scientific, Waltham, MA, USA) with a T3 column (2.1 × 100 mm, 1.8 μm) at an operating temperature of 30 °C according to the methods previously reported [21] (Figure S1). Liquid paraffin and oleic acid (OA) were purchased from Sinopharm Chemical Reagent Co., Ltd. (Shanghai, China). Other reagents were purchased from Sinopharm Chemical Reagent Co., Ltd. (Shanghai, China). Enzyme-linked immunosorbent assay (ELISA) kits were provided by Enzyme-linked Biotech Co., Ltd. (Shanghai, China).

2.2. Animal Experimental Design

C57BL/6 mice (male, 6 weeks old, 20 ± 2 g) were purchased from Charles River Laboratory Animal Technology Co., Ltd. (Jiaxing, China) and fed with commercial chow and sterile water. After 7 days of adaptive feeding, all mice were randomly grouped into 7 groups (n = 6 for each group). Naringenin and apigenin were dissolved in normal saline with Tween-80 (0.5%, *w/v*). The treatment of mice in each group was shown in Table S1. The mice in the control and model groups were orally administrated normal saline with Tween-80 (0.5%, *w/v*). After administration via gavage for a week, 0.1 mL of oleic acid

(80% *w/v*) was smeared on the mice's back once a day to build a skin injury model, which lasted for two weeks. The primary lipophilic extract derived from *Salvia miltiorrhiza*, known as CRY, is a main ingredient in ameliorating skin damage and regulating skin lipids balance and was used as the positive control (10 mg/kg/d). The animal experiments were performed with the approval of the Animal Experimentation Ethics Committee of Jiangnan University (JN.No20220717c0959628[021]).

2.3. Observation Index and Skin Damage Scores

Skin damage scores were calculated based on a previous study [22]. The main symptoms included erythema, scales, erosion, and pruritus. Scores of (0–3) were given to evaluate skin damage. A score of 0 was given when there were no symptoms. A score of 1 (mild) was given where there were symptoms including minor erythema, erosion, and pruritus, as well as sparse scales. A score of 2 (moderate) was given to symptoms including major erythema, erosion, and pruritus, as well as a wide range of scales and even scabs. A score of 3 (severe) was given to symptoms including full-scale erythema, scales (scabs), erosion, and pruritus.

2.4. Histological Analysis of Skin Tissue

Skin tissues were stained using hematoxylin and eosin (H&E) for histological analysis. The staining process included collection, dehydration, embedment, slicing, and staining. The tissues were observed according to a previous report with minor revisions [23].

2.5. Determination of Skin Lipids

The levels of skin total triglycerides (TG), non-esterified fatty acids (NEFA), and cholesterol (CHO) in skin tissues were determined according to the instructions of the kits used (Nanjing Jiancheng Bioengineering Institute, Nanjing, China). Absorption values were obtained using a microplate readers with the skanit software 4.1 (Thermo, Waltham, MA, USA).

2.6. Assessment of Antioxidant Activities in Skin Tissue

Antioxidant activities in skin lesions were characterized via the determination of the levels of malondialdehyde (MDA), superoxide dismutase (SOD), catalase (CAT), lipid peroxides (LPO), total antioxidant capacity (T-AOC) and aspartate aminotransferase (AST). The determination was performed according to the instructions of the kits used (Nanjing Jiancheng Bioengineering Institute, Nanjing, China). The obtained absorption values were processed using a one-way analysis of variance (ANOVA) via the SPSS software ver. 22.0 (IBM, Armonk, NY, USA).

2.7. Cytokine Measurements Using ELISA Assay

The IL-1β, IL-6, TNF-α, and IL-10 levels of the supernatants from skin lesions were analyzed using ELISA kits (R&D Systems, Minneapolis, MN, USA) according to the manufacturer's instructions.

2.8. Quantitative RT-PCR Analysis

Total RNA was extracted from cells and then reversely transcribed to cDNA using commercial kits (Vazyme, Nanjing, China). RT-PCR amplifications were carried out in a StepOnePlus RT-PCR system (Thermo, Waltham, MA, USA) using the qPCR kit (Vazyme, Nanjing, China). The predesigned gene-specific primers are listed in Table S2.

2.9. Immunohistochemical Analysis

Paraffin sections were heated in an oven at 90 °C until the wax fused and then successfully immersed in xylene and gradient ethanol. The tissue sections were hydrated and then incubated with 3% hydrogen peroxide for 25 min, rinsed subsequently with phosphate-buffered saline (PBS) three times, and sealed with 10% goat serum for 45 min. The sections were then incubated overnight with antibodies against NF-κB (1:200, Arigo,

Shanghai, China) at 4 °C. After hematoxylin staining, they were sealed with neutral resin and observed under a microscope. The obtained pictures were analyzed using the Image J software, and the expression of NF-κB was analyzed statically using GraphPad Prism 7.0 (GraphPad, La Jolla, CA, USA).

2.10. Statistical Analysis

Data analysis was performed to indicate significant difference using a one-way analysis of variance (ANOVA) via the SPSS software ver. 22.0 (IBM, Armonk, NY, USA). The Student–Newman–Keuls test was used for multiple comparisons using GraphPad Prism 7.0 (GraphPad, La Jolla, CA, USA). All data were presented as mean ± standard deviation (SD). $p < 0.05$ was considered statistically significant.

3. Results

3.1. Determination of Naringenin and Apigenin Based on HPLC-MS/MS

HPLC-MS/MS was used to determine the purity of naringenin and apigenin, and the chromatogram is presented in Figure S1. The peak times of naringenin and apigenin were 6.42 and 6.40 min, respectively, with obvious peak heights. The m/z values of naringenin and apigenin (273.07 and 271.05, respectively) were obtained after the MS analysis.

3.2. Effects of Naringenin and Apigenin on Body Weight and Food and Water Intake of Mice

The body weight changes in the mice with the pretreatments of naringenin (Figure 1A) and apigenin (Figure 1B) were determined after 21 days of feeding. The body weight gain in the OA-treated mice slightly increased with the intake of naringenin and apigenin compared to that in the model group, whereas the body weight gain in all other groups showed no significant differences ($p > 0.05$; Figure 1C). The food intake and water intake of the mice in all groups were also determined, and the results are shown in (Figure 1D,E) with no significant difference ($p > 0.05$). These results obtained for food and water intake were not related to the interventions with naringenin and apigenin.

3.3. Pathological Morphology of Skin Tissues in Mice

The biochemical parameters of the skin tissues from the different groups used to determine injury and inflammation are shown in Figure 2. The hyperkeratosis of skin tissues in the mice treated with OA was characterized by a thickened epidermis, blurred boundaries between epidermis and dermis, collection in hair follicle pores, infiltration of neutrophils in the dermis, and proliferation of sebaceous ducts. The damage scores for the groups are listed in Table 1. However, these pathological conditions were improved in the OA-treated mice with the pretreatments of naringenin and apigenin. The CRY pretreatment improved skin lesions in the OA-treated mice, and a better effect was shown in the naringenin and apigenin groups. The evaluation of epidermal thickness obtained consistent results (Figure 2B and Table 2), and significant differences were detected in the L-NAR and L-API groups compared to the model group ($p < 0.05$). Although there were no differences in hair follicle diameter among the NAR and API groups compared to the model group ($p > 0.05$), the OA treatment resulted in enlarged pores in the model group compared to the control group ($p > 0.05$; Figure 2C and Table 2).

Table 1. Assessment of skin damage among groups.

Symptoms	Control	Model	CRY	L-NAR	H-NAR	L-API	H-API
Erythema	0	1.38 ± 0.52 ^{a}	0.43 ± 0.53 ^{a}	0.25 ± 0.46 ^{a}	0.75 ± 0.46 ^{a}	0.25 ± 0.46 ^{a}	0.25 ± 0.46 ^{a}
Scales	0	2.25 ± 0.71 ^{a}	2.14 ± 0.69 ^{a}	1.13 ± 0.64 ^{a}	1.50 ± 0.83 ^{a}	1.00 ± 0.53 ^{a}	1.50 ± 0.53 ^{a}
Erosion	0	0.50 ± 0.53 ^{a}	0.29 ± 0.49 ^{a}	0.13 ± 0.35 ^{a}	0.38 ± 0.74 ^{a}	0.13 ± 0.35 ^{a}	0.25 ± 0.46 ^{a}
Pruritus	0	1.50 ± 0.53 ^{a}	1.29 ± 0.49 ^{a}	1.13 ± 0.35 ^{a}	1.25 ± 0.46 ^{a}	1.13 ± 0.35 ^{a}	1.13 ± 0.35 ^{a}
Total	0	5.63 ± 1.41 ^{a}	4.14 ± 1.77 ^{b}	2.63 ± 0.74 ^{bc}	3.88 ± 1.13 ^{bc}	2.38 ± 0.92 ^{c}	3.25 ± 1.04 ^{bc}

All data are presented as mean ± standard deviation (SD). "a, b, c": different letters indicate significant differences ($p < 0.05$) among different groups.

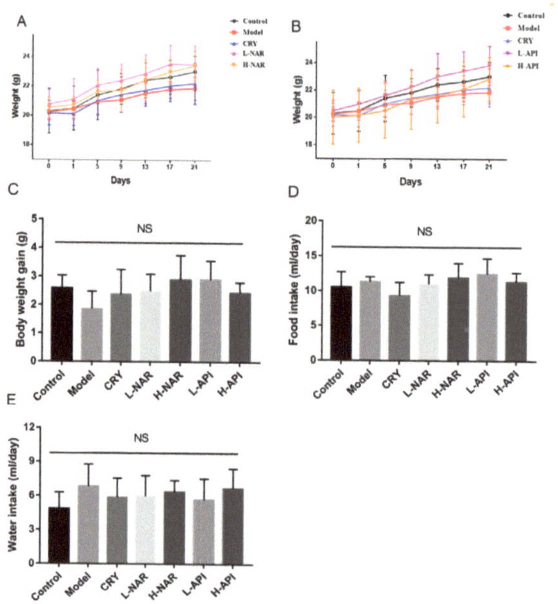

Figure 1. Influences of naringenin and apigenin pretreatments on body weight and food and water intake in mice of different groups. Body weight gain of mice with naringenin pretreatment (**A**) and apigenin pretreatment (**B**); body weight gain of mice in different groups (**C**); and food intake (**D**) and water intake (**E**) of mice in different groups. CRY: cryptotanshinone; L-NAR: naringenin with a dose of 5 mg/kg/d; H-NAR: naringenin with a dose of 10 mg/kg/d; L-API: apigenin with a dose of 5 mg/kg/d; H-API: apigenin with a dose of 10 mg/kg/d; "NS" means no significant differences ($p > 0.05$) among different groups.

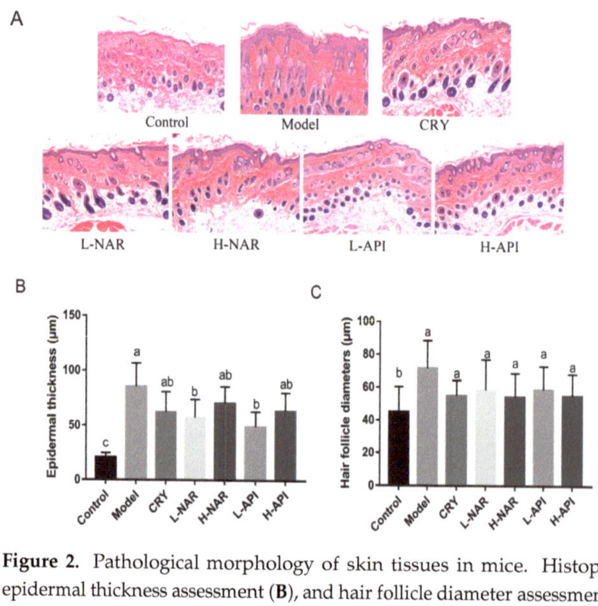

Figure 2. Pathological morphology of skin tissues in mice. Histopathological observation (**A**), epidermal thickness assessment (**B**), and hair follicle diameter assessment (**C**). CRY: cryptotanshinone; L-NAR: naringenin with a dose of 5 mg/kg/d; H-NAR: naringenin with a dose of 10 mg/kg/d; L-API: apigenin with a dose of 5 mg/kg/d; H-API: apigenin with a dose of 10 mg/kg/d. Significant differences ($p < 0.05$) among different groups were presented as "a, b, c".

Table 2. The results and significance of the main parameters among the control, model, CRY, L-NAR, H-NAR, L-API, and H-API groups.

Paramete	Group	Control	Model	CRY	L-NAR	H-NAR	L-API	H-NAR
Apparent symptoms	Damage scores	0	5.63 ± 1.41 ^{a}	4.14 ± 1.77 ^{b}	2.63 ± 0.74 ^{bc}	3.88 ± 1.13 ^{bc}	2.38 ± 0.92 ^{c}	3.25 ± 1.04 ^{bc}
	Epidermal thickness (μm)	20.5 ± 1.44 ^{c}	85.68 ± 6.55 ^{a}	62.37 ± 7.50 ^{ab}	57.67 ± 5.45 ^{b}	70.45 ± 5.84 ^{ab}	63.85 ± 5.91 ^{b}	49.24 ± 4.83 ^{ab}
	Hair follicle diameter (μm)	45.09 ± 6.23 ^{b}	71.73 ± 6.84 ^{a}	55.07 ± 3.79 ^{a}	58.63 ± 7.50 ^{a}	50.97 ± 5.89 ^{a}	51.63 ± 5.86 ^{a}	58.79 ± 5.45 ^{a}
Skin lipid expression	TG level	0.38 ± 0.08 ^{b}	0.61 ± 0.06 ^{a}	0.47 ± 0.07 ^{ab}	0.43 ± 0.05 ^{b}	0.51 ± 0.06 ^{ab}	0.39 ± 0.05 ^{b}	0.33 ± 0.03 ^{b}
	NEFA level	0.069 ± 0.008 ^{b}	0.098 ± 0.009 ^{a}	0.095 ± 0.005 ^{a}	0.090 ± 0.004 ^{a}	0.075 ± 0.003 ^{b}	0.076 ± 0.007 ^{ab}	0.089 ± 0.007 ^{ab}
	T-CHO level	0.57 ± 0.10 ^{a}	0.67 ± 0.07 ^{a}	0.43 ± 0.07 ^{a}	0.47 ± 0.09 ^{a}	0.50 ± 0.08 ^{a}	0.44 ± 0.11 ^{a}	0.50 ± 0.10 ^{a}
Antioxidative expression	CAT level	38.43 ± 4.08 ^{a}	31 ± 1.84 ^{b}	35.91 ± 2.94 ^{ab}	34.09 ± 1.55 ^{ab}	39.34 ± 2.01 ^{a}	39.21 ± 1.62 ^{a}	44.7 ± 2.47 ^{a}
	SOD level	3948 ± 541.2 ^{a}	2839 ± 380.4 ^{a}	3273 ± 195 ^{a}	3701 ± 585.1 ^{a}	3954 ± 605.3 ^{a}	3641 ± 452.3 ^{a}	3577 ± 666.9 ^{a}
	T-AOC level	0.27 ± 0.03 ^{a}	0.16 ± 0.03 ^{b}	0.27 ± 0.03 ^{a}	0.25 ± 0.04 ^{a}	0.28 ± 0.04 ^{a}	0.29 ± 0.02 ^{a}	0.26 ± 0.04 ^{a}
	MDA level	11.48 ± 0.49 ^{b}	14.02 ± 0.52 ^{a}	11.31 ± 0.77 ^{b}	11.49 ± 1.10 ^{ab}	11.68 ± 0.90 ^{b}	8.23 ± 0.28 ^{c}	11.75 ± 0.44 ^{b}
	LPO level	0.44 ± 0.03 ^{b}	1.06 ± 0.06 ^{a}	0.68 ± 0.08 ^{b}	0.61 ± 0.07 ^{b}	0.51 ± 0.06 ^{b}	0.59 ± 0.09 ^{b}	0.63 ± 0.06 ^{b}
	AST level	9.22 ± 1.81 ^{a}	7.49 ± 2.83 ^{a}	7.49 ± 1.49 ^{a}	7.49 ± 1.49 ^{a}	8.85 ± 1.18 ^{a}	8.87 ± 1.41 ^{a}	10.28 ± 1.06 ^{a}
Inflammatory cytokines	Skin IL-6 (pg/mg protein)	59.45 ± 5.43 ^{c}	98.01 ± 5.50 ^{a}	74.26 ± 6.55 ^{b}	66.37 ± 6.74 ^{a}	51.57 ± 2.68 ^{bc}	23.24 ± 6.76 ^{d}	43.87 ± 6.77 ^{c}
	Skin IL-1β (pg/mg protein)	179.8 ± 19.01 ^{c}	328.2 ± 21.85 ^{a}	250.7 ± 20.42 ^{b}	298.2 ± 23.24 ^{a}	176.4 ± 28.44 ^{bc}	58.7 ± 6.36 ^{d}	130.1 ± 21.26 ^{c}
	Skin TNF-α (pg/mg protein)	72.49 ± 8.45 ^{b}	116.8 ± 5.32 ^{a}	88.58 ± 8.44 ^{b}	77.96 ± 8.70 ^{b}	78.63 ± 9.01 ^{b}	36.7 ± 5.76 ^{c}	63.57 ± 6.40 ^{bc}
	Skin IL-10 (pg/mg protein)	60.22 ± 3.79 ^{b}	36.76 ± 3.80 ^{c}	62.90 ± 4.16 ^{b}	95.04 ± 7.32 ^{a}	60.43 ± 6.47 ^{b}	43.83 ± 3.14 ^{bc}	47.25 ± 3.41 ^{bc}
Oxidative and lipid metabolism	SREBP-1 mRNA expression	0.16 ± 0.05 ^{b}	1.10 ± 0.24 ^{a}	0.79 ± 0.06 ^{ab}	0.77 ± 0.07 ^{ab}	0.38 ± 0.20 ^{b}	0.66 ± 0.15 ^{ab}	0.57 ± 0.21 ^{ab}
	Atgl mRNA expression	0.46 ± 0.12 ^{c}	1.16 ± 0.20 ^{b}	2.10 ± 0.60 ^{a}	2.21 ± 0.31 ^{a}	2.16 ± 0.50 ^{a}	1.88 ± 0.24 ^{a}	1.27 ± 0.24 ^{b}
	PPARα mRNA expression	0.42 ± 0.07 ^{c}	1.24 ± 0.45 ^{ab}	1.23 ± 0.18 ^{b}	1.64 ± 0.65 ^{ab}	2.53 ± 0.66 ^{ab}	3.17 ± 0.55 ^{a}	3.30 ± 0.72 ^{a}
	FAS mRNA expression	1.05 ± 0.27 ^{a}	2.07 ± 0.60 ^{a}	1.05 ± 0.24 ^{a}	2.07 ± 0.52 ^{a}	2.24 ± 0.40 ^{a}	2.16 ± 0.45 ^{a}	1.69 ± 0.45 ^{a}
	Nrf2 mRNA expression	1.38 ± 0.13 ^{a}	0.62 ± 0.18 ^{b}	1.34 ± 0.13 ^{a}	0.87 ± 0.25 ^{ab}	1.18 ± 0.24 ^{ab}	1.75 ± 0.25 ^{a}	1.57 ± 0.15 ^{a}
Inflammation signaling pathway	Relative Akt mRNA expression	1.46 ± 0.47 ^{a}	1.06 ± 0.20 ^{b}	1.30 ± 0.14 ^{a}	1.83 ± 0.77 ^{a}	1.76 ± 0.38 ^{a}	1.36 ± 0.29 ^{a}	1.36 ± 0.23 ^{a}
	Relative COX-2 mRNA expression	0.15 ± 0.08 ^{c}	1.11 ± 0.02 ^{a}	0.73 ± 0.14 ^{ab}	0.45 ± 0.10 ^{b}	0.53 ± 0.14 ^{b}	0.45 ± 0.14 ^{bc}	0.60 ± 0.11 ^{ab}
	Relative NF-κB mRNA expression	0.11 ± 0.03 ^{c}	1.08 ± 0.20 ^{a}	0.90 ± 0.25 ^{ab}	0.48 ± 0.11 ^{b}	0.70 ± 0.09 ^{b}	0.24 ± 0.07 ^{bc}	0.39 ± 0.12 ^{bc}
	Relative IL-1β mRNA expression	0.01 ± 0.003 ^{c}	1.05 ± 0.27 ^{a}	0.33 ± 0.16 ^{ab}	0.36 ± 0.14 ^{ab}	0.41 ± 0.15 ^{ab}	0.11 ± 0.01 ^{b}	0.26 ± 0.17 ^{ab}
	Relative IL-6 mRNA expression	0.67 ± 0.14 ^{ab}	1.06 ± 0.18 ^{a}	0.64 ± 0.22 ^{ab}	0.74 ± 0.24 ^{a}	0.51 ± 0.17 ^{ab}	0.26 ± 0.24 ^{b}	0.74 ± 0.11 ^{ab}
	Relative IL-10 mRNA expression	1.64 ± 0.07 ^{a}	0.96 ± 0.16 ^{b}	1.00 ± 0.26 ^{ab}	0.65 ± 0.36 ^{b}	0.51 ± 0.23 ^{b}	1.42 ± 0.45 ^{ab}	1.03 ± 0.33 ^{ab}
	Relative TNFα mRNA expression	0.04 ± 0.01 ^{b}	1.04 ± 0.16 ^{a}	0.58 ± 0.21 ^{ab}	0.32 ± 0.12 ^{b}	0.40 ± 0.19 ^{ab}	0.53 ± 0.20 ^{ab}	0.50 ± 0.10 ^{ab}
Immunohisto-chemical analysis	NF-κB expression in skin issues	9.806 ± 0.76 ^{b}	16.56 ± 0.62 ^{a}	14.3 ± 0.65 ^{b}	12.48 ± 0.84 ^{b}	13.02 ± 0.84 ^{b}	11.18 ± 0.88 ^{b}	11.32 ± 1.07 ^{b}
	NF-κB expression in the epidermal layer	9.47 ± 0.46 ^{b}	12.76 ± 1.37 ^{a}	8.51 ± 1.10 ^{b}	11.32 ± 1.42 ^{a}	11.7 ± 1.61 ^{a}	9.18 ± 0.71 ^{b}	7.48 ± 0.74 ^{b}
	NF-κB expression in the sebaceous gland	6.41 ± 0.60 ^{b}	10.41 ± 0.98 ^{a}	7.71 ± 1.38 ^{a}	7.23 ± 1.16 ^{a}	6.39 ± 0.68 ^{b}	7.02 ± 0.95 ^{b}	7.59 ± 0.52 ^{b}

All data are presented as mean ± standard deviation (SD). "a, b, c": different letters indicate significant differences ($p < 0.05$) among different groups.

3.4. Decreased Skin Sebum Levels with Pretreatment with Naringenin and Apigenin

The skin TG, NEFA, and T-CHO levels were determined based on pretreatment with naringenin and apigenin, respectively (Figure 3 and Table 2). Apigenin significantly inhibited the TG levels ($p < 0.05$). The TG levels in skin lesions significantly decreased in the L-NAR group ($p < 0.05$), whereas there was no significant difference between the H-NAR and model groups ($p > 0.05$). Although there were no significant differences in the T-CHO levels among the groups ($p > 0.05$), a decreasing trend was observed after the pretreatments with naringenin and apigenin (Figure 3C and Table 2). However, the content of NEFA decreased significantly in the H-NAR group when compared to the model group ($p < 0.05$; Figure 3B and Table 2).

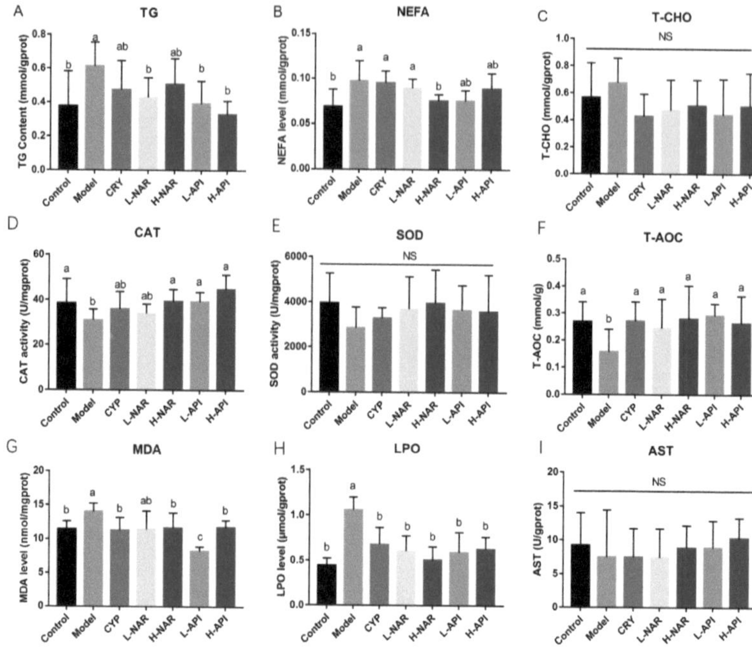

Figure 3. Influences of naringenin and apigenin pretreatments on skin lipids and antioxidative abilities. Skin lipids include total triglyceride (TG; **A**), non-esterified fatty acid (NEFA; **B**), and cholesterol (CHO; **C**) levels in mice skin after the interventions with naringenin and apigenin. Assessment of antioxidative abilities via determining the levels of catalase (CAT; **D**), superoxide dismutase (SOD; **E**), total antioxidant capacity (T-AOC; **F**), malondialdehyde (MDA; **G**), lipid peroxide (LPO; **H**), and aspartate aminotransferase (AST; **I**). CRY: cryptotanshinone; L-NAR: naringenin with a dose of 5 mg/kg/d; H-NAR: naringenin with a dose of 10 mg/kg/d; L-API: apigenin with a dose of 5 mg/kg/d; H-API: apigenin with a dose of 10 mg/kg/d. Significant differences ($p < 0.05$) among different groups were presented as "a, b, c". Non-significant differences ($p > 0.05$) were presented as "NS".

3.5. Improved Antioxidative System after Naringenin and Apigenin Supplementation

The increase in the CAT levels is shown in Figure 3D, and apigenin promoted CAT expression when compared to the value in the model group ($p < 0.05$). The SOD levels maintained a similar level in the skin after the naringenin and apigenin pretreatments ($p > 0.05$; Figure 3E and Table 2). The total antioxidative ability was assessed after the T-AOC determination, and a significant increase was observed with the intake of naringenin and apigenin ($p < 0.05$; Figure 3F and Table 2). A remarkable decrease in the MDA level was observed in the H-ANR, L-API, and H-API groups compared to that in the model group ($p < 0.05$), and the MDA level in the L-API group was significantly lower than that in the H-ANR and H-API groups ($p < 0.05$; Figure 3G and Table 2). LPO was determined, and obvious decreases are evident in Figure 3H ($p < 0.05$). Figure 3I exhibits no difference in the AST expression among the groups ($p > 0.05$).

3.6. Prevention of Inflammatory Response with Naringenin and Apigenin Pretreatments

The levels of IL-6, IL-1β, TNF-α, and IL-10 in the OA-treated mice's skin were evaluated (Figure 4 and Table 2). The levels of IL-6, IL-1β, and TNF-α significantly decreased with naringenin and apigenin pretreatments ($p < 0.05$). The decreases in the levels of these proinflammatory factors appeared to be dose-dependent with the pretreatment of naringenin; on the contrary, the OA-treated mice in the L-API group were more efficient at alleviating inflammation than those in the H-API group (Figure 4 and Table 2). Naringenin induced an increase in IL-10 expression ($p < 0.05$) compared to that in the model group, and naringenin (10 mg/kg) was thought to be better for IL-10 expression (Figure 4D and Table 2). Apigenin at 5 mg/kg and 10 mg/kg doses significantly reduced IL-10 expression in the skin lesions of the OA-treated mice (Figure 4D and Table 2). These results suggested that naringenin and apigenin alleviated inflammatory response in the OA-treated mice.

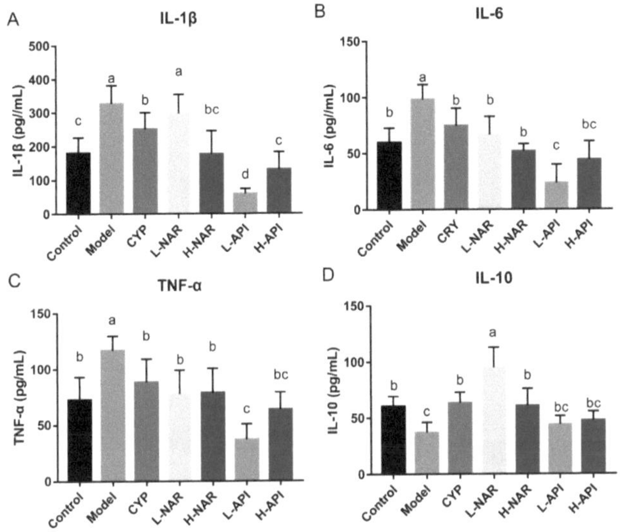

Figure 4. Influences of naringenin and apigenin pretreatments on skin interleukin (IL)-1β, IL-6, tumor necrosis factor (TNF)-α, and IL-10 levels in OA-treated mice. IL-1β expression (**A**), IL-6 expression (**B**), TNF-α expression (**C**), IL-10 expression (**D**). CRY: cryptotanshinone; L-NAR: naringenin with a dose of 5 mg/kg/d; H-NAR: naringenin with a dose of 10 mg/kg/d; L-API: apigenin with a dose of 5 mg/kg/d; H-API: apigenin with a dose of 10 mg/kg/d. Significant differences ($p < 0.05$) among different groups were presented as "a, b, c".

3.7. Effects of Naringenin and Apigenin on mRNA Levels Related to Antioxidation and Inflammation

To further evaluate the influence of naringenin and apigenin pretreatments on the OA-treated mice, the mRNA expression levels of genes related to skin sebum and inflammation were analyzed (Figure 5 and Table 2). The level of sterol response element-binding protein (SREBP) decreased in the naringenin and apigenin pretreatment groups compared to that in the model group (Figure 5A), and similar levels of SREBP were observed in the H-NAR and control groups. Naringenin and apigenin promoted peroxisome proliferator-activated receptor α (PPARα) and adipose triacylglyceride lipase (Atgl) expression, and it seemed that the PPAR levels were dose-dependent (Figure 5B,C and Table 2). However, fatty acid synthetase (FAS) gene expression did not vary significantly among the groups (Figure 5D and Table 2).

Compared to the model group, apigenin increased Nrf2 expression (Figure 5E), but no difference was shown in the NAR group. The mRNA expression of proinflammatory cytokines, including IL-6, TNF-α, and IL-1β, reduced after naringenin and apigenin pretreatments compared to that in the model group (Figure 5F–H and Table 2). A clear reduction in IL-1β and IL-6 was presented in the L-API group, and a significant decrease in TNF-α was shown in the L-NAR group. The naringenin and apigenin pretreatment groups showed no differences in the expression of IL-10 level compared to that in the model group (Figure 5I, Table 2). The mRNA expression of inflammatory factors indicated the importance of relieving inflammation to cure sebum metabolism disorders. Furthermore, protein kinase B (Akt), NF-κB, and COX-2 expressions were determined. Increased Akt level and decreased NF-κB and COX-2 levels were observed after naringenin and apigenin pretreatments in the OA-treated mice (Figure 5J–L and Table 2).

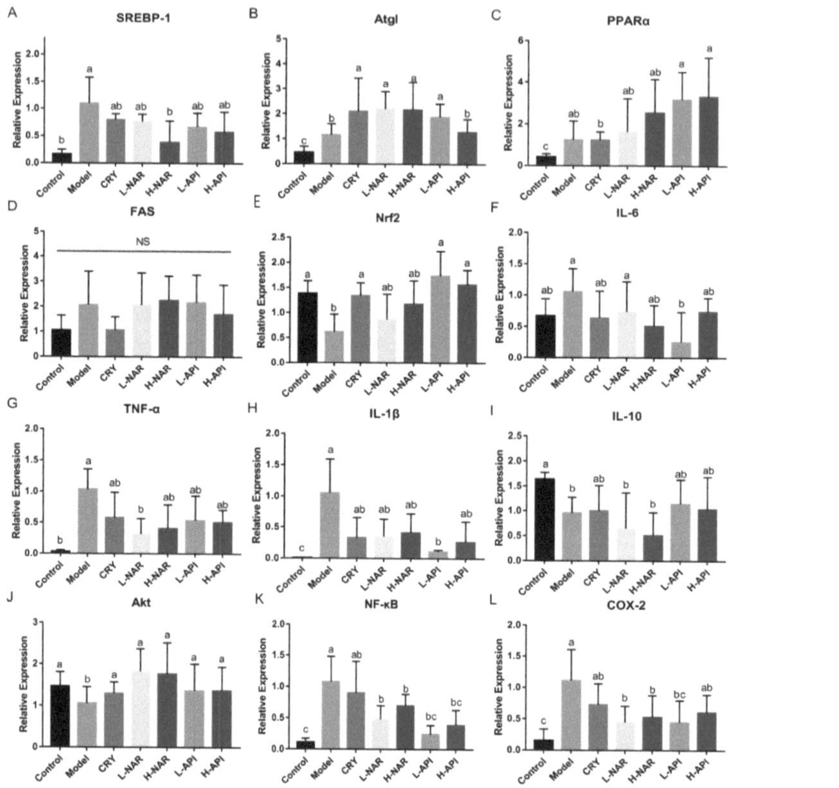

Figure 5. Influences of naringenin and apigenin pretreatments on lipid metabolism and skin inflammation. SREBP-1 expression (**A**), Atgl expression (**B**), PPARα expression (**C**), FAS expression (**D**), Nrf2 expression (**E**), IL-6 expression (**F**), TNF-α expression (**G**), IL-1β expression (**H**), IL-10 expression (**I**), Akt expression (**J**), NF-κB expression (**K**), COX-2 expression (**L**). CRY: cryptotanshinone; L-NAR: naringenin with a dose of 5 mg/kg/d; H-NAR: naringenin with a dose of 10 mg/kg/d; L-API: apigenin with a dose of 5 mg/kg/d; H-API: apigenin with a dose of 10 mg/kg/d; SREBP-1: sterol response element-binding protein-1; Atgl: adipose triacylglyceride lipase; PPARα: peroxisome proliferator-activated receptor α; FAS: fatty acid synthetase; Nrf2: transcription factor nuclear factor (erythroid-derived 2)-like 2; AKT: protein kinase B; NF-κB: nuclear factor-kappa B; COX-2: cyclooxygenase-2. Significant differences ($p < 0.05$) among different groups were presented as "a, b, c". Non-significant differences ($p > 0.05$) were presented as "NS".

3.8. *Effects of Naringenin and Apigenin on NF-κB Expression*

To further confirm whether the expression of NF-κB played an important role in improving skin lesions in the OA-treated mice after being administered naringenin and apigenin, immunohistochemical analysis was performed (Figure 6). NF-κB was expressed in skin issues, mainly in the epidermis and sebaceous glands. A high expression of inflammatory cells in the dermis was also observed in the OA-treated mice. Based on the quantitative analysis, apigenin induced a significantly lower expression of NF-κB in skin issues ($p < 0.05$) than that in the model group (Figures 6 and S3 and Table 2). The epidermal layer appeared to exhibit higher NF-κB expression, which was related to its thickness. The number of sebaceous glands limited the content of NF-κB, although NF-κB levels were presented in all groups.

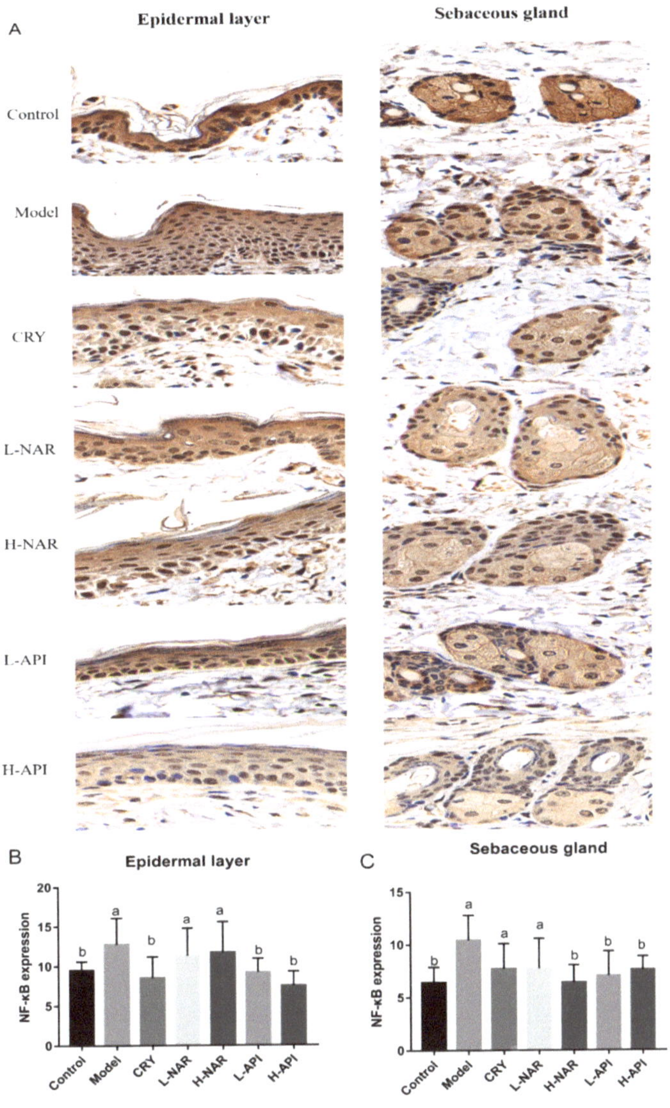

Figure 6. NF-κB expression based on immunohistochemical analysis in the epidermal layer and sebaceous gland. Immunohistochemical photographs of the epidermal layer and sebaceous gland (**A**), and quantitative analysis of NF-κB expression in the epidermal layer (**B**) and in the sebaceous gland (**C**). CRY: cryptotanshinone; L-NAR: naringenin with a dose of 5 mg/kg/d; H-NAR: naringenin with a dose of 10 mg/kg/d; L-API: apigenin with a dose of 5 mg/kg/d; H-API: apigenin with a dose of 10 mg/kg/d. Significant differences ($p < 0.05$) among the different groups were presented as "a, b". Non-significant differences ($p > 0.05$) were presented as "NS".

4. Discussion

Ingredients obtained from herbs and edible plants have medicinal and nutritional value and are widely used in medicine, foods, cosmetics, and other fields. Naringenin and apigenin are regarded as two universal molecules that have gained considerable attention; they were used in the present study to explore their effects on the maintenance of skin barriers. The back of mice was smeared with OA after shaving, which is a well-known model for causing multiple layers of tight keratinized cells in the epithelium of hair follicles to block hair follicle pores and promote inflammation [24,25]. Changes in skin lipids affect skin barrier functions, such as palmitic acid, stearic acid, and OA, which penetrate the dermo-epidermal barrier to promote the secretion of inflammatory factors and activate macrophages and related inflammatory and oxidative systems. Thus, it would be beneficial to explore the protective properties of naringenin and apigenin pretreatments against sebum metabolism disorders and related inflammation in OA-treated mice.

The skin barrier was broken and symptoms of erythema, scales, erosion, and pruritus gradually emerged after the OA application in mice. The overall assessment of skin damage showed conspicuous improvements in the skin condition with the naringenin and apigenin pretreatments ($p < 0.05$; Table 1 and Figure S2). Pathological tissue analysis indicated the apigenin pretreatment decreased epidermal thickness and reduced inflammatory invasion. Previous studies have presented that apigenin can repair skin and relieve inflammation [11,12]. In addition, fewer lipid droplets in skin tissues were observed in the API group than in the NAR group, suggesting apigenin may affect sebum balance via regulating skin fat metabolism.

To further determine whether apigenin was more effective than naringin in improving the skin barrier and alleviating inflammation, the main lipid components related to antioxidative and inflammatory factors in the skin were determined. The lipid-lowering ability of apigenin has been reported in vivo and in vitro studies [26–28], which is also confirmed in our results showing that apigenin has better effects on reducing TG and NEFA contents. Changes in sebum levels in terms of quality and quantity can induce keratinization abnormality in ductal epithelial cells of hair follicle sebaceous glands, followed inflammation [29]. Fatty acids are highly expressed in the skin lesions of patients with sebum disorders, resulting in increased secretion of inflammatory factors, especially IL-1β, IL-6, and TNF-α [9,24]. The results of this study showed that naringenin and apigenin decreased the NEFA levels and reduced the levels of IL-1β, IL-6, and TNF-α. The evidence suggested that pretreatment with naringenin and apigenin might improve skin lipids issues by inhibiting proinflammatory cytokines. The disruption of sebum balance affects the activities of lipases, such as FAS, hormone-sensitive lipase (Hsl), and Atgl [30–33]. FAS, Atgl, and Hsl are responsible for fatty acid synthesis, TG hydrolysis, and lipolysis [25,33]. FAS is also related to immune activation, and blocked FAS expression can attenuate obesity-induced adipose tissue inflammation by inhibiting pro-inflammatory pathways [34]. Atgl is a major enzyme involved in adipose tissue TG catabolism, and ATGL is downregulated in mouse models of obesity [30]. There were no significant changes in the FAS levels; however, the mRNA level of Atgl significantly increased, explaining the lower levels of TG contents after the naringenin and apigenin pretreatments in the OA-treated mice. To understand the increases in Atgl levels, especially in the NAR group, related key proteins were analyzed. Sebum improvement has been reported to be associated with SREBP-1 and PPAR expressions [35,36]. SREBP-1 is a lipogenesis-related factor that maintains the balance in glucose metabolism and cholesterol and lipid metabolism in vivo [33]. PPARs participate in various processes, including improving endothelial function, promoting CHO transfer, regulating glucose and lipid metabolism, and inhibiting inflammation [37–39]. Some reports have illustrated that decreased SREBP-1 and PPARα expressions regulate a low level of lipids in human sebocytes [33]. For the naringenin pretreatment, the lower mRNA expression of SERBP-1 than that in the model group indicated naringenin might regulate sebum depending on fat metabolism via the SREBP-1/ATGL pathway.

In addition, a trial demonstrated that the skin surface of patients with acne presented a mass of lipoperoxides owing to squalene peroxidation and a reduction in antioxidant sebum (vitamin E) [40]. This indicates that oxidized lipids play an important role in the development of skin lesions. Naringenin reduces UVB-induced oxidative stress in the skin of hairless mice [6]. A study demonstrated that T-AOC and NRF were improved in apigenin-treated aged mice, thereby inhibiting oxidative stress [41]. Apigenin restores the viability of UV-treated normal human dermal fibroblasts by altering their anti-inflammatory and antioxidant properties [26]. Naringenin and apigenin improved the antioxidative system in the skin, decreased MDA and LPO levels, and increased CAT and T-AOC levels. Significantly increased expression of Nrf2 level was observed after the naringenin and apigenin pretreatments. The Nrf2 pathway mediates the oxidative system and ultimately exerts anti-inflammatory effects. The transcription factor Nrf2 and its principal negative regulator, the E3 ligase adaptor Kelch-like ECH-associated protein 1, play a pivotal role in maintaining intracellular redox homeostasis and regulating inflammation [42]. In addition, the crosstalk between the Nrf2 pathway and NF-κB expression has been demonstrated. NF-κB is a key biomarker involved in various inflammatory diseases, such as arthritis, carditis, folliculitis, enteritis, and pneumonia. We focused on the exploration of the NF-κB pathway, and NF-κB pathway is a common, central pathway involved in promoting the formation of pro-inflammatory and pro-labor mediators in tissues [43,44]. NF-κB protein is located in the cytoplasm and can be activated by several different stimuli. The activation of NF-κB is expressed via the positive and negative regulations of many important genes involved in the essential inflammatory process, including chemokines and proinflammatory cytokines. PPARs block this activation and reduce the synthesis and secretion of inflammatory molecules. IL-1β, IL-6, and TNF-α are triggers of NF-κB [43]. Skin inflammation is suppressed via the baicalin-regulated NF-κB/MAPK signaling pathway [45]. The migration of neutrophils and lymphocytes and the activation of NF-κB were found to be present in patients with seborrheic dermatitis [46]. The Akt pathway also plays an important role in the development of inflammation, and Akt activation downregulates NF-κB expression, thus affecting the NF-κB pathway [47]. Previous studies demonstrated that inflammatory diseases were improved by mediating Akt/NF-κB pathway signaling [43]. Consistently, downregulation of NF-κB and upregulation of Akt were observed after the naringenin and apigenin pretreatments compared to the levels in the model group. Furthermore, the decreased mRNA levels of IL-1β, IL-6, and TNF-α were consistent. COX-2 expression was also downregulated after the naringenin and apigenin pretreatments in the OA-treated mice. COX-2 is thought to be an inflammatory inducer, with high expression of COX-2 found in inflammatory regions [48]. In general, COX-2 is not expressed in most tissues but is highly expressed in damaged and inflamed tissues [49,50]. Previous studies showed that COX-2 expression was regulated by not only proinflammatory cytokines but also NF-κB levels [50]. Notably, NF-κB has two binding sites in the promoter of COX-2, and increases in COX-2 are induced after specific changes in these binding sites, indicating that the transcriptional activation of COX-2 requires the participation of NF-κB [50]. When drugs were applied in acute lung injury, the conditions were improved following decreased expression of COX-2 and NF-κB levels [51]. The decreases in COX-2 level reflected the reduction in NF-κB expression, suggesting the potential therapeutic effects of naringenin and apigenin in regulating NF-κB signaling.

The present study demonstrated that naringenin and apigenin can ameliorate skin damage by regulating the Nrf2 and NF-κB pathways to improve antioxidative stress and anti-inflammatory response in OA-induced mice, indicating naringenin and apigenin as potential dietary supplements to protect against skin damage. In addition, fruits and vegetables rich in apigenin and naringin could be consumed in a more appropriate way. The information provided in this study is useful to the development of new health-promoting diets and can play an important role in the improvement of human health. However, more studies on phytochemicals and their healthy components are required to further explore their benefits for health.

5. Conclusions

Pretreatments with naringenin and apigenin alleviated skin damage in the OA-induced mice by decreasing skin lipid levels, improving keratinization and antioxidation, and reducing the release of proinflammatory factors. The improvement in antioxidative abilities was also mediated by the activation of the Nrf2 pathway, and the reduction in inflammation was regulated by naringenin and apigenin via NF-κB signaling.

Supplementary Materials: The following supporting information can be downloaded at https://www.mdpi.com/article/10.3390/foods12112120/s1. Figure S1: Chromatogram of naringenin and apigenin based on LC-MS/MS: chromatogram of apigenin (A) and chromatogram of naringenin (B). Figure S2: Overall assessment score of skin injury in mice with OA treatment. Scores were given by evaluating symptoms such as skin erythema, scales, erosion, and pruritus. Single symptoms were divided into 4 levels: a 0 score for asymptomatic effects, 1 for mild symptoms, 2 for moderate symptoms, and 4 for severe symptoms. Figure S3: (A,B) NF-κB expression based on immunohistochemical analysis in skin issues. Significant differences ($p < 0.05$) among different groups were presented as "a, b, c". Non-significant differences ($p > 0.05$) were presented as "NS". Table S1: List of treatments in different groups. Table S2: Primer sequences for the quantitative real-time PCR (qRT-PCR) analysis used in the present study.

Author Contributions: Conceptualization, J.L. and B.M.; methodology, X.T.; validation, X.T., Q.Z. and H.Z.; formal analysis, J.Z.; investigation, J.L. and Q.Z.; resources, J.L. and J.Z.; writing—original draft preparation, J.L.; writing—review and editing, J.L., B.M., H.Z. and S.C.; visualization, B.M.; supervision, B.M., H.Z. and S.C.; project administration, H.Z.; funding acquisition, S.C. All authors have read and agreed to the published version of the manuscript.

Funding: This work was supported by the National Natural Science Foundation of China (Grant No. 32172173) and the Collaborative Innovation Center of Food Safety and Quality Control in Jiangsu Province.

Data Availability Statement: Data are contained within the article.

Conflicts of Interest: The authors declare no conflict of interest.

References

1. Lu, C.L.; Li, X.F. A Review of *Oenanthe javanica* (Blume) DC. as Traditional Medicinal Plant and Its Therapeutic Potential. *Evid.-Based Complement. Altern. Med.* **2019**, *2019*, 6495819. [CrossRef]
2. Proteggente, A.R.; Saija, A.; De Pasquale, A.; Rice-Evans, C.A. The compositional characterisation and antioxidant activity of fresh juices from sicilian sweet orange (*Citrus sinensis* L. Osbeck) varieties. *Free Radic. Res.* **2003**, *37*, 681–687. [CrossRef]
3. Lee, D.H.; Lee, J.S.; Lee, I.H.; Hong, J.T. Therapeutic potency of fermented field water-dropwort (*Oenanthe javanica* (Blume) DC.) in ethanol-induced liver injury. *RSC Adv.* **2020**, *10*, 1544–1551. [CrossRef] [PubMed]
4. Kelebek, H.; Selli, S. Determination of volatile, phenolic, organic acid and sugar components in a Turkish cv. Dortyol (*Citrus sinensis* L. Osbeck) orange juice. *J. Sci. Food Agric.* **2011**, *91*, 1855–1862. [CrossRef] [PubMed]
5. Al-Roujayee, A.S. Naringenin improves the healing process of thermally-induced skin damage in rats. *J. Int. Med. Res.* **2017**, *45*, 570–582. [CrossRef] [PubMed]
6. Martinez, R.M.; Pinho-Ribeiro, F.A.; Steffen, V.S.; Caviglione, C.V.; Vignoli, J.A.; Barbosa, D.S.; Baracat, M.M.; Georgetti, S.R.; Verri, W.A., Jr.; Casagrande, R. Naringenin Inhibits UVB Irradiation-Induced Inflammation and Oxidative Stress in the Skin of Hairless Mice. *J. Nat. Prod.* **2015**, *78*, 1647–1655. [CrossRef]
7. Martinez, R.M.; Pinho-Ribeiro, F.A.; Steffen, V.S.; Silva, T.C.; Caviglione, C.V.; Bottura, C.; Fonseca, M.J.; Vicentini, F.T.; Vignoli, J.A.; Baracat, M.M.; et al. Topical Formulation Containing Naringenin: Efficacy against Ultraviolet B Irradiation-Induced Skin Inflammation and Oxidative Stress in Mice. *PLoS ONE* **2016**, *11*, e0146296. [CrossRef] [PubMed]
8. Kumar, R.; Bhan Tiku, A. Naringenin Suppresses Chemically Induced Skin Cancer in Two-Stage Skin Carcinogenesis Mouse Model. *Nutr. Cancer* **2020**, *72*, 976–983. [CrossRef]
9. Hernandez-Aquino, E.; Muriel, P. Beneficial effects of naringenin in liver diseases: Molecular mechanisms. *World J. Gastroenterol.* **2018**, *24*, 1679–1707. [CrossRef]
10. Sanaye, P.M.; Mojaveri, M.R.; Ahmadian, R.; Jahromi, M.S.; Bahramsoltani, R.J.P. Different mechanisms by which apigenin improves treatment of dermatological disorders Apigenin and its dermatological applications: A comprehensive review. *Phytochemistry* **2022**, *203*, 113390. [CrossRef]
11. Park, C.H.; Min, S.Y.; Yu, H.W.; Kim, K.; Kim, S.; Lee, H.J.; Kim, J.H.; Park, Y.J. Effects of Apigenin on RBL-2H3, RAW264.7, and HaCaT Cells: Anti-Allergic, Anti-Inflammatory, and Skin-Protective Activities. *Int. J. Mol. Sci.* **2020**, *21*, 4620. [CrossRef] [PubMed]
12. Ma, X.; Lin, Y.; Liu, y.; Li, W.; He, J.; Fang, M.; Lin, D. Effects of Apigenin Treatment on Random Skin Flap Survival in Rats. *Front. Pharmacol.* **2021**, *12*, 625733. [CrossRef] [PubMed]
13. Cao, C.; Xiao, Z.; Wu, Y.; Ge, C. Diet and Skin Aging-From the Perspective of Food Nutrition. *Nutrients* **2020**, *12*, 870. [CrossRef]

14. Naidoo, K.; Birch-Machin, M. Oxidative Stress and Ageing: The Influence of Environmental Pollution, Sunlight and Diet on Skin. *Cosmetics* **2017**, *4*, 4. [CrossRef]
15. Williams, H.C.; Dellavalle, R.P.; Garner, S. Acne vulgaris. *Lancet* **2012**, *379*, 361–372. [CrossRef] [PubMed]
16. Darvin, M.E.; Patzelt, A.; Knorr, F.; Blume-Peytavi, U.; Sterry, W.; Lademann, J. One-year study on the variation of carotenoid antioxidant substances in living human skin: Influence of dietary supplementation and stress factors. *J. Biomed. Opt.* **2008**, *13*, 044028. [CrossRef] [PubMed]
17. Shin, E.J.; Lee, J.S.; Hong, S.; Lim, T.G.; Byun, S. Quercetin Directly Targets JAK2 and PKCdelta and Prevents UV-Induced Photoaging in Human Skin. *Int. J. Mol. Sci.* **2019**, *20*, 5262. [CrossRef]
18. Katsuta, Y.; Iida, T.; Hasegawa, K.; Inomata, S.; Denda, M. Function of oleic acid on epidermal barrier and calcium influx into keratinocytes is associated with N-methyl D-aspartate-type glutamate receptors. *Br. J. Dermatol.* **2009**, *160*, 69–74. [CrossRef]
19. Yang, Y.; Trevethan, M.; Wang, S.; Zhao, L. Beneficial Effects of Citrus Flavanones Naringin and Naringenin and Their Food Sources on Lipid Metabolism: An Update on Bioavailability, Pharmacokinetics, and Mechanisms. *J. Nutr. Biochem.* **2022**, *104*, 108967. [CrossRef]
20. Ganai, S.A. Plant-derived flavone Apigenin: The small-molecule with promising activity against therapeutically resistant prostate cancer. *Biomed. Pharmacother.* **2017**, *85*, 47–56. [CrossRef]
21. Guo, W.; Cui, S.; Tang, X.; Zhang, Q.; Zhao, J.; Mao, B.; Zhang, H. Intestinal Microbiomics and Metabolomics Insights into the Hepatoprotective Effects of Lactobacillus paracasei CCFM1222 Against the Acute Liver Injury in Mice. *Probiotics Antimicrob. Proteins* **2022**, 1–15. [CrossRef]
22. Kim, Y.R.; Kim, J.H.; Shin, H.J.; Choe, Y.B.; Ahn, K.J.; Lee, Y.W. Clinical Evaluation of a New-Formula Shampoo for Scalp Seborrheic Dermatitis Containing Extract of Rosa centifolia Petals and Epigallocatechin Gallate: A Randomized, Double-Blind, Controlled Study. *Ann. Dermatol.* **2014**, *26*, 733–738. [CrossRef]
23. Su, L.; Su, Y.; An, Z.; Zhang, P.; Yue, Q.; Zhao, C.; Sun, X.; Zhang, S.; Liu, X.; Li, K.; et al. Fermentation products of Danshen relieved dextran sulfate sodium-induced experimental ulcerative colitis in mice. *Sci. Rep.* **2021**, *11*, 16210. [CrossRef]
24. Ryu, A.R.; Kim, Y.-W.; Lee, M.-Y. Chlorin e6 and halogen light as a sebostatic photomedicine modulates linoleic acid-induced lipogenesis. *Mol. Cell. Toxicol.* **2018**, *15*, 49–56. [CrossRef]
25. Torocsik, D.; Fazekas, F.; Poliska, S.; Gregus, A.; Janka, E.A.; Dull, K.; Szegedi, A.; Zouboulis, C.C.; Kovacs, D. Epidermal Growth Factor Modulates Palmitic Acid-Induced Inflammatory and Lipid Signaling Pathways in SZ95 Sebocytes. *Front. Immunol.* **2021**, *12*, 600017. [CrossRef] [PubMed]
26. Hsu, M.C.; Guo, B.C.; Chen, C.H.; Hu, P.A.; Lee, T.S. Apigenin ameliorates hepatic lipid accumulation by activating the autophagy-mitochondria pathway. *J. Food Drug Anal.* **2021**, *29*, 240–254. [CrossRef]
27. Kang, S.R.; Park, K.I.; Park, H.S.; Lee, D.H.; Kim, J.A.; Nagappan, A.; Kim, E.H.; Lee, W.S.; Shin, S.C.; Park, M.K.; et al. Anti-inflammatory effect of flavonoids isolated from Korea *Citrus aurantium* L. on lipopolysaccharide-induced mouse macrophage RAW 264.7 cells by blocking of nuclear factor-kappa B (NF-κB) and mitogen-activated protein kinase (MAPK) signalling pathways. *Food Chem.* **2011**, *129*, 1721–1728. [CrossRef]
28. Wang, Z.; Zeng, M.; Wang, Z.; Qin, F.; Chen, J.; He, Z. Dietary Luteolin: A Narrative Review Focusing on Its Pharmacokinetic Properties and Effects on Glycolipid Metabolism. *J. Agric. Food Chem.* **2021**, *69*, 1441–1454. [CrossRef] [PubMed]
29. Shin, H.S.; Lee, Y.; Shin, M.H.; Cho, S.I.; Zouboulis, C.C.; Kim, M.K.; Lee, D.H.; Chung, J.H. Histone Deacetylase 1 Reduces Lipogenesis by Suppressing SREBP1 Transcription in Human Sebocyte Cell Line SZ95. *Int. J. Mol. Sci.* **2021**, *22*, 4477. [CrossRef]
30. Ghosh, M.; Niyogi, S.; Bhattacharyya, M.; Adak, M.; Nayak, D.K.; Chakrabarti, S.; Chakrabarti, P. Ubiquitin Ligase COP1 Controls Hepatic Fat Metabolism by Targeting ATGL for Degradation. *Diabetes* **2016**, *65*, 3561–3572. [CrossRef] [PubMed]
31. Fang, X.; Zhao, Z.; Jiang, P.; Yu, H.; Xiao, H.; Yang, R. Identification of the bovine HSL gene expression profiles and its association with fatty acid composition and fat deposition traits. *Meat Sci.* **2017**, *131*, 107–118. [CrossRef] [PubMed]
32. Zhang, Z.; Guo, M.; Shen, M.; Li, Y.; Tan, S.; Shao, J.; Zhang, F.; Chen, A.; Wang, S.; Zheng, S. Oroxylin A regulates the turnover of lipid droplet via downregulating adipose triglyceride lipase (ATGL) in hepatic stellate cells. *Life Sci.* **2019**, *238*, 116934. [CrossRef]
33. Shin, J.; Kim, K.P.; Ahn, H.Y.; Kim, B.; Cho, Y. Alterations in IL-4, IL-10 and IFN-gamma levels synergistically decrease lipid content and protein expression of FAS and mature SREBP-1 in human sebocytes. *Arch. Dermatol. Res.* **2019**, *311*, 563–571. [CrossRef] [PubMed]
34. Nyambuya, T.M.; Dludla, P.V.; Nkambule, B.B. Diet-Induced Obesity Promotes the Upregulation of Fas Expression on T-cells. *Biology* **2021**, *10*, 217. [CrossRef]
35. Lovaszi, M.; Mattii, M.; Eyerich, K.; Gacsi, A.; Csanyi, E.; Kovacs, D.; Ruhl, R.; Szegedi, A.; Kemeny, L.; Stahle, M.; et al. Sebum lipids influence macrophage polarization and activation. *Br. J. Dermatol.* **2017**, *177*, 1671–1682. [CrossRef]
36. Borda, L.J.; Perper, M.; Keri, J.E. Treatment of seborrheic dermatitis: A comprehensive review. *J. Dermatolog. Treat.* **2019**, *30*, 158–169. [CrossRef]
37. Gupta, M.; Mahajan, V.K.; Mehta, K.S.; Chauhan, P.S.; Rawat, R. Peroxisome proliferator-activated receptors (PPARs) and PPAR agonists: The 'future' in dermatology therapeutics? *Arch. Dermatol. Res.* **2015**, *307*, 767–780. [CrossRef] [PubMed]
38. Jia, Y.; Kim, J.Y.; Jun, H.J.; Kim, S.J.; Lee, J.H.; Hoang, M.H.; Hwang, K.Y.; Um, S.J.; Chang, H.I.; Lee, S.J. The natural carotenoid astaxanthin, a PPAR-alpha agonist and PPAR-gamma antagonist, reduces hepatic lipid accumulation by rewiring the transcriptome in lipid-loaded hepatocytes. *Mol. Nutr. Food Res.* **2012**, *56*, 878–888. [CrossRef]

39. Capitao, A.; Lopes-Marques, M.; Pascoa, I.; Ruivo, R.; Mendiratta, N.; Fonseca, E.; Castro, L.F.C.; Santos, M.M. The Echinodermata PPAR: Functional characterization and exploitation by the model lipid homeostasis regulator tributyltin. *Environ. Pollut.* **2020**, *263*, 114467. [CrossRef]
40. Capitanio, B.; Lora, V.; Ludovici, M.; Sinagra, J.L.; Ottaviani, M.; Mastrofrancesco, A.; Ardigo, M.; Camera, E. Modulation of sebum oxidation and interleukin-1alpha levels associates with clinical improvement of mild comedonal acne. *J. Eur. Acad. Dermatol. Venereol.* **2014**, *28*, 1792–1797. [CrossRef]
41. Wang, D.; Yang, Y.; Zou, X.; Zhang, J.; Zheng, Z.; Wang, Z. Antioxidant Apigenin Relieves Age-Related Muscle Atrophy by Inhibiting Oxidative Stress and Hyperactive Mitophagy and Apoptosis in Skeletal Muscle of Mice. *J. Gerontol. A Biol. Sci. Med. Sci.* **2020**, *75*, 2081–2088. [CrossRef] [PubMed]
42. Ahmed, S.M.; Luo, L.; Namani, A.; Wang, X.J.; Tang, X. Nrf2 signaling pathway: Pivotal roles in inflammation. *Biochim. Biophys. Acta Mol. Basis Dis.* **2017**, *1863*, 585–597. [CrossRef]
43. Zinatizadeh, M.R.; Schock, B.; Chalbatani, G.M.; Zarandi, P.K.; Jalali, S.A.; Miri, S.R. The Nuclear Factor Kappa B (NF-kB) signaling in cancer development and immune diseases. *Genes Dis.* **2021**, *8*, 287–297. [CrossRef]
44. Zhao, W.; Yan, Y.; Xiao, Z.; Wang, M.; Xu, M.; Wang, Z.; Wang, Y.; Zhuang, Z.; Yang, D.; Chen, G.; et al. Bicyclol ameliorates nonalcoholic fatty liver disease in mice via inhibiting MAPKs and NF-kappaB signaling pathways. *Biomed. Pharmacother.* **2021**, *141*, 111874. [CrossRef]
45. Fang, F.; Xie, Z.; Quan, J.; Wei, X.; Wang, L.; Yang, L. Baicalin suppresses Propionibacterium acnes-induced skin inflammation by downregulating the NF-kappaB/MAPK signaling pathway and inhibiting activation of NLRP3 inflammasome. *Braz. J. Med. Biol. Res.* **2020**, *53*, e9949. [CrossRef]
46. Adalsteinsson, J.A.; Kaushik, S.; Muzumdar, S.; Guttman-Yassky, E.; Ungar, J. An update on the microbiology, immunology and genetics of seborrheic dermatitis. *Exp. Dermatol.* **2020**, *29*, 481–489. [CrossRef]
47. Caporali, S.; Levati, L.; Graziani, G.; Muzi, A.; Atzori, M.G.; Bonmassar, E.; Palmieri, G.; Ascierto, P.A.; D'Atri, S. NF-kappa B is activated in response to temozolomide in an AKT-dependent manner and confers protection against the growth suppressive effect of the drug. *J. Transl. Med.* **2012**, *10*, 252. [CrossRef]
48. Khan, A.A.; Iadarola, M.; Yang, H.Y.; Dionne, R.A. Expression of COX-1 and COX-2 in a clinical model of acute inflammation. *J. Pain.* **2007**, *8*, 349–354. [CrossRef]
49. Ren, Q.; Guo, F.; Tao, S.; Huang, R.; Ma, L.; Fu, P. Flavonoid fisetin alleviates kidney inflammation and apoptosis via inhibiting Src-mediated NF-kappaB p65 and MAPK signaling pathways in septic AKI mice. *Biomed. Pharmacother.* **2020**, *122*, 109772. [CrossRef] [PubMed]
50. Poligone, B.; Baldwin, A.S. Positive and negative regulation of NF-kappaB by COX-2: Roles of different prostaglandins. *J. Biol. Chem.* **2001**, *276*, 38658–38664. [CrossRef] [PubMed]
51. Wang, J.; Liu, Y.T.; Xiao, L.; Zhu, L.; Wang, Q.; Yan, T. Anti-inflammatory effects of apigenin in lipopolysaccharide-induced inflammatory in acute lung injury by suppressing COX-2 and NF-kB pathway. *Inflammation* **2014**, *37*, 2085–2090. [CrossRef] [PubMed]

foods

MDPI

Article

Effective Therapeutic Verification of Crocin I, Geniposide, and Gardenia (*Gardenia jasminoides* Ellis) on Type 2 Diabetes Mellitus In Vivo and In Vitro

Haibo Zhou [1], Sen Zhang [1], Lianghua Chen [2], Yimei Liu [1], Luhong Shen [1] and Jiuliang Zhang [1,3,*]

1 College of Food Science and Technology, Huazhong Agricultural University, No.1, Shizishan Street, Hongshan District, Wuhan 430070, China
2 Key Laboratory of Fujian Province for Physiology and Biochemistry of Subtropical Plant, Fujian Institute of Subtropical Botany, Xiamen 361006, China
3 Key Laboratory of Environment Correlative Dietology, Ministry of Education, Wuhan 430070, China
* Correspondence: zjl_ljz@mail.hzau.edu.cn; Tel./Fax: +86-027-87282111

Abstract: For many centuries, Gardenia (*Gardenia jasminoides* Ellis) was highly valued as a food homologous Chinese herbal medicine with various bioactive compounds, including crocin I and geniposide. However, the functional mechanism underlying the hypoglycemic effect of gardenia is absent in the literature. To evaluate the effect of gardenia and its different extracts on type 2 diabetes mellitus (T2DM) in in vivo and in vitro experiments, the dried gardenia powder was extracted using 60% ethanol and eluted at different ethanol concentrations to obtain the corresponding purified fragments. After that, the active chemical compositions of the different purified gardenia fragments were analyzed using HPLC. Then, the hypoglycemic effects of the different purified gardenia fragments were compared using in vitro and in vivo experiments. Finally, the different extracts were characterized using UPLC-ESI-QTOF-MS/MS and the mass spectrometric fragmentation pathway of the two main compounds, geniposide and crocin I, were identified. The experimental results indicated that the inhibitory effect of the 40% EGJ (crocin I) on the α-glucosidase was better than the 20% EGJ (geniposide) in vitro. However, the inhibitory effect of geniposide on T2DM was better than crocin I in the animal experiments. The different results in vivo and in vitro presumed potentially different mechanisms between crocin I and geniposide on T2DM. This research demonstrated that the mechanism of hypoglycemia in vivo from geniposide is not only one target of the α-glucosidase but provides the experimental background for crocin I and the geniposide deep processing and utilization.

Keywords: gardenia; α-glucosidase; crocin I; geniposide; type 2 diabetes mellitus

Citation: Zhou, H.; Zhang, S.; Chen, L.; Liu, Y.; Shen, L.; Zhang, J. Effective Therapeutic Verification of Crocin I, Geniposide, and Gardenia (*Gardenia jasminoides* Ellis) on Type 2 Diabetes Mellitus In Vivo and In Vitro. *Foods* **2023**, *12*, 1668. https://doi.org/10.3390/foods12081668

Academic Editor: Jee-Young Imm

Received: 7 March 2023
Revised: 10 April 2023
Accepted: 14 April 2023
Published: 17 April 2023

1. Introduction

T2DM is a metabolic disorder disease that is closely linked to an insufficient or defective insulin secretion. The main manifestation of T2DM is postprandial hyperglycemia [1]. Currently, there are more than 400 million people with diabetes worldwide and these records will continue to increase in the future [2]. The long-term vascular complications are the leading causes of an increased need for medical care, a reduced quality of life, and an increased risk of early death [3,4]. Amidst the strategies being developed for the management of T2DM with the established therapies, the α-glucosidase inhibitors have been approved in several treatment protocols [5]. The α-glucosidase is a kind of glucosidase that has a significant ability for digestion and hydrolysis starching in the small intestine [6]. The hypoglycemic principle inhibits the starch digestion in the small intestine and delays the absorption of glucose by inhibiting the α-glucosidase, thereby preventing the rate of abnormal postprandial glucose production [7]. Although hypoglycemic drugs, such as acarbose, metformin, and thiazolidinediones, clinically restrict the glucose absorption, adverse events to the undesirable gastrointestinal symptoms impede their application [8].

As a result, natural and harmless remedies are constantly being sought, and the search for natural products to treat diabetes is a viable and valuable approach.

The typical clinical symptoms of type 2 diabetes are polydipsia, polyuria, polyphagia, and weight loss. Gardenia can quench thirst and is widely used alone or in combination with other medicines for the treatment of T2DM in China and throughout Asia (The Pharmacopoeia Commission of PRC, 2015). Gardenia is a kind of homologous food medicine that is rich in various bioactive active compounds, such as geniposide and crocetin (including crocin I and crocin II) [9]. Geniposide is the most abundant iridoid and is the material basis of its biological activity. Geniposide has a significant effect on the digestive, cardiovascular, and central nervous systems [10–12]. Crocetin, also known as crocin, is divided into crocin I and crocin II, according to the type of the connection disaccharides at both ends of the crocin acid [13]. Crocin I accounts for 70% of the crocetin [14]. As crocetin is rare and precious, it has become the main trend to acquire it from gardenia. In recent years, gardenia was found to have many pharmacological benefits, such as anti-inflammatory effects, tumor prevention, cardiovascular disease prevention, and water-soluble nutrients that can be easily absorbed in the body [15–17]. One study showed that the *gardenia latifolia* extract (GLE) had an anti-diabetic potential in rats with a high-fat diet (HFD) + streptozotocin (STZ) induced type 2 diabetes mellitus (T2DM) [18]. However, there was less information about the functional mechanism of the hypoglycemic effects of crocin I, geniposide, from gardenia on T2DM in vivo and in vitro. Therefore, it was necessary to investigate and compare the intervention of geniposide, crocin I, and the extract of gardenia on hyperglycemic mice.

The objective of the present study was to compare the hypoglycemic effects of the bioactive compounds from gardenia in in vivo and in vitro experiments and to identify their structures. First, the major bioactive components of the extract fractions were identified using analytical methods. Next the activity of the major bioactive components against the enzyme and on the mice were compared. Then, different concentrations of alcohol were used to isolate and purify the crude extracts of gardenia (EGJ). Finally, geniposide and crocin I were obtained. This paper explored the hypoglycemic role-playing effects of the geniposide, crocin I, and EGJ intervention on T2DM. This research can also enrich the potential use of gardenia as a hypoglycemic functional food and lay the foundation for a new high added value product development of gardenia.

2. Materials and Methods

2.1. Materials and Reagents

Gardenia (*G. jasminoides* Ellis) from the Hubei province in China was provided and characterized by Hubei HaoYSJ Biotechnology Co. Limited (Yichang, China). The voucher specimens of gardenia (No. 190925) were deposited in the College of Food Science and Technology, Huazhong Agricultural University. The α-glucosidase from *Saccharomyces cerevisiae* (EC 3.2.1.20) was obtained from Sigma-Aldrich (St. Louis, MO, USA). The *p*-nitrophenyl (*p*NPG), α-D-glucosidase (purity > 99%), and acarbose (purity > 95%) were purchased from Yuanye Biological Technology (Shanghai, China). The AB-8 macroporous adsorption resin (0.3–1.25 mm particle size) was purchased from Nankai Hecheng Science & Technology Co. (Tianjin, China). The crocin I and geniposide (HPLC grade) were purchased from Shanghai Yuanye Biotechnology Co. (Shanghai, China). The methanol and acetic acid (HPLC grade) were obtained from J.T. Baker Co. (Phillipsburg, NJ, USA) and Aladdin Industrial Co. (Shanghai, China), respectively. The water was purified using a Milli-Q system supplied by Millipore (Billerica, MA, USA). All the other reagents were of an analytical grade.

2.2. Isolation and Identification of Geniposide and Crocetin

The purified and modified fractions of gardenia were mainly prepared according to the preliminary laboratory experiments [17]. The powder of gardenia was extracted using 60% ethanol at a solid–liquid ratio of 1:12 (g mL^{-1}) at 60 °C for 300 min. The supernatant was then concentrated using a rotary evaporator to obtain the viscous infusion. The 1:7 volume ratio of the petroleum ether was added to the viscous infusion, and the crude extract of gardenia (EGJ) was acquired after the concentration and vacuum freeze-drying processes.

According to the results of the preliminary experiments, the elution effect of the AB-8 macroporous adsorption resin was the best [19]. The crude EGJ was loaded onto AB-8 resin columns (25 mm × 500 mm) and eluted with 2 BV of water (1 BV = 1500 mL) and 20%, 40%, and 60% ethanol successively. Next, the samples were obtained by rinsing them with a 20% ethanol solution, 40% ethanol solution, and 60% ethanol solution, respectively. Subsequently, the 20% EGJ and 40% EGJ eluents were separately collected at 500 mL intervals for the relatively single components, which were numbered in the order of the sample from No.1 to No.30 and analyzed using the absorbance value measurement. According to the characteristic wavelength, the No.4, No.8, No.9, and No.14 samples were analyzed using HPLC. The Waters module 2695 was equipped with a 2998 photodiode array detector (Milford, MA, USA) and a Diamonsil Plus C18-A column (250 mm × 4.6 mm, 5 μm, Dikma, Shanghai, China). The column temperature was kept at 30 °C and the injection volume was 20 μL. An elution using solvent A (0.5% acetic acid) and solvent B (methanol) in a step gradient manner at a flow rate of 1 mL min^{-1} was carried out as follows: 0–10 min, 10–20% B; 10–20 min, 20–50% B; 20–30 min, 50–50% B; 30–40 min, 50–70% B; 40–50 min, 70–95% B; 50–60 min, 95–95% B; 60–65 min, 95–10% B; 65–75 min, 10–10% B. The detection wavelength was set to 238 and 440 nm.

In addition, the structures of the geniposide and crocetin were identified following an optimized method [7,20,21]. To obtain the structure of the geniposide and crocetin in detail, the UPLC-ESI-QTOF-MS/MS system was performed on an ACQUITY UPLC instrument connected to a Xevo G2Q-TOF mass spectrometer via an ESI interface (Waters Corp, Milford, MA, USA). For the qualitative analysis, an analytical column (100 mm × 2.1 mm, 1.7 μm) was used to separate the complex fraction compounds. The injection volume was 5 μL, consisting of a 0.1% formic acid solution (A) and acetonitrile (B) as the mobile phase. The flow rate was 0.4 mL/min and the gradient elution conditions were 0 min, 1% B; 1 min, 1% B; 25 min, 99% B; 27 min, 99% B; and 30 min, 1% B. In the meantime, the column temperature was 30 °C, the voltage was 550 V, the dry gas temperature was 500 °C, the atomizing gas was N_2, the pressure was 50 psi, and the mass-to-charge ratio was 50–1500 m/z.

2.3. Inhibition of Enzyme Activity In Vitro

Based on the characteristic wavelengths of the samples, the No. 4, No. 8, No. 9, and No. 14 samples were used for the in vitro α-glucosidase activity inhibition assays. The α-glucosidase inhibition assays were carried out following the method established by Mojica et al. (2015) [22]. In brief, 40 μL of the samples were added to 20 μL of a 0.25 unit/mL yeast α-glucosidase solution. The α-glucosidase was incubated in 0.1 M of a sodium phosphate buffer (PBS) (pH 6.9) for 15 min at 37 °C in a 96-well plate. Then, 40 μL of the 0.5 mM *p*NPG in a PBS was added to each well and incubated at 37 °C for 20 min. The reaction was stopped using 100 μL of 0.1 M Na_2CO_3. Acarbose was used as the positive control and the PBS was used as the blank control. In the reaction system, a half-maximal inhibitory concentration (IC_{50}) of the No.8 and No.9 samples was performed by fixing the concentration of the enzyme (0.25 U mL^{-1}) and substrate *p*NPG (0.5 mol L^{-1}). The inhibitory activity was expressed as the percentage inhibition of the enzyme activity and was calculated using Equation (1).

$$Inhibitory\ activity\ (\%) = \frac{A_{sample} - A_{k2}}{A_{control} - A_{k1}} \times 100\% \quad (1)$$

where $A_{control}$ is the absorbance of the PBS that replaces the samples, A_{sample} is the absorbance of the samples, A_{k1} is the absorbance of the PBS that replaces the samples and the enzyme, and A_{k2} is the absorbance of the PBS that replaces the enzyme [23].

2.4. Animal Experiment

The 8-week-old SPF male Kunming mice were purchased from the Experimental Animal Center of Huazhong Agricultural University (Certificate Number: SCXK (Hubei) 2015-0019, Wuhan, China). The animal experiment was carried out at the Experimental Animal Center of Huazhong Agricultural University with the approval of the Scientific Ethics Committee (Permission No. HZAUMO-2018-068). All the experimental operations were performed in strict accordance with the experimental animal standards established by the People's Republic of China (GB 14925-2010) and the guidelines for animal feeding and use during the experiment of the Experimental Animal Center of Huazhong Agricultural University (license number: No. SYXK (Hubei) 2015-0084). This study was conducted according to the standards in the Guiding Principles in the Care and Use of Animals. During the experimental period, the mice were housed in each cage under external environmental conditions of a temperature of 24 ± 1 °C, a humidity of $40 \pm 10\%$, and a normal light/dark cycle (12 h/12 h). Throughout the experiment, the animals had free access to distilled water and standard laboratory pellets. They were acclimatized and fed for one week before the start of the experiment.

2.4.1. Establishment of the Type 2 Diabetes Mellitus (T2DM) Mouse Model and Drug Administration

According to the adaptive feeding situation and the weight of the mice, nine mice were chosen as the normal control group and continued to be fed using an ordinary diet. The other mice were fed using a high-fat diet [24]. Each group consisted of nine mice. Four weeks later, the mice in the high-fat diet group were intraperitoneally injected with streptozotocin (STZ, 65 mg $kg^{-1}bw^{-1}$) for three consecutive days [6]. Then, the mice continued to be fed using a high-fat diet for one week. The fasting blood glucose levels of the mice were measured after fasting for 12 h. The group with a fasting blood glucose value ≥ 7.8 mmol/L could be regarded as the successful T2DM mice. They were divided into a normal control group (NC group), a T2DM group (DM group), a metformin positive control group (M group, 200 mg kg^{-1}), a 20% EGJ group (100 mg kg^{-1}), a 40% EGJ group (200 mg kg^{-1}), and an EGJ group (600 mg kg^{-1}). The mice in each group were fed for 5 weeks. The flow diagram of the animal experiments is shown in Scheme 1.

2.4.2. Determination of the Body Weight, Food Intake, Water Intake, Blood Glucose, and Oral Glucose Tolerance

The mice were weighed daily, and their food and water intake were recorded. The fasting blood glucose (FBG) was measured once a week until the blood glucose level of the experimental group showed a downward trend. The oral glucose tolerance test (OGTT) was performed 2 days before the end of the animal experiment. The test procedure was as follows. After fasting for 12 h, the first FBG value was determined, recorded as BG_0, and then the glucose solution was perfused at a dose of 1 g $kg^{-1}bw^{-1}$. The test time for each mouse was strictly controlled, and the glucose was measured at 15 min, 30 min, 60 min, and 120 min after the gavage, respectively. The values were recorded as BG_{15}, BG_{30}, BG_{60}, and BG_{120}. The blood glucose curve was drawn for each group and then the area under the curve (AUC) for the glucose was calculated.

2.4.3. Biochemical Assays and the Detection of the Organ Indexes

The retro-orbital site was used for the blood sampling. The mice were euthanized using cervical dislocation and were quickly dissected to obtain the livers and the kidneys. Within 24 h, the blood was centrifuged at 3000 r/min for 15 min. The serum was sub-packaged and stored at −20 °C for the detection of the glycosylated serum protein (GSP), total cholesterol (TC), triglyceride (TG), high-density lipoprotein cholesterol (HDL-C), and low-density lipoprotein cholesterol (LDL-C). The atherosclerosis index (AI) and cardiovascular risk index (CVRI) of the lipid metabolism were calculated using Equations (2) and (3) [24]. SOD and CAT belong to the antioxidant enzymes, which play a key role in the oxidation and antioxidant balance of the body and can protect the integrity of the β cell structure and function. MDA is the final peroxidation product of the lipids, and its expression level can reflect the level of oxygen free radicals in the body and measure the damage degree of the oxidative stress, which has been proven to be an important cause of inducing apoptosis of the pancreatic β cells and inhibiting the insulin secretion. For the assays of the total protein, hepatic malondialdehyde (MDA), catalase (CAT), and superoxide dismutase (SOD), the liver tissues were homogenized to a 10% liver homogenate and centrifuged at 4000 r/min for 10 min at 4 °C. Then, the supernatant was obtained. All the biochemical indexes were determined using the detection kits. The liver and kidney indexes were calculated according to Equation (4).

$$AI = \frac{LDL - C}{HDL - C} \tag{2}$$

$$CVRI = \frac{TG}{HDL - C} \tag{3}$$

$$\text{Organ index } (\%) = \frac{\text{Orgain weight (mg)}}{\text{Weight of mice (g)}} \times 100\% \tag{4}$$

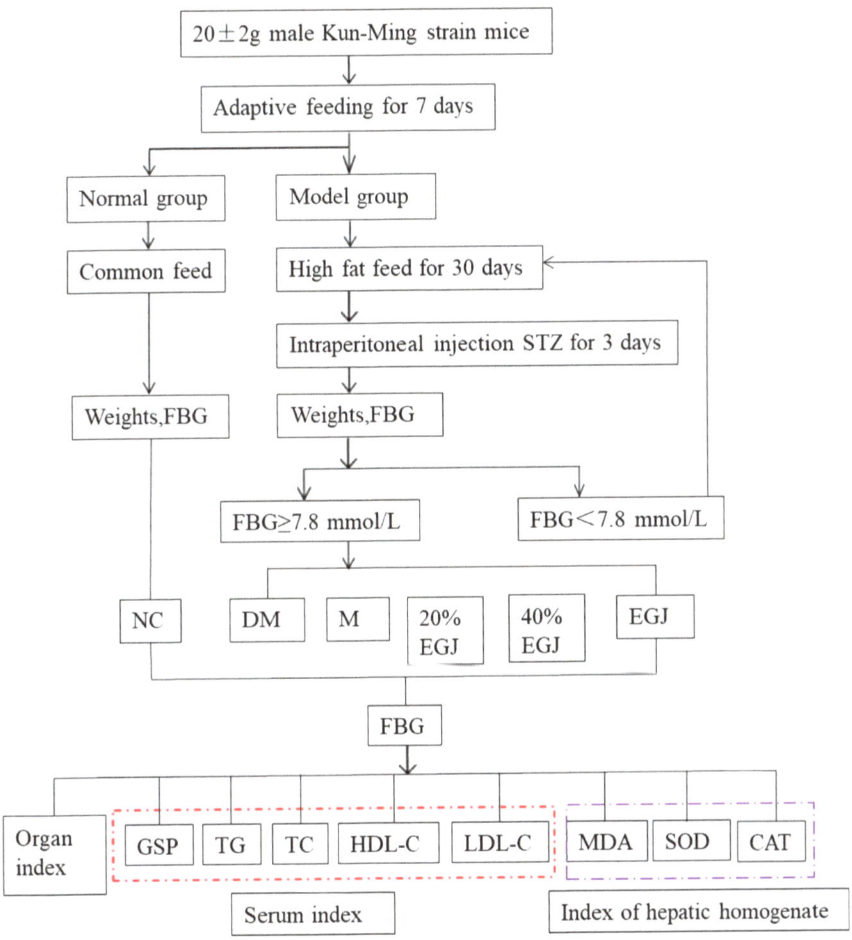

Scheme 1. Flow diagram of the animal experiments.

2.5. Statistical Analysis

The data were expressed as the mean ± the standard deviation (SD). The SPSS Statistics 22.0 was used to analyze the significant difference between the logarithm data of the one-way analysis of variance (ANOVA) and Duncan's multiple range (DMRT). When the value of $p < 0.05$, there was a statistical difference between the judgment data. The GraphPad Prism 5 software was used to draw the corresponding charts for the data analysis results.

3. Results and Discussion

3.1. HPLC Analysis of Geniposide and Crocetin

As shown in Figures 1 and 2, the HPLC chromatograms of the crude EGJ and its fractions were recorded at $\lambda = 238$ nm or 440 nm. The samples were prepared according to the above method. The peak of the geniposide standard was recorded at 238 nm and its retention time was 20.72 min. The characteristic wavelength of the No.4 sample was 238 nm and the retention time was as same as the geniposide standard. Therefore, the No.4 sample mainly included geniposide. By the same argument, the retention time of crocin I was 28.67 min with a corresponding wavelength of 440 nm. Thus, the No.8 and No.9 samples mainly contained crocin I. The peak of the 20% EGJ fraction was recorded at 238 nm, which implied that the 20% EGJ fraction mainly contained geniposide. In addition, the crocin I standard and geniposide standard were taken as the references for the quantitative analysis. The two linear regression equations for geniposide and crocin I were $y = 1.0664x + 1.4066$ ($R^2 = 0.9994$) and $y = 1.1889x - 10.92$ ($R^2 = 0.9959$). The peak areas of geniposide and crocin I from gardenia were 57.603 ± 0.93 and 58.481 ± 1.25. According to the above formula, the contents of geniposide and crocin I in gardenia were 53.541 ± 1.17 mg g^{-1} and 57.679 ± 0.95 mg g^{-1}. This result laid the groundwork for the further screening of the four fractions on the α-glucosidase activity in vitro.

3.2. Identification of Geniposide and Crocetin Using UPLC-ESI-QTOF-MS/MS

Figure 3A–C show the pattern of total ion chromatogram (TIC) of the UPLC-ESI-MS for the geniposide and crocin I standards, the EGJ, and the No.9 sample. As shown in Figure 3A, the peak times for geniposide and crocin I was recorded at 6.76 min and 8.60 min, respectively. As shown in Figure 3B, the two peaks of crocin I were recorded at 8.54 min and 10.81 min. Since the peak of the crocin I standard was recorded at 8.60 min, the former should be crocin I, while the latter might be the isomer $C_{44}H_{64}O_{24}$. The crude EGJ was captured with 12 peaks, including geniposide and crocin I. As shown in Figure 3C, the No. 9 sample was recorded with 18 peaks, including geniposide and crocin I. Meanwhile, the size of the peak indicated that it contained a large amount of crocin I and a very small amount of geniposide. The mass cracking pathways of these two main compounds could be inferred based on the MS and MS–MS spectrum information.

According to the primary and secondary mass spectra of geniposide, the mass-to-charge ratio (m/z) of geniposide with sodium ($[M+Na]^+$) was 411.1267, the two geniposide couples ($[2M+H]^+$) corresponded to 777.2850, the geniposide ($[M+H]^+$) was 389.1268; the geniposide stripped of one molecule of monosaccharide ($[M\text{-glc}+H]^+$) was the fragment ion 227.0921, and the geniposide stripped of one molecule of monosaccharide and one molecule of water ($[M\text{-glc-}H_2O+H]^+$) was 209.0805, as shown in Figure 4A. It was presumed that it was relatively easy for geniposide to remove one molecule of monosaccharide, followed by a further removal of one molecule of water from the parent nucleus of the cyclic enol ether terpene. Its mass spectrometry cleavage pathway is shown in Figure 4A.

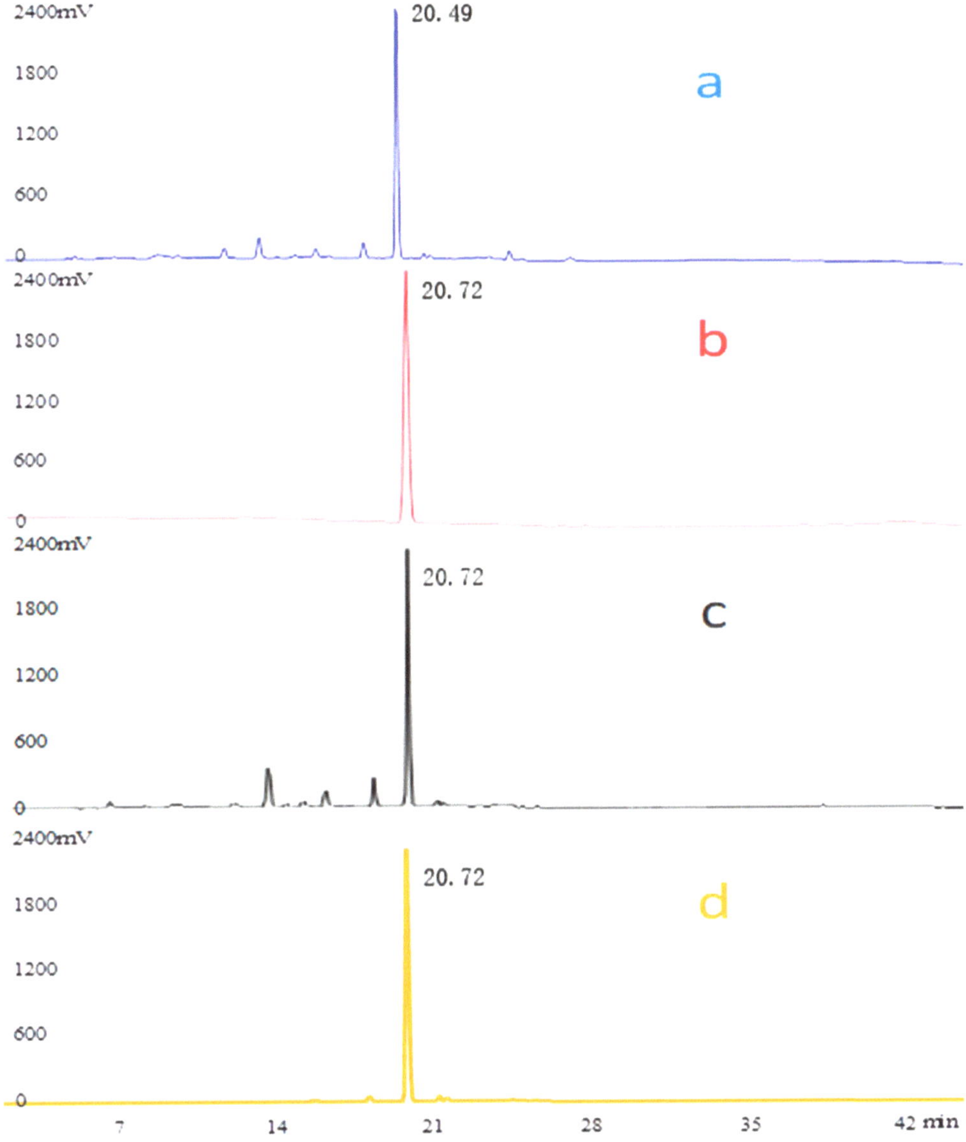

Figure 1. HPLC chromatograms. λ = 238 nm of the crude EGJ (**a**), the standard compound of geniposide (**b**), the 20% EGJ (**c**), and the No.4 sample (**d**).

On the basis of the primary and secondary mass spectra of crocin I, the mass-to-charge ratio (m/z) of crocin I ($[M+Na]^+$) with sodium ions was 999.3741, crocin I ($[M+H]^+$) corresponded to 976.3801; crocin I with the sodium ions stripped of the disaccharides ($[M-2glc+Na]^+$) was the fragment ion 675.2639, crocin I stripped of two disaccharides ($[M-4glc]^+$) was 329.1755, and one molecule of disaccharide with the sodium ions ($[M-2glc-C_{20}H_{20}O_3]^+$) was 347.0957, as shown in Figure 4B. It was presumed that it was relatively easy to break the glycosidic bond attached to the parent chain from the disaccharide at both ends of crocin I. Its mass spectral cleavage pathway is shown in Figure 4B.

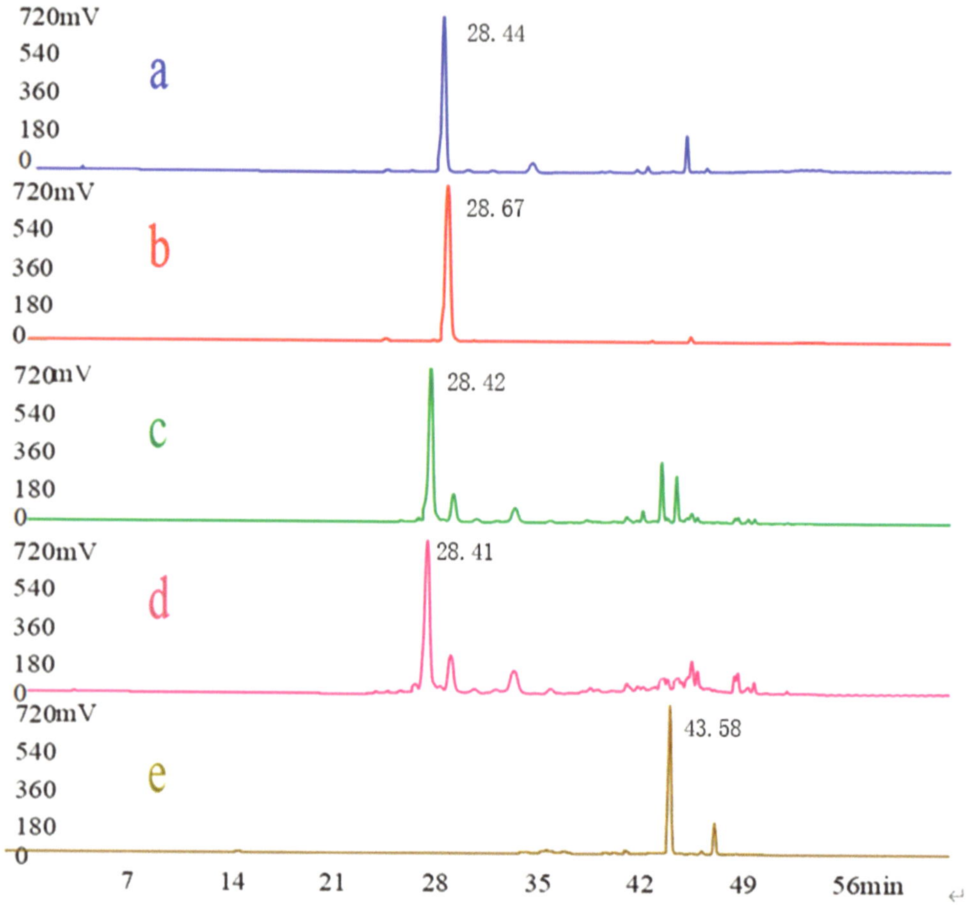

Figure 2. HPLC chromatograms. λ = 440 nm of the crude EGJ (**a**), the standard compound of crocin I (**b**), the No.8 sample (**c**), the No.9 sample (**d**), and the No.14 sample (**e**).

Based on the time of each peak and their mass spectrometry information, sixteen compounds were identified, consisting of flavonoids, iridoid, organic acids, and diterpene, as shown in Table 1.

3.3. *Screening of Geniposide and Crocetin on the α-Glucosidase Activity In Vitro*

As shown in Figure 5A, the different compounds had different inhibitory effects on the α-glucosidase at the same concentration. The concentration of the No. 8 and No. 9 samples was 0.5 mg mL^{-1} and inhibition rates were 35.74 ± 4.86% and 97.61 ± 8.12%, respectively. At the same concentration as the No. 8 and No. 9 samples, the No.14 sample (21.79 ± 4.27%) had a slightly weaker inhibition ability than the acarbose (35.99 ± 3.28%). Therefore, both the No. 8 and No. 9 samples had a better ability of inhibiting the enzyme. The IC_{50} of the No. 8 and No. 9 samples were tested to further screen their inhibition ability in vitro. As shown in Figure 5B, the IC_{50} of the No.8 and No.9 samples were 0.577 ± 0.041 mg mL^{-1} and 0.204 ± 0.007 mg mL^{-1}, respectively. This result indicated that the enzyme inhibitory effect of the No.9 sample was the best. Since the No. 4 sample was obtained through the elution of a 20% ethanol concentration and the No. 9 sample was obtained through the elution of the 40% ethanol concentration, the No.4 and No.9 samples as well as the crude

EGJ were named as the 20% EGJ, 40% EGJ, and EGJ group, respectively, and used for the subsequent animal experiments.

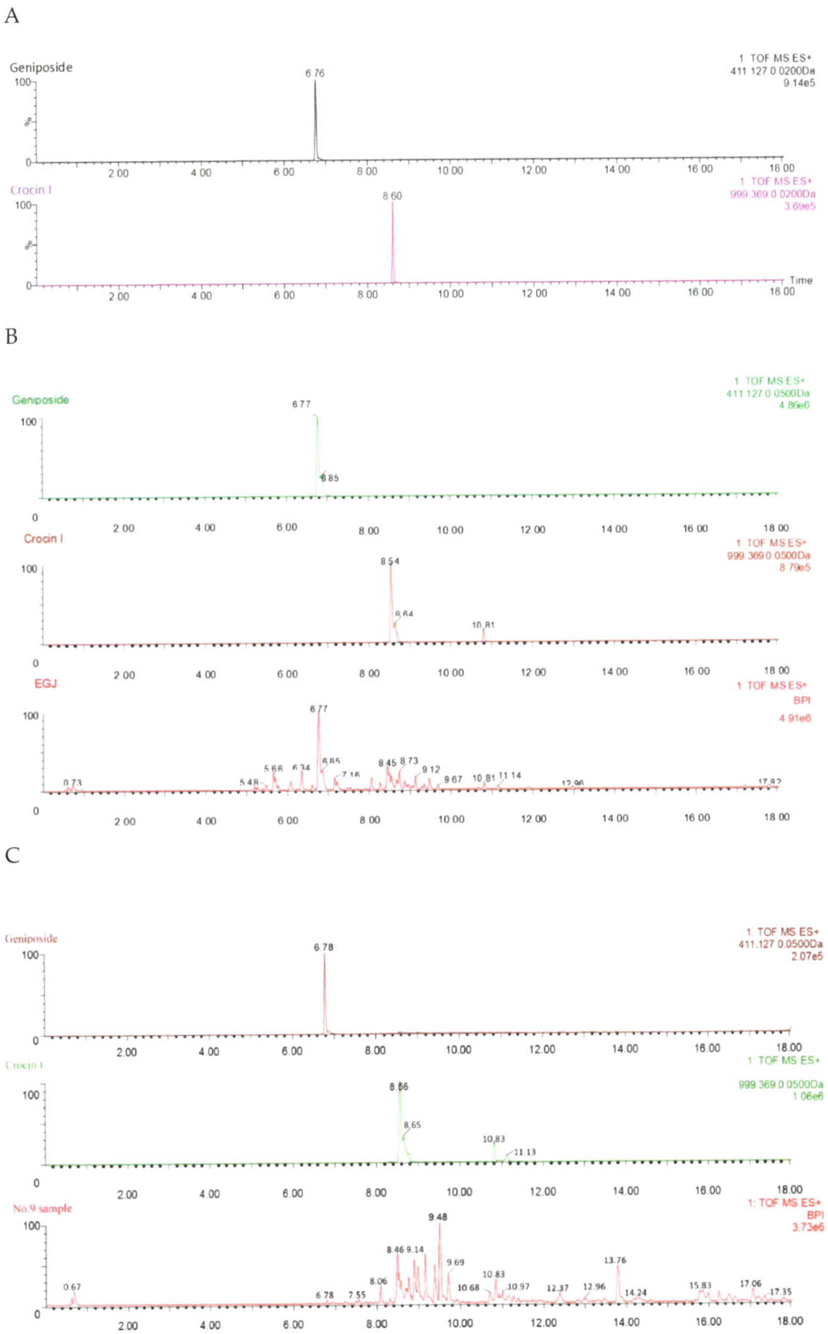

Figure 3. (**A**) The geniposide and crocin I standards, (**B**) the crude EGJ, and (**C**) the No.9 sample.

3.3.1. Type of Inhibition for the No.9 Sample

As mentioned before, the No. 9 sample showed the best inhibition of the enzyme, so this sample was used to study the type of inhibition. Using the enzyme concentration as the horizontal coordinate and the reaction rate as the vertical coordinate, the straight lines were obtained for the different concentrations of the No. 9 sample (0.18–0.22 mg/mL). The results are shown in Figure 6, where each straight line passed through the origin and the slope was negatively correlated with the concentration of the sample, i.e., the type of inhibition of the α-glucosidase for the No. 9 sample was reversible and the enzyme inhibited by the sample could be revived using physical methods, such as dialysis and ultrafiltration [25].

A

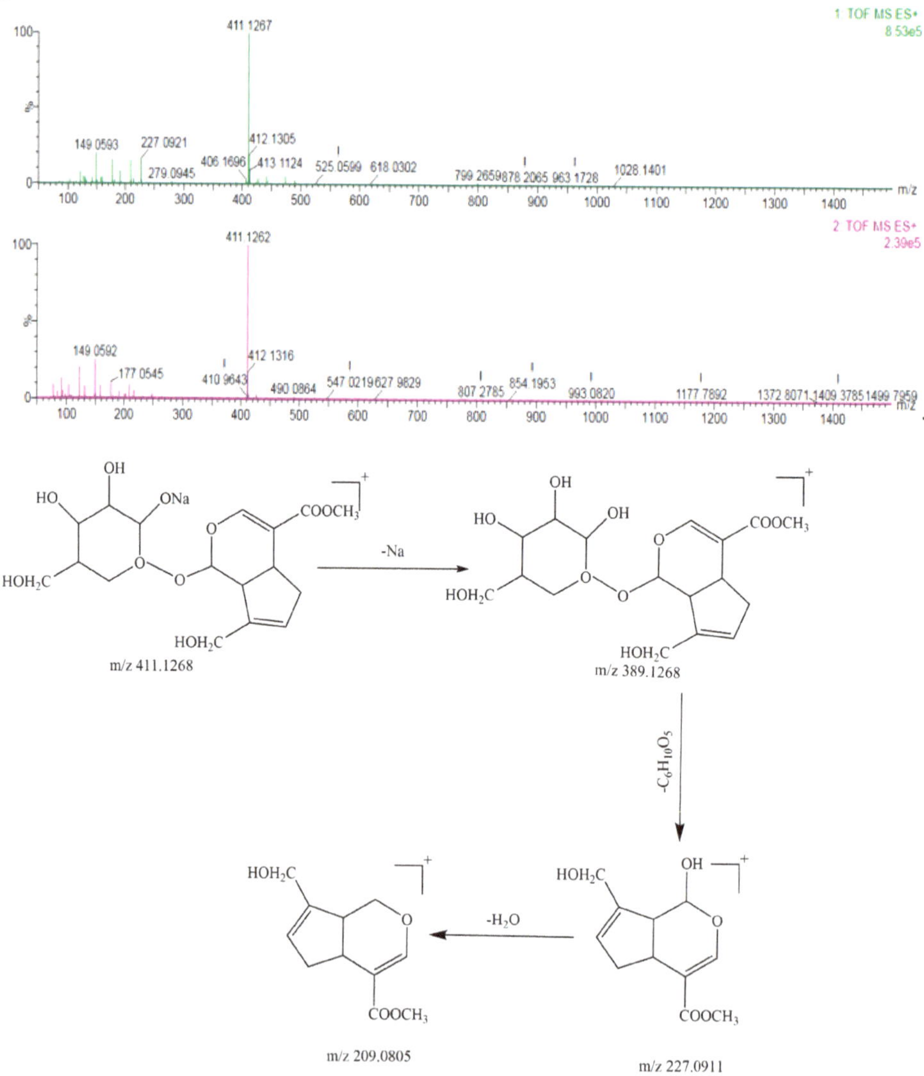

Figure 4. *Cont.*

B

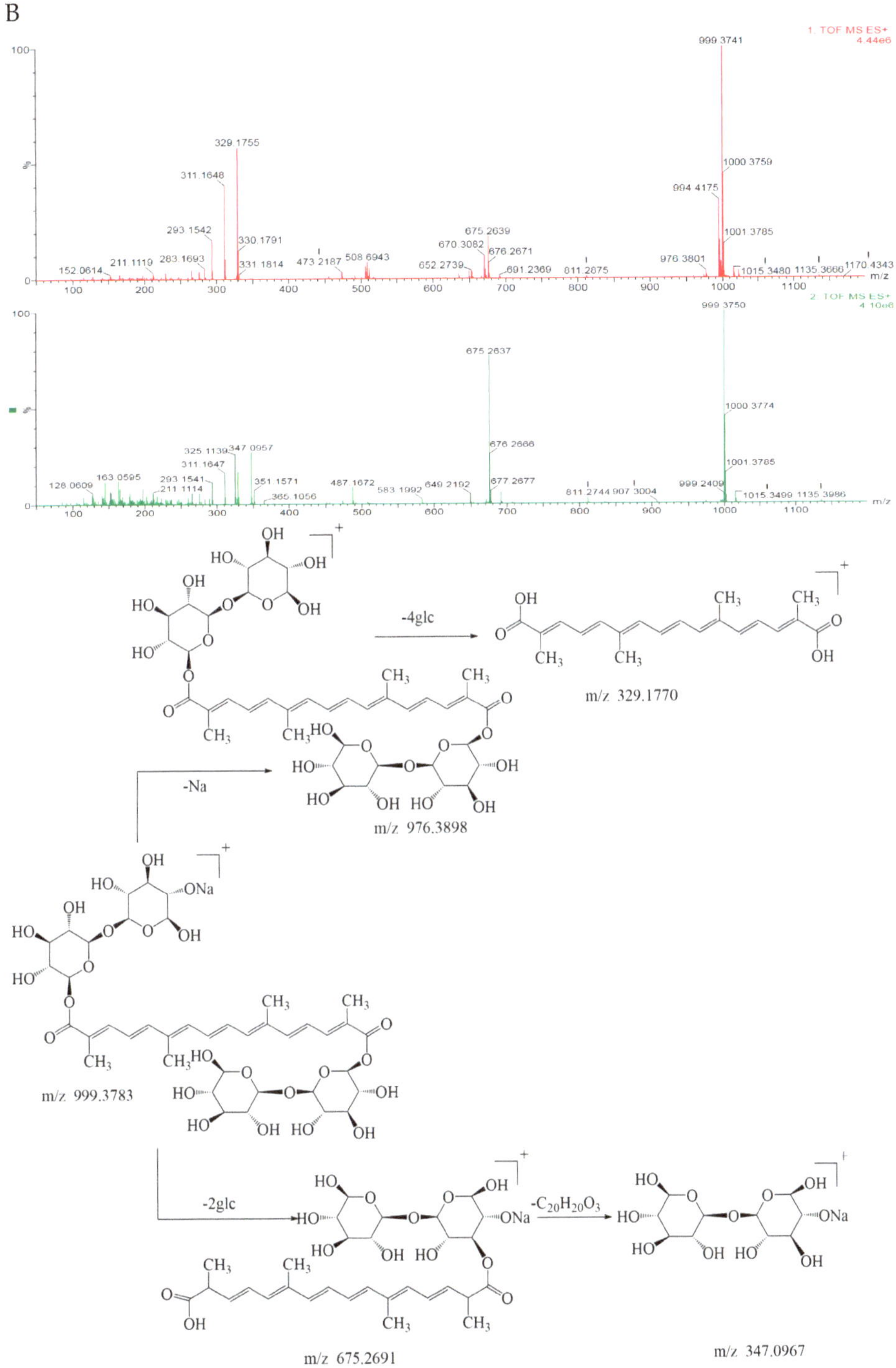

Figure 4. The TOF−MS, TOF−MS/MS spectra and the mass spectrometric fragmentation pathway of (**A**) geniposide and (**B**) crocin I.

Table 1. Identification of the compounds in geniposide and crocetin using UPLC−ESI−QTOF−MS/MS in the positive ion mode.

No.	Rt/min	Proposed Formula	Relative Molecular Mass/Da	$[M+H]^+$ Measured Value	Theoretical Value	Fragment Ions ESI (+)	ppm	Tentative Identification	Crude EGJ	No.9 Sample
1	5.265	$C_{16}H_{24}O_{11}$	392.1385	393.1385	393.1397	415.1232[M+Na] 431.0829[M+K] 397.1082[M+Na-H_2O]	−1.2	Shanzhiside	+	−
2	5.481	$C_{17}H_{24}O_{11}$	404.1319			375.1460[M+H-H_2O] 427.1217[M+Na] 443.0999[M+K] 409.2032[M+Na-H_2O]		6-alpha-Hydroxy geniposide	+	−
3	5.666	$C_{17}H_{24}O_{11}$	404.1319	405.1559	405.1563	427.1218[M+Na] 443.0805[M+K] 409.1115[M+Na-H_2O]	−0.4	Feretoside	+	−
4	5.760	$C_{17}H_{24}O_{11}$	404.1319	405.1554	405.1563	427.1212[M+Na] 443.1036[M+K]	−0.9	Feretoside	+	−
5	6.034	$C_{16}H_{18}O_9$	354.0951	355.1003	355.1029	731.1856[2M+Na] 393.0562[M+K] 377.0871[M+Na] 163.0306[M+H-QA] 135.0393[M+H-QA-CO]	−2.6	Chlorogenic acid	+	−
6	6.339	$C_{23}H_{34}O_{15}$	550.1898	551.1967	551.1976	573.1804[M+Na] 589.1512[M+K] 389.1396[M+H-glc] 227.0854[M+H-2glc]	−0.9	Genipin-1-O-β-D-gentian glycosides	+	−
7	6.817	$C_{17}H_{24}O_{10}$	388.1369			411.1262[M+Na] 777.2932[M+2Na]		Geniposide	+	+
8	8.278	$C_{27}H_{30}O_{14}$	578.1636	579.3737	579.3714	271.0556[M-rha-glc]	2.3	Apigenin-7-O-rutinoside	+	+
9	8.494	$C_{32}H_{40}O_{17}$	696.2265	697.2177	697.2144	719.2186[M+Na] 735.1713[M+K] 309.0976[M+H-gen]	3.3	6″-O-P-coumaroyl genipingentiobioside	+	+

Table 1. *Cont.*

No.	Rt/min	Proposed Formula	Relative Molecular Mass/Da	$[M+H]^+$ Measured Value	Theoretical Value	Fragment Ions ESI (+)	ppm	Tentative Identification	Crude EGJ	No.9 Sample
10	8.668	$C_{44}H_{64}O_{24}$	976.5788	977.3801	977.3786	999.3741[M+Na] 675.2639[M+Na-2glc] 347.0957[M-2glc-$C_{20}H_{20}O_3$] 329.1755[M-4glc]	1.5	Corcin I	+	+
11	9.031	$C_{38}H_{54}O_{19}$	814.3174	815.3259	815.3255	675.2301[M+Na-2glc] 347.0930[M-2glc-$C_{20}H_{20}O_3$] 329.1750[M-4glc]	0.4	Crocin II	+	+
12	9.110	$C_{22}H_{22}O_{11}$	462.1162	463.1225	463.124	947.2373[2M+Na] 485.2435[M+Na]	−1.5	2-O-Cinnamoyl-glucogallin	+	+
13	11.134	$C_{21}H_{18}O_{11}$	446.0849	447.0956	447.0927	469.1128[M+Na] 271.0462[M-glc]	2.9	Baicalin	+	+
14	12.374	$C_{21}H_{21}NO_4$	351.1471	352.2179	352.1849	337.1254[M-CH_3] 308.1244[M-CH_3-H-CO]	33	Palmatine	–	+

Note 1: glc–glucose; rha–rhamnose; gen–gentian disaccharide; QA–quinic acid. Note 2: "+" and "–" refer to "detected" and "not detected or content too low".

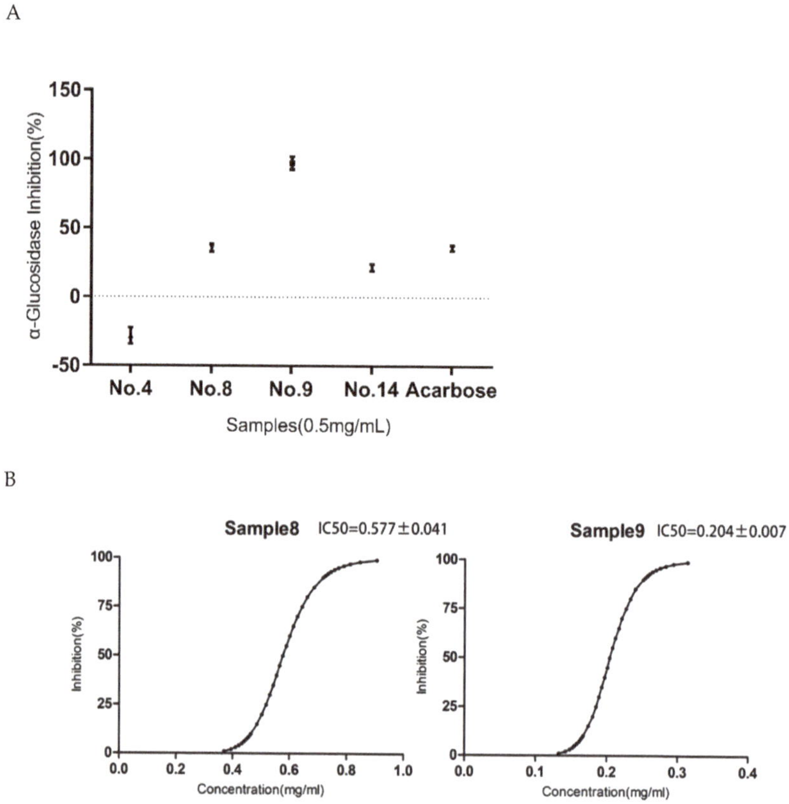

Figure 5. (**A**): Inhibitory effects of the different samples (No.4, 8, 9, and 14) on the α−glucosidase in vitro at the same concentrations (T = 37 °C, pH 6.9); c (α−glucosidase) = 0.25 U/mL, c (pNPG) = 0.5 mmol/L. (**B**): The value of the half maximal inhibitory concentration (IC_{50}) of the No.8 and No.9 samples was performed by fixing the concentration of the enzyme (0.25 U mL^{-1}) and the substrate pNPG where c (α−glucosidase) = 0.25 U/mL, c (*p*NPG) = 0.5 mmol/L, c (No.8 sample) = 0.30 to 1.00 mg/mL, and c (No.9 sample) = 0.10 to 0.35 mg/mL, respectively. The data were expressed as mean ± the SD (*n* = 3).

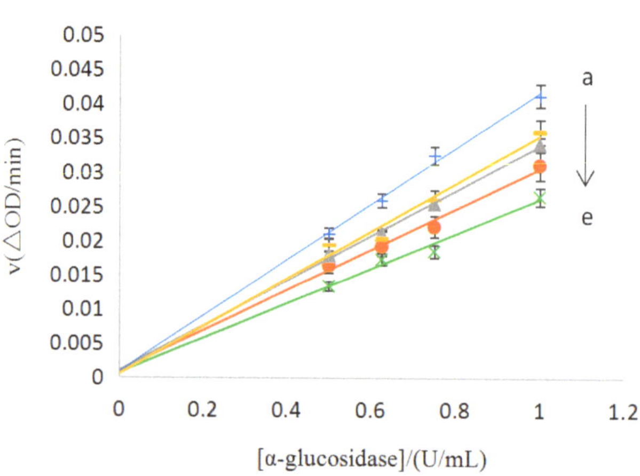

Figure 6. The inhibition of the reversibility of the No.9 sample on the α-glucosidase [C(No.9) = 0.18, 0.19, 0.20, 0.21, and 0.22 mg/mL for the curves a→e, respectively].

3.3.2. Mechanism of Inhibition for the No.9 Sample

As stated before, the type of inhibition of the No.9 sample was reversible, and this sample was then subjected to a kinetic study of the inhibition. Using the reciprocal of the pNPG concentration as the horizontal coordinate and the reciprocal of the reaction rate as the vertical coordinate, the Lineweaver–Burk plot was used to obtain a straight line for the different concentrations of the No.9 sample (0.06–0.14 mg/mL), which was judged by the size of the intercept and the position of the coordinate. The results are shown in Figure 7. When the sample concentration increased, the vertical coordinate of the corresponding line grew, indicating that the 1/V increased and the V_{max} decreased. A decrease in the horizontal coordinate indicated a decrease in $-1/K_m$ and an increase in the Mie's constant K_m. According to the principle of the enzymatic reaction, it was clear that the No. 9 sample was anti-competitive for the α-glucosidase inhibition [26].

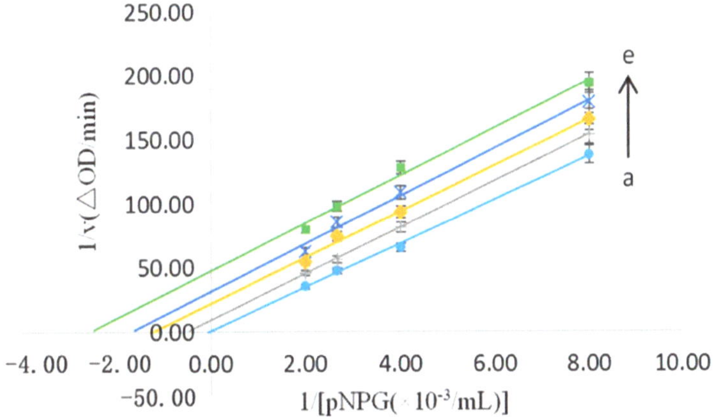

Figure 7. The enzyme kinetics of the reversibility of the No.9 Sample on the α−glucosidase [C(No.9) = 0.06, 0.08, 0.10, 0.12, and 0.14 mg/mL for the curves a→e, respectively].

3.4. *Effects of the Crude EGJ, 20% EGJ, and 40% EGJ in the DM Male Mice*

3.4.1. Effects of the Crude EGJ, 20% EGJ, and 40% EGJ on the Fasting Blood Glucose, Body Weight, and Water Intake

The weight changes of the mice after three days of the STZ injection were shown in Figure 8. The weight of the mice in the general group increased slowly while the weight of the mice in the model group increased on the second day and then decreased continuously. Until the eighth day, the weight of the model group dropped to 37.3 ± 1.6 g, while the weight in the general group increased to 44.0 ± 1.1 g. At the same time, the level of the blood glucose in the different groups was measured. The level of the blood glucose in the general group and the model group fluctuated between 4.67 mmol/L and 19.14 mmol/L, respectively. In addition, it was found that the hair of the mice in the general group was normal white and the mice were active, while the hair of mice in the model group w yellow, wet, and the mice behaved lazily, indicating that the DM modeling was successful.

The weight changes of the mice in each group during the administration period is shown in Figure 9A. The weight of the mice in the NC group always maintained the highest level and constantly increased. The initial body weights of the mice in the other groups were lower and later changed to varying degrees. The body weight of the mice in the DM group decreased, while the mice in the M group did not change much, indicating that the weight of the T2DM mice could be controlled using metformin. Compared to the EGJ group and the 40% EGJ group, the weight in the 20% EGJ group maintained an upward trend and increased to approx. 42 g at the end of the fifth week, indicating that the 20% EGJ (geniposide) had a certain effect on controlling the weight loss caused by diabetes.

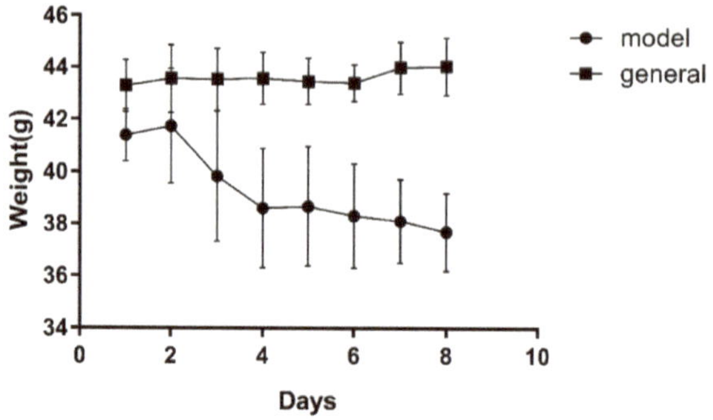

Figure 8. Body weight changes during the modeling. Note: the data were expressed as the mean ± the SD (n = 9).

A

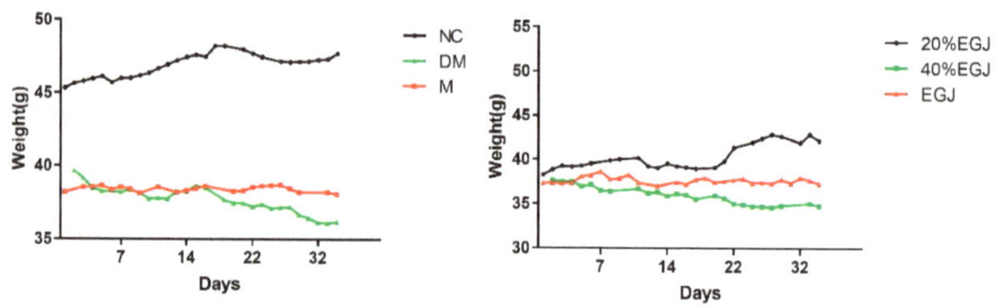

B

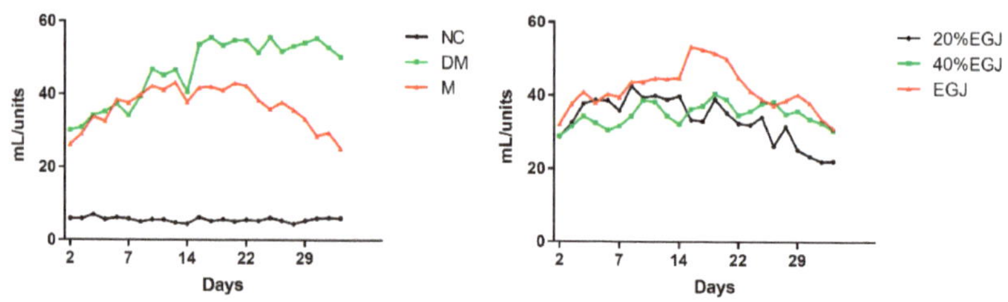

C

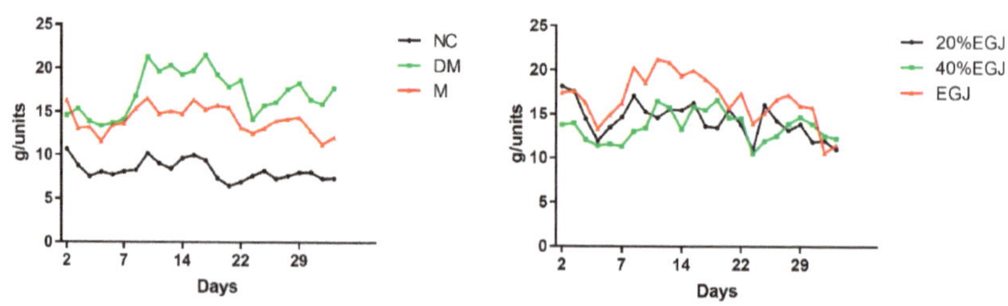

Figure 9. The changes of (**A**) the body weight (**B**) the water intake, and (**C**) the food intake.

The quantity of the water and food intake of the mice in each group during the administration period are shown in Figure 9B,C. The water intake in the 20% EGJ group also decreased, and the quantity of the food intake was kept at a low level at the end of the experiment, at approx. 11.01 g/unit. The results suggested that metformin and geniposide had a good effect on adjusting the quantity of the water and food intakes of the T2DM mice, and that geniposide had the highest efficacy in this study.

3.4.2. Effects of the Crude EGJ, 20% EGJ, and 40% EGJ on the FBG, OGTT, and AUC

As shown in Figure 10A, during the period of administration, the level of the FBG in the NC group was basically stable at the normal level ($\leq$7.8 mmol/L). Compared to the NC group, the FBG in the other groups were significantly higher. The FBG in the DM group was the highest, indicating that the hyperglycemia symptoms were the most severe [27]. In the M group, the level of the FBG reached the highest level in the first week, and then decreased, indicating that the metformin could restore the FBG of the T2DM to fluctuate within the normal range. There was no significant difference between the 20% EGJ group and the EGJ, for which the FBG levels were separately 11.06 $\pm$ 2.62 and 17.33 $\pm$ 2.25 mmol/L at the fifth week. However, the 40% EGJ group did not show a significant downward trend. These results indicated that the metformin, 20% EGJ (geniposide), and EGJ could effectively alleviate the level of the FBG in the T2DM mice.

The OGTT experiment was carried out two days before the end of the experiment. As shown in Figure 10B, the level of the blood glucose of the mice in each group increased after 15 min caused by the glucose intake on fasting. Except for the DM group, the level of the blood glucose in the other groups recovered at 2 h in varying degrees, and better recoveries were provided in the metformin and 20% EGJ groups. According to the AUC curve (Figure 10C), geniposide and metformin had the same effect for impairing the glucose tolerance, which was 63.52% lower than the DM group. The value of the AUC in the 40% EGJ and EGJ groups decreased by 39.17% and 50.06%, respectively. However, the results of the postprandial blood glucose were unstable, and more results were needed.

3.4.3. Effects of the Crude EGJ, 20% EGJ, and 40% EGJ on the Organ Indexes and GSP

The organ indexes in the NC group were the lowest, and the indexes in the DM group were the largest, indicating that diabetes caused hepatomegaly and liver injury in the mice. Metformin could alleviate the liver injury better than the 20% EGJ, and there was no significant difference between metformin and the 20% EGJ (Table 2). The GSP reflected the blood glucose concentration in the first two weeks. The GSP index of the mice in each group after 5 weeks of administration is shown in Figure 11. The level of the GSP in the NC, M, and 20% EGJ groups were significantly different compared to the DM and 40% EGJ groups ($p < 0.05$). The results showed that the 40% EGJ (crocin I) and 20% EGJ (geniposide) groups both had a certain effect for reducing the GSP index in vivo, while the geniposide fraction was better.

3.4.4. Effects of the Crude EGJ, 20% EGJ, and 40% EGJ on the Serum TC, TG, HDL-C, LDL-C, AI, and CVRI

After the drug administration, the level of the TC of the mice in the NC group, the DM group and the sample group altered remarkably with significant differences ($p < 0.05$), as shown in Figure 12A. The TC levels of the sample groups all decreased by approx. 32% compared to the DM group. As shown in Figure 12B, the decrease in the levels of the TC was similar in the M and the 20% EGJ groups, which was 15% lower than the NC group. Compared to the DM group, the EGJ group showed a significant decrease ($p < 0.05$). However, the 40% EGJ group failed to increase the level of the TC. As shown in Figure 12C, the HDL-C levels were significantly increased in all the sample groups, except for the 40% EGJ group. The HDL-C levels in both the M and the 20% EGJ groups were approximately 56% higher than those in the DM group. The abnormal lipid metabolism symptoms of the sample group mice were improved since the HDL-C contributed to the removal of the cholesterol. As shown in Figure 12D, the level of the LDL-C in the M group returned to normal, and the decrease in the LDL-C in the 20% EGJ group was the most obvious; the value was 17.55% lower than NC group. Since an excessive LDL-C can lead to cholesterol accumulation, the administration group could improve the abnormal lipid metabolism of the mice. In addition, the level of the LDL-C in the 40% EGJ group was similar to the M group. As shown in Figure 12E,F, the AI and CVRI indexes were the highest in the DM group, indicating the highest risk index for atherosclerosis and cardiovascular disease. The index was significantly lower in the M group, 20% EGJ group, 40% EGJ and crude EGJ group. The above results showed that the 20% EGJ (geniposide) group showed an obvious improvement effect in the T2DM mice.

A

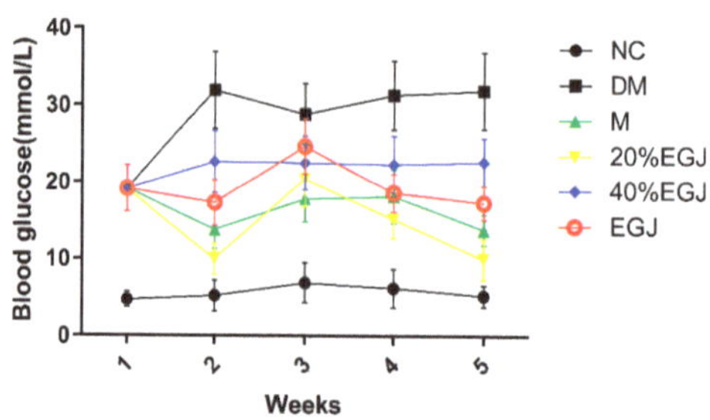

B

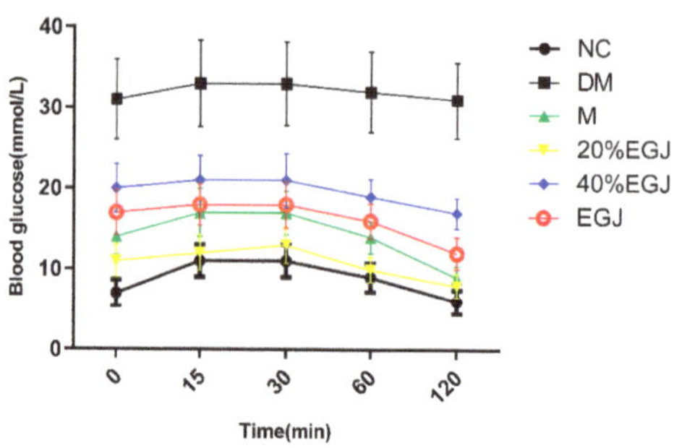

C

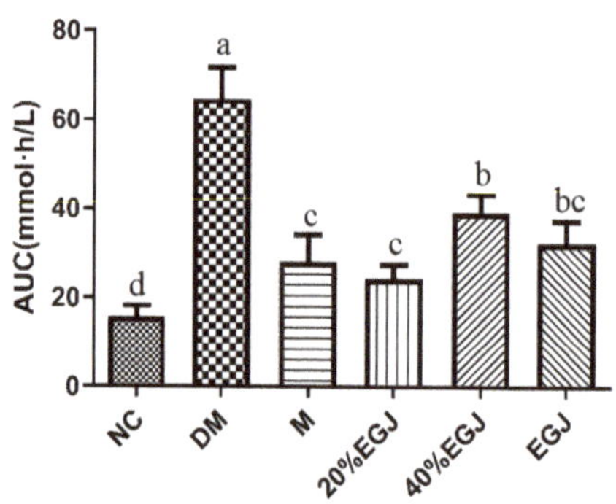

Figure 10. (**A–C**) represent the changes of the FBG, OGTT, and AUC, respectively. Note: the data were expressed as the mean ± the SD (n = 9) and the different letters marked above the bars showed the significant differences (p < 0.05).

Table 2. Effect of the samples on the organ indexes in the diabetes mice.

Group	Liver (g)	Kidney (g)	Liver Index (%)	Kidney Index (%)
NC	1.533 ± 0.266 [b]	0.553 ± 0.0482 [b]	3.166 ± 0.546 [d]	1.141 ± 0.099 [d]
DM	1.780 ± 0.327 [a]	0.642 ± 0.058 [a]	5.021 ± 0.932 [a]	1.812 ± 0.134 [a]
M	1.493 ± 0.156 [b]	0.513 ± 0.074 [b]	3.6137 ± 0.571 [cd]	1.304 ± 0.311 [cd]
20%EGJ	1.674 ± 0.279 [ab]	0.573 ± 0.097 [b]	3.876 ± 0.649 [bc]	1.326 ± 0.226 [bc]
40%EGJ	1.585 ± 0.369 [ab]	0.556 ± 0.116 [b]	4.259 ± 0.9931 [b]	1.494 ± 0.401 [b]
EGJ	1.496 ± 0.2632 [b]	0.531 ± 0.0301 [b]	4.187 ± 0.7367 [bc]	1.485 ± 0.089 [b]

Abbreviations: NC, normal control group; DM, type 2 diabetes mellitus model group; M, metformin positive control group. The EGJ, 20% EGJ, and 40% EGJ are the groups with the purified components of gardenia. The values were expressed as the means ± the SD ($n = 9$). The different letters indicate the significant difference between the groups, and the same superscript indicates no significant difference.

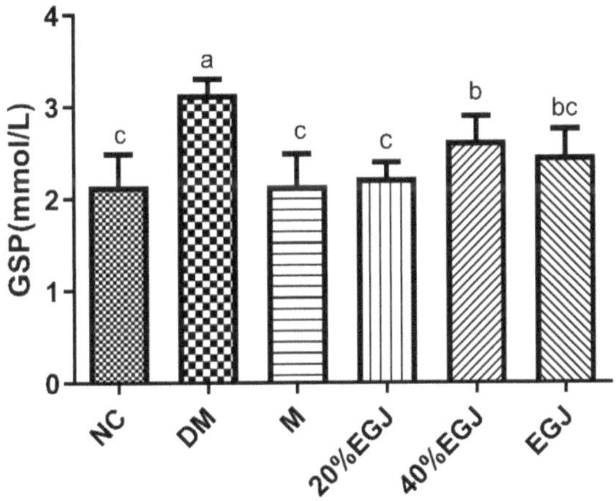

Figure 11. GSP of the mice in each group. Note: the data were expressed as the mean ± the SD ($n = 9$) and the different letters marked above the bars showed the significant differences ($p < 0.05$).

3.4.5. Effects of the Crude EGJ and Its 20% and 40% Fractions on Hepatic MDA, CAT, and SOD

As shown in Figure 13A, the MDA levels in the M and 20% EGJ groups were reduced. However, there was no significant difference between the MDA levels in the 40% EGJ group and the model group, indicating that the 40% EGJ (crocin I) group did not show an ability to eliminate the free radicals in the DM mice, whereas the 20% EGJ (geniposide) group could scavenge the free radicals in the DM mice. Then, we further measured the level of CAT. As shown in Figure 13B, there was no significant difference between the 40% EGJ group and the DM group. The level of CAT in the other groups could not reach the level in the NC group but was higher than the DM group, indicating that metformin and the 20% EGJ could improve the level of CAT in the T2DM mice. Consequently, metformin and the 20% EGJ (geniposide) could accelerate the decomposition of hydrogen oxide to produce oxygen and water in the body and reduce the damage to the body. SOD could increase the active substances in the body and reduce the harmful substances produced by the metabolism. The level of SOD was similar to the level of MDA and CAT. The 40% EGJ group failed to improve the level of SOD the mice, but the level of SOD in the other groups of mice were increased [28], and the effect of the 20% EGJ (geniposide) was better than the M group. The above showed that geniposide could improve the oxidative stress damage in the T2DM mice and reduce its damage to the pancreatic islet β cells. The geniposide has the best effect and the extent of the compounds on T2DM followed the order of geniposide > crocin I > crude EGJ.

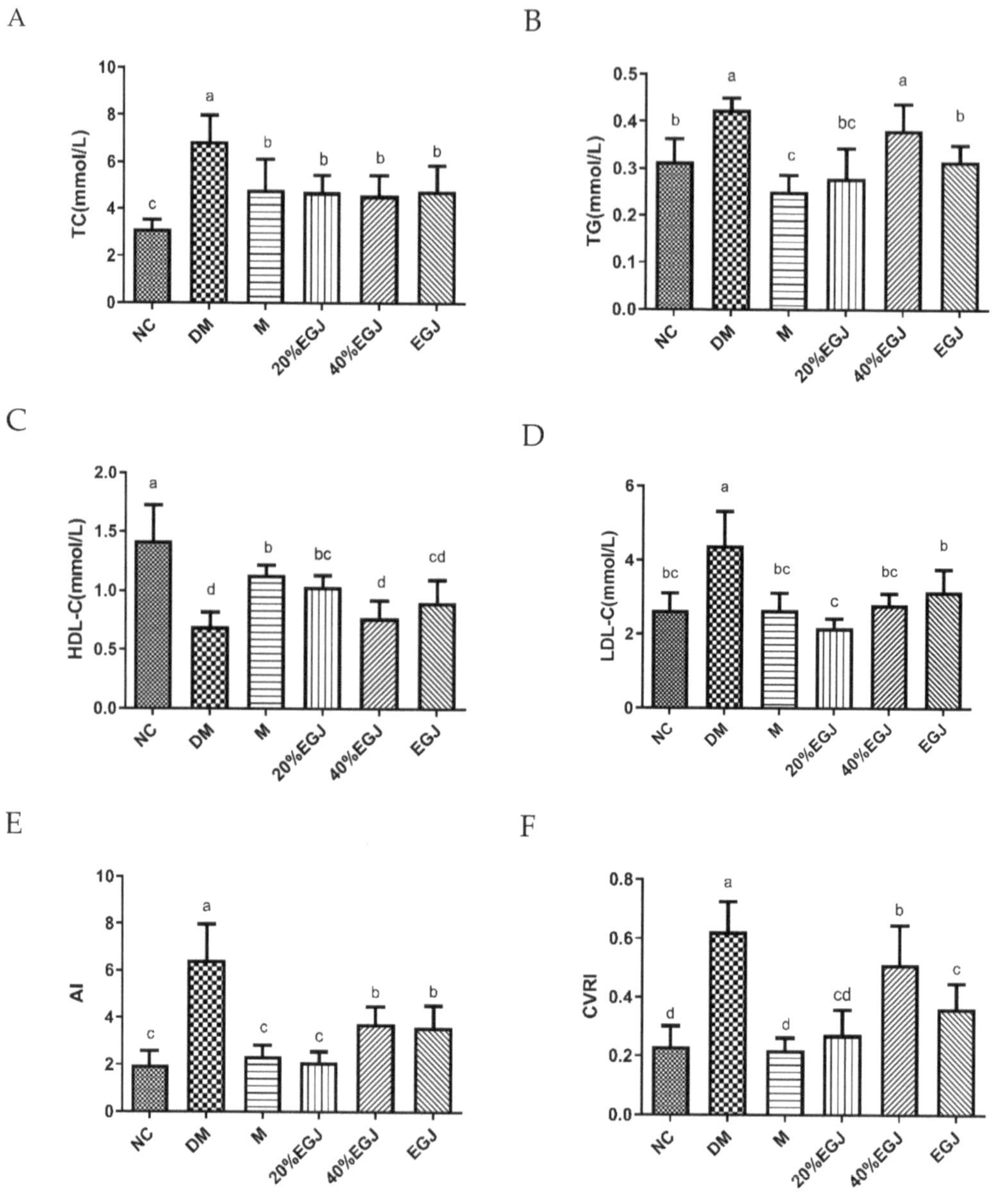

Figure 12. (**A–F**) stands for the TC, TG, HDL-C, LDL-C, AI, and CVRI of the mice in each group, respectively. Note: the data were expressed as the mean ± the SD (n = 9) and the different letters marked above the bars showed the significant differences ($p < 0.05$).

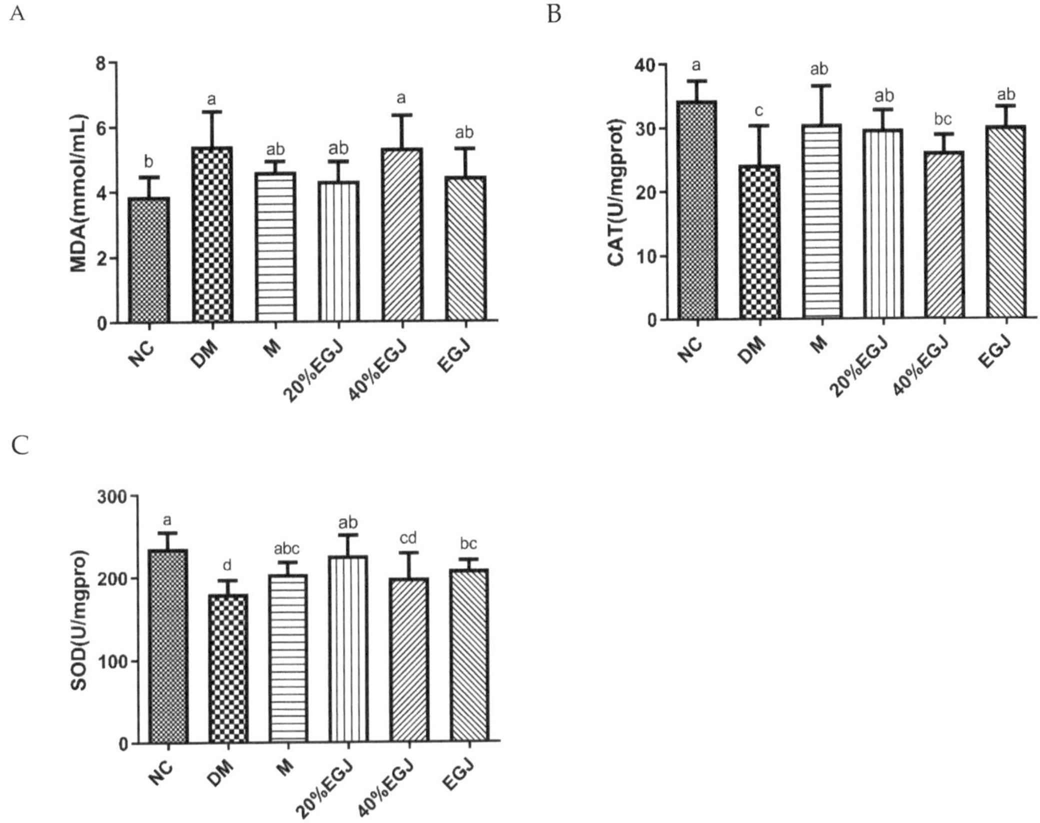

Figure 13. (**A–C**) stands for the MDA, CAT, and SOD of the mice in each group, respectively. Note: the data were expressed as the mean ± the SD (n = 9) and the different letters marked above the bars showed the significant differences from the ANOVA multiple test ($p < 0.05$).

4. Conclusions

The hypoglycemic mechanism of the α-glucosidase inhibitors showed that, by inhibiting α-glucosidase in the intestinal mucosa, the breakdown of starches into glucose was slowed, reducing and delaying the absorption of glucose in the small intestine to lower the blood glucose, which was more obvious for postprandial hyperglycemia [29,30]. The main α-glucosidase inhibitors commonly used were acarbose and voglibose [31,32]. Related studies showed that gardenia can effectively inhibit the activity of the α-glucosidase [33]. However, there has been little comprehensive comparison between the hypoglycemic effects of geniposide, crocin I and crocin II or in vitro and in vivo evaluation. In this chapter, a type 2 diabetes mellitus (T2DM) mice model was induced using a high-fat diet plus an intraperitoneal injection of streptozocin (STZ). A crude extract of gardenia (EGJ), geniposide, and crocin I was administrated to the mice. The high-fat diet plus the intraperitoneal injection of STZ induced the rapid decreases in the body weights with a yellow and wet hair color. Approx. 10 days after the establishment of the diabetes model, the body weight of the mice in the model group was as low as 37.3 ± 1.6 g, the water intake and food intake were as high as 44.73 mL/units and 22.47 g/units, respectively, and the blood glucose level fluctuated around 19.14 mm/L. The mice in the model group showed the lowest body weight and the highest water and food intakes and blood sugar levels compared to the mice in the normal group, proving that the diabetes model was successful. The mice in the administration metformin positive drug and administration 20% EGJ groups were accompanied by a higher body weight gain, a lower water and food intake, and a lower blood glucose level the T2DM group. The protective effect of the 20% EGJ was similar to metformin in the diabetic mice from the liver injury. Compared to the model group, the 20% EGJ, 40% EGJ, and crude EGJ all could decrease the serum GSP, TC, TG, HDL-C, and other indicator concentrations, among which the 20% EGJ had the most significant effect for regulating the glucolipid metabolism in the T2DM mice. The main content of the 20% EGJ was geniposide, indicating that geniposide had a certain effect on the glucose metabolism and the lipid metabolism in the diabetic mice. The effect of geniposide for treating T2DM was much higher than that of crocin I in vivo. However, the main component of the No. 8 sample (IC_{50} = 0.577 $\pm$ 0.041 mg ml^{-1}) was geniposide, which was a weak inhibitor of the α-glucosidase in vitro. The component of the No. 9 sample (IC_{50} = 0.204 $\pm$ 0.007 mg ml^{-1}) was crocin I, which had a stronger α-glucosidase inhibitory activity in vitro compared to the No.8 sample. This different result was not contradictory to the in vivo results. The studies suggested that the hypoglycemic effect of the alleviated geniposide might have been mediated by alleviating the glycogen phosphorylase and glucose-6-phosphatase activities [34]. It was also found that geniposide could significantly upregulate the protein levels of Glc NAc T-IV, a glycosyltransferase, but had no significant effect on the expression of clathrin [35]. Some studies found that gardeniside could stimulate the glycogen synthesis in the mice with a high-fat diet and a streptozotocin injection [36]. Therefore, in this research, the result of geniposide was effective in vivo but not in vitro, which also complemented the mechanism that the primary target of the geniposide participating in the treatment of T2DM was not α-glucosidase. Additionally, the action target of crocin I was α-glucosidase in vitro. The evidence reported that crocin I had benefits on the glycemic parameters and the type 2 diabetes-related complications in vivo [37,38]. However, there has been minimal research on its targets. This paper added that the target for the treatment of crocin I was α-glucosidase. At the same time, this experiment proved that geniposide was better than crocin I for the regulation of the blood glucose and blood lipids in the diabetic mice.

In conclusion, the separated and purified monomers of the geniposide and crocetin compounds from the crude extract of gardenia (EGJ) were collected and analyzed using HPLC. The monomers of the geniposide and crocin I compounds were compared using their hypoglycemic effects in vitro and in vivo. The extent of the compounds on T2DM followed the order of geniposide > crocin I > crude EGJ. The findings identified the different mechanism between crocin I and geniposide. Crocin I was effective on T2DM by inhibiting the α-glucosidase, geniposide played a major role on the hypoglycemic effect in the apparent levels of glucose, the lipid metabolism, and the apparent level of oxidative stress in the mice. This paper laid a further experimental foundation for the development of the crude EGJ as a new nutritional supplement and provided a theoretical reference for the high value utilization of the crude EGJ. Further studies are needed to compare the hypoglycemic mechanisms between the geniposide, crocin I, and EGJ.

Author Contributions: H.Z.: conceptualization, methodology, writing—original draft. S.Z.: methodology, investigation, software, writing—original draft. L.C.: methodology, investigation, funding acquisition. Y.L.: methodology, resources. L.S.: methodology, writing—review and editing. J.Z.: writing—review and editing, project administration, funding acquisition. All authors have read and agreed to the published version of the manuscript.

Funding: This research was supported by the Fundamental Research Funds for the Central Universities of China (No. 2662018PY022) and the Xiamen Major Science and Technology Plan Project (No. 3502Z2021005).

Data Availability Statement: Data is contained within the article.

Conflicts of Interest: The authors declare no conflict of interest.

Abbreviations

AI	Atherosclerosis index
AUC	Area under the curve
CAT	Catalase
CVRI	Cardiovascular risk index
EGJ	Extract of gardenia (*Gardenia jasminoides* Ellis)
FBG	Fasting blood glucose
GSP	Glycosylated serum protein
HDL-C	High-density lipoprotein cholesterol
LDL-C	Low-density lipoprotein cholesterol
MDA	Malondialdehyde
OGTT	Oral glucose tolerance test
*p*NPG	The *p*-nitrophenyl α-D-glucosidase
SOD	Superoxide dismutase
TC	Total cholesterol
TG	Triglyceride
T2DM	Type 2 diabetes mellitus

References

1. Scully, T. Diabetes in Numbers. *Nature* **2012**, *485*, S2–S3. [CrossRef] [PubMed]
2. Tonyan, Z.N.; Nasykhova, Y.A.; Danilova, M.M.; Glotov, A.S. Genetics of Macrovascular Complications in Type 2 Diabetes. *World J. Diabetes* **2021**, *12*, 1200–1219. [CrossRef] [PubMed]
3. Samarghandian, S.; Azimi-Nezhad, M.; Farkhondeh, T. Immunomodulatory and Antioxidant Effects of Saffron Aqueous Extract (*Crocus sativus* L.) on Streptozotocin-Induced Diabetes in Rats. *Indian Heart J.* **2017**, *69*, 151–159. [CrossRef] [PubMed]
4. Papademetriou, V.; Lovato, L.; Tsioufis, C.; Cushman, W.; Applegate, W.B.; Mottle, A.; Punthakee, Z.; Nylen, E.; Doumas, M.; Accord Study Group. Effects of High Density Lipoprotein Raising Therapies on Cardiovascular Outcomes in Patients with Type 2 Diabetes Mellitus, with or without Renal Impairment: The Action to Control Cardiovascular Risk in Diabetes Study. *Am. J. Nephrol.* **2017**, *45*, 136–145. [CrossRef] [PubMed]
5. Dash, R.P.; Babu, R.J.; Srinivas, N.R. Reappraisal and Perspectives of Clinical Drug-Drug Interaction Potential of Alpha-Glucosidase Inhibitors Such as Acarbose, Voglibose and Miglitol in the Treatment of Type 2 Diabetes Mellitus. *Xenobiotica* **2018**, *48*, 89–108. [CrossRef]
6. Addy, C.; Tatosian, D.; Glasgow, X.S.; Gendrano, I.N., III; Kauh, E.; Martucci, A.; Johnson-Levonas, A.O.; Selverian, D.; Matthews, C.Z.; Gutierrez, M.; et al. Pharmacokinetic and Pharmacodynamic Effects of Multiple-Dose Administration of Omarigliptin, a Once-Weekly Dipeptidyl Peptidase-4 Inhibitor, in Obese Participants with and without Type 2 Diabetes Mellitus. *Clin. Ther.* **2016**, *38*, 516–530. [CrossRef]
7. Wu, Y.; Zhou, Q.; Chen, X.-Y.; Li, X.; Wang, Y.; Zhang, J.-L. Comparison and Screening of Bioactive Phenolic Compounds in Different Blueberry Cultivars: Evaluation of Anti-Oxidation and Alpha-Glucosidase Inhibition Effect. *Food Res. Int.* **2017**, *100 Pt 1*, 312–324. [CrossRef]
8. Abou-Hany, H.O.; Atef, H.; Said, E.; Elkashef, H.A.; Salem, H.A. Crocin Mediated Amelioration of Oxidative Burden and Inflammatory Cascade Suppresses Diabetic Nephropathy Progression in Diabetic Rats. *Chem.-Biol. Interact.* **2018**, *284*, 90–100. [CrossRef]
9. Lee, S.W.; Lim, J.M.; Bhoo, S.H.; Paik, Y.S.; Hahn, T.R. Colorimetric Determination of Amino Acids Using Genipin from *Gardenia jasminoides*. *Anal. Chim. Acta* **2003**, *480*, 267–274. [CrossRef]
10. Chang, W.L.; Wang, H.Y.; Shi, L.S.; Lai, J.H.; Lin, H.C. Immunosuppressive Iridoids from the Fruits of *Gardenia jasminoides*. *J. Nat. Prod.* **2005**, *68*, 1683–1685. [CrossRef]
11. Qi, Q.; Mao, Y.; Tian, Y.; Zhu, K.; Cha, X.; Wu, M.; Zhou, X. Geniposide Inhibited Endothelial-Mesenchymal Transition Via the Mtor Signaling Pathway in a Bleomycin-Induced Scleroderma Mouse Model. *Am. J. Transl. Res.* **2017**, *9*, 1025–1036. [PubMed]
12. Lu, W.; Zhao, Y.; Kong, Y.; Zhang, W.; Ma, W.; Li, W.; Wang, K. Geniposide Prevents H_2O_2-Induced Oxidative Damage in Melanocytes by Activating the Pi3k-Akt Signalling Pathway. *Clin. Exp. Dermatol.* **2018**, *43*, 667–674. [CrossRef] [PubMed]
13. Chen, C.-C.; Hsu, C.-Y.; Chen, C.-Y.; Liu, H.-K. Fructus Corni Suppresses Hepatic Gluconeogenesis Related Gene Transcription, Enhances Glucose Responsiveness of Pancreatic Beta-Cells, and Prevents Toxin Induced Beta-Cell Death. *J. Ethnopharmacol.* **2008**, *117*, 483–490. [CrossRef] [PubMed]
14. Chen, Y.-I.; Cheng, Y.-W.; Tzeng, C.-Y.; Lee, Y.-C.; Chang, Y.-N.; Lee, S.-C.; Tsai, C.-C.; Chen, J.-C.; Tzen, J.T.-C.; Chang, S.-L. Peroxisome Proliferator-Activated Receptor Activating Hypoglycemic Effect of *Gardenia jasminoides* Ellis Aqueous Extract and Improvement of Insulin Sensitivity in Steroid Induced Insulin Resistant Rats. *BMC Complement. Altern. Med.* **2014**, *14*, 30. [CrossRef]

15. Li, S.; Liu, X.; Lei, J.; Yang, J.; Tian, P.; Gao, Y. Crocin Protects Podocytes against Oxidative Stress and Inflammation Induced by High Glucose through Inhibition of Nf-Kappa B. *Cell. Physiol. Biochem.* **2017**, *42*, 1481–1492. [CrossRef]
16. Yin, F.; Liu, J.-H. Research and Application Progress of *Gardenia jasminoides*. *Chin. Herb. Med.* **2018**, *10*, 362–370. [CrossRef]
17. Chen, T.; Kang, B.; Ren, L.; Chen, S.; Lin, H. The Optimized Parameters of Extracting Yellow Pigment from Gardenia (*Gardenia jasminoides* Ellis) Fruit. *Chin. J. Trop. Crops* **2010**, *31*, 1398–1402.
18. Alshabi, A.M.; Shaikh, I.A. Antidiabetic and Antioxidant Potential of *Gardenia latifolia* in Type-2 Diabetic Rats Fed with High-Fat Diet Plus Low-Dose Streptozotocin. *Saudi Med. J.* **2022**, *43*, 881–890. [CrossRef]
19. Zhang, J.-L.; Luo, C.-L.; Zhou, Q.; Zhang, Z.-C. Isolation and Identification of Two Major Acylated Anthocyanins from Purple Sweet Potato (*Ipomoea batatas* L. Cultivar Eshu No. 8) by Uplc-Qtof-Ms/Ms and Nmr. *Int. J. Food Sci. Technol.* **2018**, *53*, 1932–1941. [CrossRef]
20. Luo, C.-L.; Zhou, Q.; Yang, Z.-W.; Wang, R.-D.; Zhang, J.-L. Evaluation of Structure and Bioprotective Activity of Key High Molecular Weight Acylated Anthocyanin Compounds Isolated from the Purple Sweet Potato (*Ipomoea batatas* L. Cultivar Eshu No.8). *Food Chem.* **2018**, *241*, 23–31. [CrossRef]
21. Moras, B.; Loffredo, L.; Rey, S. Quality Assessment of Saffron (*Crocus sativus* L.) Extracts Via Uhplc-Dad-Ms Analysis and Detection of Adulteration Using Gardenia Fruit Extract (*Gardenia jasminoides* Ellis). *Food Chem.* **2018**, *257*, 325–332. [CrossRef] [PubMed]
22. Mojica, L.; Meyer, A.; Berhow, M.A.; de Mejia, E.G. Bean Cultivars (*Phaseolus vulgaris* L.) Have Similar High Antioxidant Capacity, in Vitro Inhibition of Alpha-Amylase and Alpha-Glucosidase While Diverse Phenolic Composition and Concentration. *Food Res. Int.* **2015**, *69*, 38–48. [CrossRef]
23. Aebi, H. Catalase in Vitro. *Methods Enzymol.* **1984**, *105*, 121–126. [PubMed]
24. Erejuwa, O.O.; Nwobodo, N.N.; Akpan, J.L.; Okorie, U.A.; Ezeonu, C.T.; Ezeokpo, B.C.; Nwadike, K.I.; Erhiano, E.; Wahab, M.S.A.; Sulaiman, S.A. Nigerian Honey Ameliorates Hyperglycemia and Dyslipidemia in Alloxan-Induced Diabetic Rats. *Nutrients* **2016**, *8*, 95. [CrossRef] [PubMed]
25. Yang, Y.; Zhang, J.-L.; Shen, L.-H.; Feng, L.-J.; Zhou, Q. Inhibition Mechanism of Diacylated Anthocyanins from Purple Sweet Potato (*Ipomoea batatas* L.) against Alpha-Amylase and Alpha-Glucosidase. *Food Chem.* **2021**, *359*, 129934. [CrossRef]
26. Zhang, G.-S.; Yang, H.-D.; Yue, X.-X.; Liu, Z.-X.; Xu, C.-M. Study on Enzymatic Combined Chemical Demulsification Process of Emulsion from Enzyme-Assisted Aqueous Extraction of Pumpkin Seed Oil. *Food Mach.* **2018**, *10*, 139–178.
27. Hirano, T.; Yamashita, S.; Takahashi, M.; Hashimoto, H.; Mori, Y.; Goto, M. Anagliptin, a Dipeptidyl Peptidase-4 Inhibitor, Decreases Macrophage Infiltration and Suppresses Atherosclerosis in Aortic and Coronary Arteries in Cholesterol-Fed Rabbits. *Metab.-Clin. Exp.* **2016**, *65*, 893–903. [CrossRef]
28. Vo, A.Q.; Feng, X.; Morott, J.T.; Pimparade, M.B.; Tiwari, R.V.; Zhang, F.; Repka, M.A. A Novel Floating Controlled Release Drug Delivery System Prepared by Hot-Melt Extrusion. *Eur. J. Pharm. Biopharm.* **2016**, *98*, 108–121. [CrossRef]
29. Liu, Y.; Zhu, J.; Yu, J.; Chen, X.; Zhang, S.; Cai, Y.; Li, L. A New Functionality Study of Vanillin as the Inhibitor for Alpha-Glucosidase and Its Inhibition Kinetic Mechanism. *Food Chem.* **2021**, *353*, 129448. [CrossRef]
30. Tan, S.Y.; Wong, J.L.M.; Sim, Y.J.; Wong, S.S.; Elhassan, S.A.M.; Tan, S.H.; Lim, G.P.L.; Tay, N.W.R.; Annan, N.C.; Bhattamisra, S.K.; et al. Type 1 and 2 Diabetes Mellitus: A Review on Current Treatment Approach and Gene Therapy as Potential Intervention. *Diabetes Metab. Syndr.* **2019**, *13*, 364–372. [CrossRef]
31. Cakar, U.; Grozdanic, N.; Pejin, B.; Vasic, V.; Cakar, M.; Petrovic, A.; Djordjevic, B. Impact of Vinification Procedure on Fruit Wine Inhibitory Activity against Alpha-Glucosidase. *Food Biosci.* **2018**, *25*, 1–7. [CrossRef]
32. Cakar, U.; Grozdanic, N.; Petrovic, A.; Pejin, B.; Nastasijevic, B.; Markovic, B.; Dordevic, B. Fruit Wines Inhibitory Activity against Alpha-Glucosidase. *Curr. Pharm. Biotechnol.* **2017**, *18*, 1264–1272. [CrossRef] [PubMed]
33. Ren, S.; Wan, Y.; Li, L.; Pan, T. Studies on the Inhibition Kinetics and Interaction Mechanism of Gardenia Yellow on Starch Digestive Enzyme. *J. Chin. Inst. Food Sci. Technol.* **2021**, *21*, 38–47.
34. Wu, S.-Y.; Wang, G.-F.; Liu, Z.-Q.; Rao, J.-J.; Lue, L.; Xu, W.; Wu, S.-G.; Zhang, J.-J. Effect of Geniposide, a Hypoglycemic Glucoside, on Hepatic Regulating Enzymes in Diabetic Mice Induced by a High-Fat Diet and Streptozotocin. *Acta Pharmacol. Sin.* **2009**, *30*, 202–208. [CrossRef]
35. Jiang, X.-Q.; Shen, S.-L.; Li, W.-Z.; Xu, X.-K.; Yin, F. Molecular Mechanism of Geniposide in Regulating Glut2 Glycosylation in Pancreatic Beta Cells. *Zhongguo Zhong Yao Za Zhi = Zhongguo Zhongyao Zazhi = China J. Chin. Mater. Med.* **2021**, *46*, 3643–3649.
36. Gao, S.; Feng, Q. The Beneficial Effects of Geniposide on Glucose and Lipid Metabolism: A Review. *Drug Des. Dev. Ther.* **2022**, *16*, 3365–3383. [CrossRef] [PubMed]
37. Khorasani, M.K.; Ahangarpour, A.; Khorsandi, L. Effects of Crocin and Metformin on Methylglyoxal-Induced Reproductive System Dysfunction in Diabetic Male Mice. *Clin. Exp. Reprod. Med.* **2021**, *48*, 221–228. [CrossRef] [PubMed]
38. Fang, K.; Gu, M. Crocin Improves Insulin Sensitivity and Ameliorates Adiposity by Regulating Ampk-Cdk5-Ppar Gamma Signaling. *Biomed Res. Int.* **2020**, *2020*, 9136282. [CrossRef]

foods

Article

The Effect and Mechanism of Corilagin from *Euryale Ferox* Salisb Shell on LPS-Induced Inflammation in Raw264.7 Cells

Minrui Wu [1], Yuhan Jiang [1], Junnan Wang [1], Ting Luo [1], Yang Yi [2], Hongxun Wang [1] and Limei Wang [1,*]

1 College of Life Science and Technology, Wuhan Polytechnic University, Wuhan 430023, China
2 College of Food Science and Engineering, Wuhan Polytechnic University, Wuhan 430023, China
* Correspondence: wanglimeiyx@163.com

Abstract: (1) Background: *Euryale ferox* Salisb is a large aquatic plant of the water lily family and an edible economic crop with medicinal value. The annual output of *Euryale ferox* Salisb shell in China is higher than 1000 tons, often as waste or used as fuel, resulting in waste of resources and environmental pollution. We isolated and identified the corilagin monomer from *Euryale ferox* Salisb shell and discovered its potential anti-inflammatory effects. This study aimed to investigate the anti-inflammatory effect of corilagin isolated from *Euryale ferox* Salisb shell. (2) Methods: We predict the anti-inflammatory mechanism by pharmacology. LPS was added to 264.7 cell medium to induce an inflammatory state, and the safe action range of corilagin was screened using CCK-8. The Griess method was used to determine NO content. The presence of TNF-α, IL-6, IL-1β, and IL-10 was determined by ELISA to evaluate the effect of corilagin on the secretion of inflammatory factors, while that of reactive oxygen species was detected by flow cytometry. The gene expression levels of TNF-α, IL-6, COX-2, and iNOS were determined using qRT-PCR. qRT-PCR and Western blot were used to detect the mRNA and expression of target genes in the network pharmacologic prediction pathway. (3) Results: Network pharmacology analysis revealed that the anti-inflammatory effect of corilagin may be related to MAPK and TOLL-like receptor signaling pathways. The results demonstrated the presence of an anti-inflammatory effect, as indicated by the reduction in the level of NO, TNF-α, IL-6, IL-1β, IL-10, and ROS in Raw264.7 cells induced by LPS. The results suggest that corilagin reduced the expression of TNF-α, IL-6, COX-2, and iNOS genes in Raw264.7 cells induced by LPS. The downregulation of the phosphorylation of IκB-α protein related to the toll-like receptor signaling pathway and upregulation of the phosphorylation of key proteins in the MAPK signaling pathway, P65 and JNK, resulted in reduced tolerance toward lipopolysaccharide, allowing for the exertion of the immune response. (4) Conclusions: The results demonstrate the significant anti-inflammatory effect of corilagin from *Euryale ferox* Salisb shell. This compound regulates the tolerance state of macrophages toward lipopolysaccharide through the NF-κB signaling pathway and plays an immunoregulatory role. The compound also regulates the expression of iNOS through the MAPK signaling pathway, thereby alleviating the cell damage caused by excessive NO release.

Keywords: inflammatory reaction; Raw264.7 macrophage; corilagin from *Euryale ferox* Salisb shell; cell pathway

Citation: Wu, M.; Jiang, Y.; Wang, J.; Luo, T.; Yi, Y.; Wang, H.; Wang, L. The Effect and Mechanism of Corilagin from *Euryale Ferox* Salisb Shell on LPS-Induced Inflammation in Raw264.7 Cells. *Foods* **2023**, *12*, 979. https://doi.org/10.3390/foods12050979

Academic Editor: Carla Gentile

Received: 4 January 2023
Revised: 9 February 2023
Accepted: 13 February 2023
Published: 25 February 2023

1. Introduction

The immune stimulator, LPS, can activate multiple signaling pathways in macrophages, leading to a series of pathophysiological responses [1], and is often used in in vitro inflammation studies. Inhibition of excessive activation of macrophages and its mediated inflammation has been demonstrated to be beneficial in many disease models [2,3], suggesting that targeting macrophage activation is a promising strategy for preventing inflammatory diseases.

Euryale ferox Salisb is a large aquatic plant belonging to the water lily family, which has been frequently reported for its use in lowering blood sugar and blood lipid [4], in addition to its antioxidation properties [5]. Shuliang He et al. characterized the constituents of the volatile oil of *Euryale ferox* Salisb and identified its biological activity, particularly its antioxidant activity [6]. Additionally, Wen-Na Zhang et al. characterized the polysaccharide in *Euryale ferox* Salisb and investigated its hypoglycemic effect [7]. The biosynthesis mechanism of flavonoids in *Euryale ferox* Salisb was analyzed by Peng Wu et al. through metabolomic and transcriptomic analyses, which revealed the key factors involved in the biosynthesis of flavonoids in *Euryale ferox* Salisb, its main functional substances [8]. Transcriptomic analysis of *Euryale ferox* Salisb at different developmental stages was performed by Xian Liu et al. [9], knowledge of which is particularly useful for the development and utilization of *Euryale ferox* Salisb.

With an annual output of more than 1000 tons, the *Euryale ferox* Salisb shell accounts for around 40% of the seed. It is often used as fuel or transported for disposal, resulting in waste of resources and environmental pollution [10]. Cheng Ying Wu et al. studied the antioxidant and anti-fatigue properties of phenolic extracts of the *Euryale ferox* Salisb shell, which led to the discovery of potential antioxidant agents [11].

Corilagin has been reported to exhibit various pharmacological activities, including inhibition of inflammatory development [12], antiviral [13], liver protection [14], and antitumor effects [15]. Li et al. [16] found that corilagin significantly reduced the levels of IL-6 and IL-1β in the serum of cells and mice and exhibited an anti-inflammatory role by downregulating the TLR4 signaling molecules, improving the extreme inflammatory state in patients with sepsis. Additionally, Tong et al. [17] found that corilagin may inhibit the activation of the nuclear factor-κB pathway in a STAT3-related manner and reduce the secretion of IL-1β and TNF-α, thereby reducing radiation-induced brain injury in mice. Previous studies have demonstrated that corilagin mainly improves cellular inflammation through the TLR signaling pathway, but there is no report of its activity in the LPS-induced RAW264. Previously, we isolated and identified the corilagin monomer from *Euryale ferox* Salisb shell; however, the anti-inflammatory effect was not evaluated.

We applied a network pharmacology approach to predicting the anti-inflammatory mechanism of corilagin, using the LPS-stimulated Raw264.7 cells as an in vitro inflammatory model to determine the effect of corilagin from *Euryale ferox* Salisb on the expression of inflammatory and anti-inflammatory factors in macrophages and the possible molecular mechanisms at three levels, i.e., biochemical factors, transcription, and protein levels. The theoretical groundwork is provided by developing and utilizing the Euryale ferox Salisb shell.

2. Materials and Methods

2.1. Materials and Chemicals

Raw264.7 cells were purchased from Shanghai Cell Bank, Chinese Academy of Sciences. The cell viability detection kit was purchased from Japan Tongren Reagent Company, and lipopolysaccharide for inducing inflammation was purchased from Sigma Company. The ELISA kit was purchased from Beijing Sizheng Bo Bio Co., Ltd. (Beijing, China), the ROS kit was purchased from Beijing Prilai Gene Technology Co., Ltd. (Beijing, China), and the reverse transcription kit was provided by Bao Bioengineering Co., Ltd. (Dalian, China). Primers were provided by Shanghai Sangon Biology Co., Ltd. (Shanghai, China). The primary and secondary antibodies used for Western blotting were provided by CST (Pi3K and p-Pi3K, Bioss; AKT and JNK, Wuhan Miting).

2.2. Separation and Identification of Corilagin from Euryale ferox Salisb Shell

Euryale ferox Salisb shells were dried and crushed, filtered with a 200-mesh sieve, ultrasonically extracted with 70% ethanol, concentrated under reduced pressure, and freeze-dried to obtain *Euryale ferox* Salisb shell polyphenol alcohol extracts. The chitosan polyphenol extract was added with water and ultrasonicated before being fractionally extracted with petroleum ether, ethyl acetate, and n-butanol (v/v = 1:1). The extract phase of the *Euryale ferox* Salisb shell was collected and packed on a silica gel column (60 mesh) by the wet method for separation. Elution was carried out with a mixture of ethyl acetate and petroleum ether (2:1), and the elution fractions were collected. The fraction with the highest activity was concentrated and lyophilized. The eluent was petroleum ether:ethyl acetate (100:15), and the eluent was collected. The samples were separated on a Sephadex LH-20 (Hydroxypropyl Sephadex) chromatographic column with 50% methanol and water, and the monomer compound was obtained by semi-preparative liquid phase, which was identified as corilagin (Figures S1 and S2).

2.3. Corilagin and Inflammation Target Prediction and Screening Application

The 2D structure of corilagin was obtained from the PubChem database, and the sdf file of the drug structure was imported into PharmMapper and Swiss Target Prediction, the TCMSP database, to obtain drug-related targets by merging and de-weighting. The disease-related targets were obtained from the GeneCards database by setting the search keyword "inflammation" as the genus "human origin".

2.4. Construction of PPI Network and Acquisition of Crossover Genes

The species was set as the human species, and the minimum relationship score was 0.4. The key proteins with the cross-repetition of corilagin and inflammation were input into the protein interaction database (STRING), and the proteins without an interaction relationship were removed to obtain the protein interaction map.

2.5. HUB Genes' Acquisition and KEGG and GO Enrichment Analysis

Cytoscape 3.9.1 was supplied with the protein–protein interaction diagram obtained from the STRING database to obtain the top 30 central target genes for interactions with other proteins in the network diagram. GO and KEGG analysis of central target genes were performed using the R-package clusterProfiler and enrichment plot. The data with p-value < 0.05 were screened, and the relevant legends were plotted using the R package ggplot2.

2.6. Docking Analysis

The top five degrees in the PPI network were used as the receptor proteins, the top nodes in the "active ingredient-target-disease" network were used as the ligands, and the structures of the receptor proteins were downloaded from the PDB database. The proteins and ligands were pre-processed using PyMOL-2.3.4. Subsequently, AutoDockTools software was used to pre-process the protein and ligands, and Vina was used for predicting the binding energy of the ligands of small size to the proteins, with the lowest binding energy indicating the optimal conformation. The receptor–ligand docking files were processed by PyMOL and uploaded to the online website called Plip to visualize the validation results.

2.7. Cell Culture and Model Establishment

Raw264.7 cells, the mouse monocytic leukemia cells, were cultured in a DMEM medium containing 10% FBS and 1% penicillin and streptomycin dual-antibodies at 37 °C in an incubator containing 5% CO_2. The culture was passaged when grown to more than 90% in cell culture flasks, and selected experiments were performed on counted cells. The experiments were conducted by comparing different groups of experimental subjects: the control group (without LPS and corilagin intervention), the LPS stimulation group (with the addition of 1 μg/mL of LPS for intervention), the experimental group (different concentrations of corilagin were pretreated for 2 h and the final concentration of 1 μg/mL was added and LPS co-treated for 24 h), and the positive drug group (50 μmol/L of dexamethasone pretreatment for 2 h, LPS with a final concentration of 1 μg/mL added for 24 h).

2.8. Cell Morphology Observation and CCK-8 Assay to Detect the Proliferation Toxicity of Corilagin on Raw264.7 Cells

Cells were seeded into 6-well plates (at a density of 5×10^5 cells per well). The cells were divided into a normal control group, an LPS stimulation group, and a corilagin treatment group, and the experiments were conducted in replicates (2 wells for each group). The cellular morphology was observed by an inverted microscope.

To screen for the safe concentration of corilagin from the Gorgon husk source, the CCK-8 method was used to analyze the effect of corilagin on the survival rate of Raw264.7 cells.

Cells in the logarithmic growth phase were sampled for cell counting, and 100 μL of cell suspension was added to each well of a 96-well plate (density of 3000 cells per well). The surrounding wells were sealed with PBS and grown for 24 h.

The experiment was divided into the blank group (without cells and drugs), the control group (without drugs), and the experimental group. The drugs were prepared by diluting with complete medium to 2-fold gradient dilution, followed by filtration with a membrane of 0.22 μm pore size, and used immediately. The cultures were grown for 24 h before the analysis. When testing, the medium in the 96-well plate was aspirated, washed twice with PBS, and patted dry on thick paper. CCK-8 was prepared in the dark to avoid errors caused by residual CCK-8 in the pipette tip left when adding samples. A complete medium was used to dilute CCK-8, and the diluted solution was mixed well for later use. The cultures were incubated for 3 h in an incubator, and the OD value at a wavelength of 450 nm was determined.

2.9. Determination of NO Content by Griess Method

Raw264.7 cells were seeded into a 24-well plate (at a density of 2×10^5 cells per well) and placed in a cell incubator for 12 h. Different groups were cultured for 24 h according to the corresponding treatment, and the cell supernatant was collected. The NO content in the supernatant was detected by the Griess reagent method, and the amount of released NO of each group was calculated using the standard curve.

2.10. ELISA Method to Determine the Effect of Corilagin on the Secretion of Inflammatory Factors

Cell treatment was kept consistent with the pretreatment method used for cell morphology observation, and the cell supernatant was collected. The contents of TNF-α, IL-6, IL-1β, and IL-10 were determined according to the instructions of the ELISA kit.

2.11. Detection of Intracellular Reactive Oxygen Species by Flow Cytometry

The cells in the logarithmic growth phase were inoculated into 6-well plates and cultured for 12 h. Different groups were cultured for 24 h according to the corresponding treatment. The liquid in the 6-well plate was discarded and washed twice with PBS. The control group was added with 2 mL of complete medium. The base, lipopolysaccharide, and experimental groups were added with 2 mL of 20 μmol DCFH-DA diluted in a complete medium and incubated in an incubator for 2 h. After incubation, PBS was used for rinsing twice, i.e., 1 mL of PBS was added to each well and the cells were detached by pipetting, collected into a centrifuge tube, and centrifuged at 1000 r/min for 3 min, the supernatant was discarded, and 1 mL of PBS was added to each tube. The mixtures were mixed by pipetting, the cell suspension was transferred to a 1.5 mL flow centrifuge tube, and the intensity of the intracellular ROS fluorescence was measured by flow cytometry.

2.12. qRT-PCR Detection of TNF-α, IL-6, COX-2, and iNOS Gene Expression Levels

The six-well plate was taken out, and the medium discarded and rinsed twice with PBS. Then, 1 mL of RNA iso plus was added and left to stand for 1 min before pipetting. The cell suspension was collected by pipetting, transferred into a sterile tube, and added with 200 μL of chloroform, and the mixture was mixed well. Next, extraction was carried out on the ice for 15 min with inversion every 5 min. The EP tube with three layers of supernatant was carefully removed, and the supernatant was pipetted into another 1.5 mL EP tube using a 100 μL pipette. Then, 500 μL of isopropanol was added and mixed well, and the tube was placed on ice for 15 min. The mixture was then centrifuged at 12,000 rpm and at 4 °C for 15 min to recover a white precipitate. The supernatant was carefully removed to avoid disturbing the pellet. Next, 1 mL of 75% ethanol was added, and the pellet was gently lifted. The wall of the tube was washed by inversion, at 7500 rpm. The supernatant was discarded, and the pellet was air-dried with the lid opened at room temperature for 5 min before the addition of an appropriate amount of DEPC water to dissolve the pellet. The RNA concentration was measured using a UV micropipette and adjusted to obtain the RNA concentration of 1000 ng/μL per tube. The RNA was reverse-transcribed into cDNA according to the protocols for reverse transcription, and the real-time fluorescence quantitative PCR reaction system using 10 μL of SYBR Premix Ex Taq TM, 8 μL of primers, and 2 μL of cDNA was conducted.

Primer sequences (Table 1): The cDNA sequences of each gene were retrieved from NCBI, and specific primer sequences were designed and synthesized by Shanghai Sangon Bioengineering Co., Ltd.

Table 1. Primer sequences.

Gene Name	Forward Primer (5→3)	Reverse Primer (3→5)
β-actin	CTACCTCATGAAGATCCTGACC	CACAGCTTCTCTTTGATGTCAC
TNF-α	ATGTCTCAGCCTCTTCTCATTC	GCTTGTCACTCGAATTTTGAGA
IL-6	CTCCCAACAGACCTGTCTATAC	CCATTGCACAACTCTTTTCTCA
COX-2	ATTCCAAACCAGCAGACTCATA	ATTCCAAACCAGCAGACTCATA
iNOS	CGGACGAGACGGATAGGCAGAG	GGAAGGCAGCGGGCACATG

2.13. Western Blot Analysis of Key Proteins' Expression in NF-κB, MAPK, and PI3K-AKT Signaling Pathways

After preconditioning cells, the excess medium in the well plate was aspirated, and 1 mL of PBS was used for washing twice. Cells were digested with 1 mL of trypsin, transferred to a 1.5 mL EP tube, and centrifuged at 3000 rpm for 1 min. The resulting supernatant was discarded. Next, 100 μL of lysis buffer (prepared by RIPA lysis buffer and protease inhibitor 1:100) was added to each tube and evenly pipetted. The cells were lysed on ice for 30 min to ensure complete cell lysis. The cells were then centrifuged at 12,000 rpm for 10 min at 4 °C, and the supernatant was collected to obtain the total protein. A 5× protein loading buffer 4:1 was added to the protein sample, mixed by vortexing, and incubated in a water bath at 95 °C for 10 min. The treated samples were stored in a −20 °C refrigerator for later use. The treated protein samples were transferred to 10% SDS-polyacrylamide gel for electrophoretic separation. The PVDF membrane was placed on the glue and covered with wet filter paper and a sponge. The mounted membrane transfer system was secured with the membrane transfer clip, and the transfer was conducted at 200 mA for 1 h. The membrane was then blocked with 5% skimmed milk at room temperature for 1 h with gentle shaking, followed by overnight incubation with a primary antibody at 4 °C. The blocked membrane was then washed thrice on a decolorizing shaker for 5 min at room temperature. Two hours later, the membrane was then washed thrice on a decolorizing shaker for 5 min. The membrane was then exposed to ECL, and the signal was analyzed using gel imaging software.

2.14. Data Statistics and Analysis

GraphPad Prism 8 was used for statistical analysis. All data are expressed as the mean ± standard deviation unless otherwise stated. The *p*-value of <0.05 was considered significant.

3. Results

3.1. Acquisition of Corilagin and Inflammatory Targets and the "Corilagin-Target-Inflammation" Interaction Network

The 3D structure of corilagin from *Euryale ferox* Salisb shell was obtained from the PubChem database (Figure 1A). A total of 307 corilagin genes were obtained from the PharmMapper website and the Swiss Target Prediction database. Among human species, the GeneCard database showed that there were 111,109 genes related to inflammation. Then, 268 cross-repeating genes (Figure 1B) that appear in both drug targets and inflammatory targets were selected, suggesting that these genes may be the key genes in regulating inflammation of corilagin. To further investigate the relationship between corilagin and inflammation, a "Corilagin-Target-Inflammation" network has been constructed in Cytoscape 3.9.1 (Figure 2).

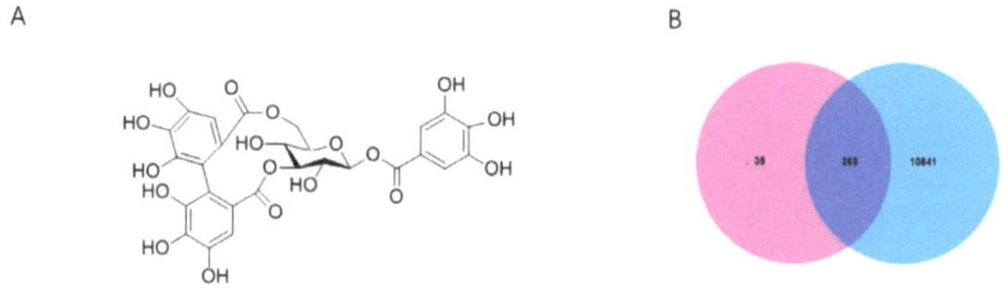

Figure 1. Corilagin and inflammation intersection targets: (**A**) structural formula of corilagin and (**B**) intersection Venn diagram.

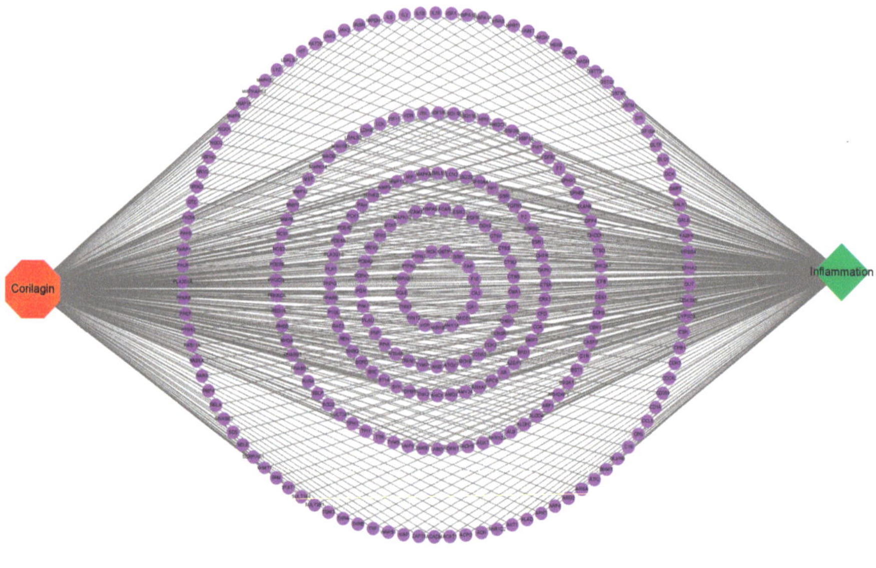

Figure 2. "Corilagin-Target-Inflammation" network.

3.2. Construction of PPI Network and Acquisition of HUB Genes

The crossover genes were imported into the STRING website for protein interactions, and a protein interaction map containing 268 nodes and 3724 edges was obtained. Cytoscape software visualized the protein interaction map, and the top 30 target proteins of the protein interaction network were calculated using the MCC calculation method in the CytoHubba plugin (Figure 3). The results predicted that the above genes and related proteins play an important role in treating hepatocellular carcinoma.

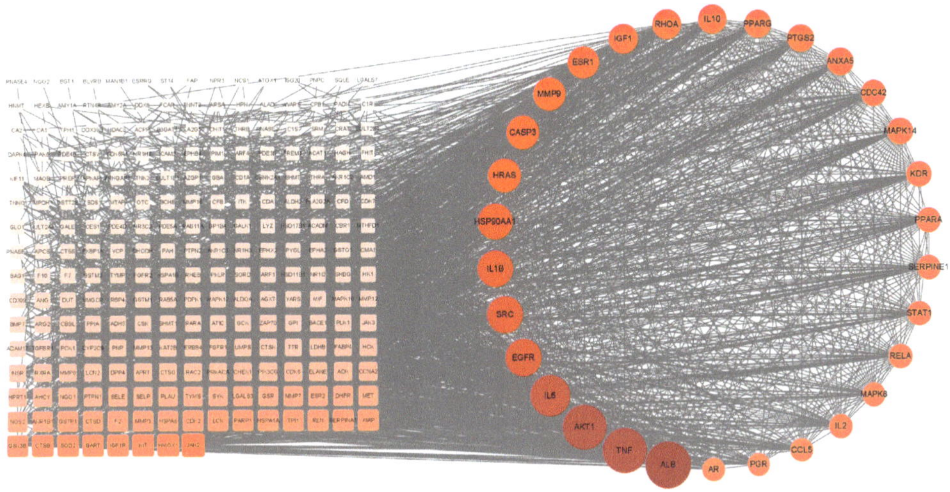

Figure 3. PPI network map and screening of HUB genes.

3.3. GO and KEGG Pathway Analysis

The GO analysis (Figure 4A) revealed the expression of genes localized in the nucleus, while the KEGG (Figure 4B) analysis revealed the MAPK and TOLL-like receptor signaling pathways. Based on findings from the GO and KEGG analyses, it was concluded that corilagin may validate protein expression by regulating the MAPK signaling pathway, which is related to the secretion of inflammatory factors by macrophages, as well as in the NF-B and PI3K signaling pathways, which are closely related to toll-like receptors that attenuate the tolerance level of macrophages.

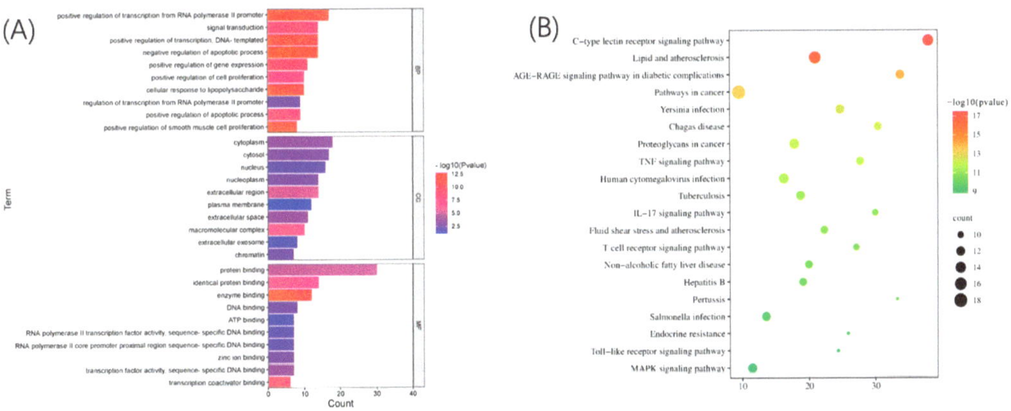

Figure 4. GO and KEGG pathway analyses of HUB genes: (**A**) GO analysis and (**B**) KEGG analysis.

3.4. Validation of Molecular Docking

The binding energy (kcal/mol) between the target and corilagin was predicted based on molecular docking (Figure 5A), whereby the negative binding energy of the ligand and receptor usually indicates the binding affinity between them. The docking revealed that the binding energy was less than −5 kcal/mol between all hub gene targets and keratine, from which only three models with good binding energy were selected for visualization (Figure 5B–D).

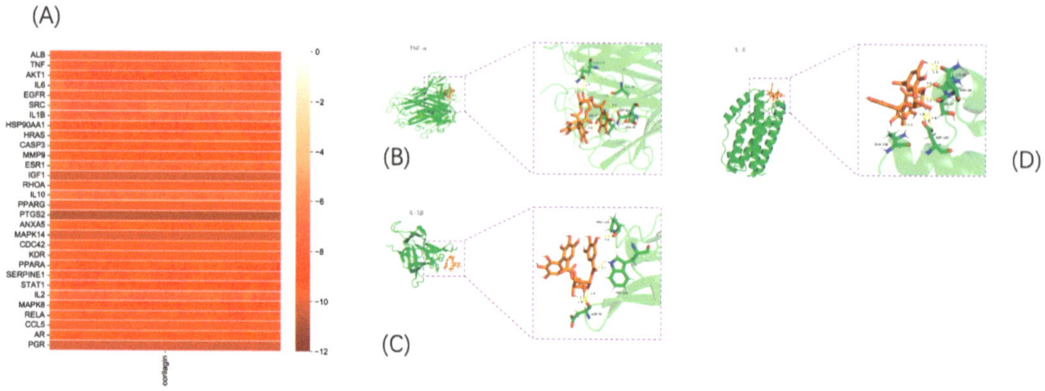

Figure 5. Validation of molecular docking: (**A**) molecular binding energy heatmap and (**B–D**) visualization of molecular docking results.

3.5. The Effect of Corilagin on the Viability of Raw264.7 Cells and Changes in Cellular Morphology

The investigation of the effect of corilagin on the viability of Raw264.7 cells demonstrated that (Figure 6) cell viability was significantly decreased after incubation for 24 h at concentrations exceeding 100 μm/L, i.e., 25, 50, and 100 μmol/L were selected as the low, medium, and high doses of corilagin from the Gorgon shell source in the experiment.

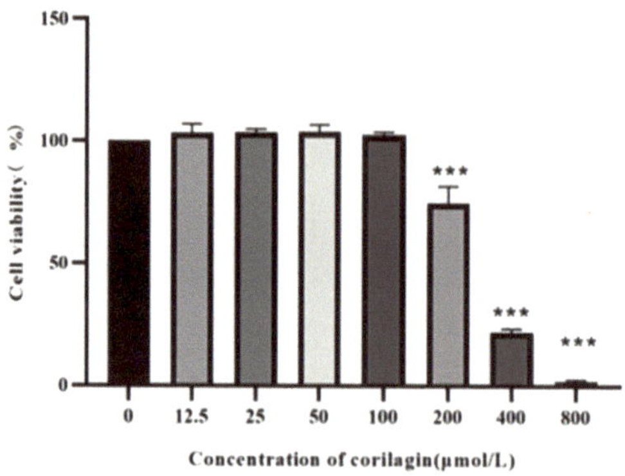

Figure 6. Effect of LPS on Raw264.7 cell viability. (Compared with control group, *** $p < 0.001$).

Raw264.7 is a mononuclear macrophage derived from the leukemia virus in Balb/c mice [18]. The health of the cells can be impacted by the state of the cells, and since Raw264.7 cells are small and bright in appearance, they are not in a good shape. After the LPS stimulation, the cells formed a long shuttle with elongated false feet [19]. Additionally, pseudopodia were reduced in cells treated with corilagin, with most of the cells being round in shape (Figure 7). The results indicated that corilagin from *Euryale ferox* Salisb could mitigate inflammation by inhibiting the differentiation of Raw264.7 cells.

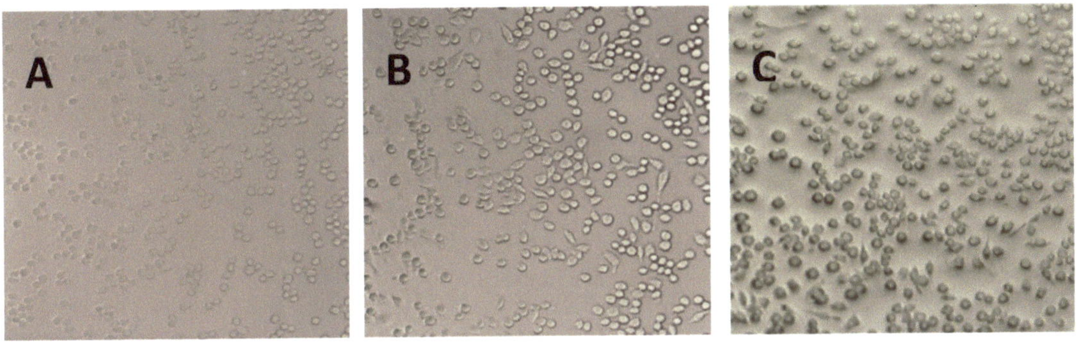

Figure 7. Morphology of Raw264.7 cells under different treatment conditions: (**A**) control cells, (**B**) lipopolysaccharide-stimulated cells, and (**C**) cells treated with corilagin from *Euryale ferox* Salisb shell.

3.6. Effects of Corilagin on LPS-Induced NO Secretion by Raw264.7 Macrophages

NO is an endogenous synthetic gas signal molecule, synthesized in the cytoplasm, which quickly diffuses through the cell membrane. The molecule rapidly reacts with other free radicals, producing high levels of active peroxidase (oxidant) and other active nitrogen derivatives. These molecules can reflect inflammation and diseases such as atherosclerosis [20]. The effects of NO secretion on the volume of Raw264.7 cells induced by LPS via the Griess method were investigated. The results (Figure 8) depict that at the concentration of higher than 25, 50, and 100 μmol/L, the corilagin intervention resulted in a significant reduction in the volume of LPS-induced Raw264.7 cells' supernatant, indicating the potential inflammation-relieving activity of corilagin, which concerns its ability to suppress the release of NO.

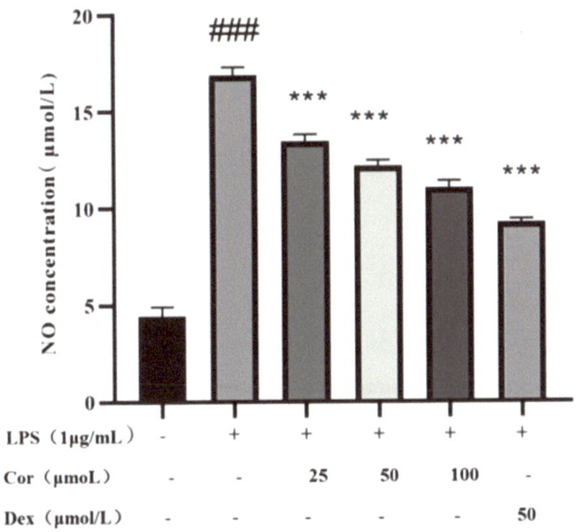

Figure 8. The effect of corilagin from *Euryale ferox* Salisb shell on NO secretion of Raw264.7 cells. (Compared with the control group, ### $p < 0.001$; Compared with LPS group, *** $p < 0.001$).

3.7. The Effect of Corilagin on the Expression of Inflammatory Factors in LPS-Induced Raw264.7 Macrophages

When macrophages are activated, inflammatory cytokines such as inflammatory enzymes and TNF-α, IL-6, IL-1β, and IL-10 are secreted. Following anti-inflammatory drug intervention, macrophages secrete anti-inflammatory factors, which are responsible for anti-inflammatory effects. The investigation of the effect of corilagin on the secretion of TNF-α, IL-6, IL-1β, and IL-10 in Raw264.7 cells (Figure 9) revealed that treatment with corilagin at 25, 50, and 100 μmol/ L significantly reduced the secretion of TNF-α and IL-6 in LPS-induced Raw264.7 cells. At 50 and 100 μmol/L, corilagin significantly reduced the secretion of IL-1β and IL-10 in LPS-induced Raw264.7 cells.

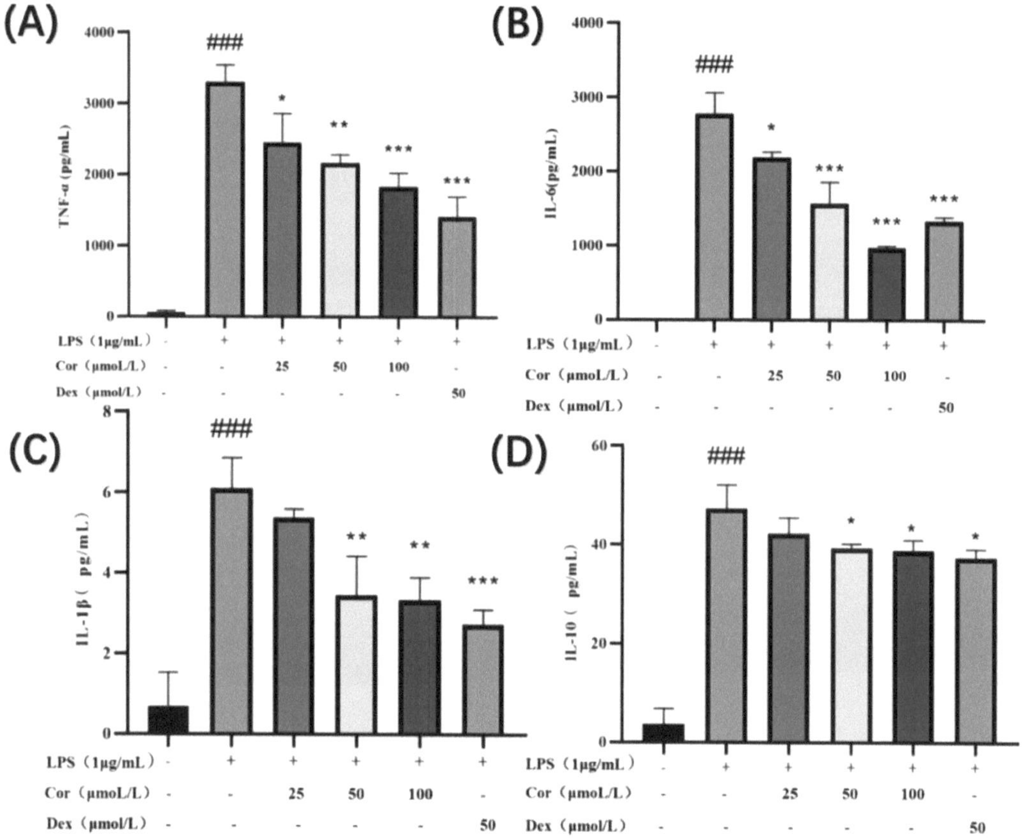

Figure 9. Effect of corilagin on TNF-α, IL-6, IL-1β, and IL-10 secretion of Raw264.7 cells: the amount of (**A**) TNF-α, (**B**) IL-6, (**C**) IL-1β, and (**D**) IL-10 secreted in the cell supernatant. (Compared with the control group, ### $p < 0.001$; Compared with LPS group, * $p < 0.05$, ** $p < 0.01$ and *** $p < 0.001$).

3.8. The Effect of Corilagin on the Content of Reactive Oxygen Species in Raw264.7 Cells

Activated oxygen is an active oxygen family, which includes superoxide and hydroxyl radicals that stimulate macrophages and neutrophils. This reactive oxygen species is produced by many biologically active media. Reactive oxygen species and other pro-inflammatory factors activate nuclear transcription factors (NF-κB) and cell nuclear binding, thereby promoting inflammatory factor transcription, sometimes resulting in inflammatory allergic reactions. These factors play a key role in the inflammation and wound healing processes, which lack specificity in bacteria. Excessive active oxygen can destroy the integrity of the mitochondrial membrane, resulting in changes in mitochondrial permeability; thus, eliminating tissue and damaging active oxygen is highly important. The effects of corilagin on the content of reactive oxygen species in Raw264.7 cells at 25, 50, 100, and 50 μmol/L of dexamethasone are presented in Figure 10. Following the intervention with corilagin at 25, 50, 100, and 50 μmol/L of dexamethasone, the intracellular ROS was reduced by 9.49%, 10.91%, 15.43%, and 8.73%, respectively. The results demonstrate that the corilagin from *Euryale ferox* Salisb shell can prevent the inflammatory reaction by reducing reactive oxygen species.

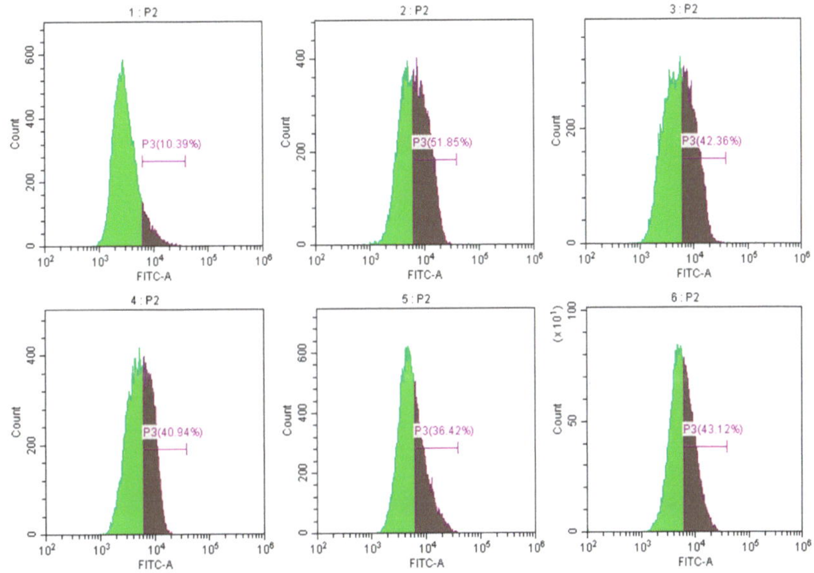

Figure 10. Effect of corilagin on ROS content of Raw264.7 cells. Pictures 1–6 correspond to the control group, LPS treatment group, 25, 50, 100 μmol/L of corilagin treatment group, and the 50 μmol/L dexamethasone treatment group, in which P3 represents intracellular reactive oxygen species content.

3.9. *Effects of Corilagin on the Gene Expression of TNF-α, IL-6, iNOS, and COX-2 in LPS-Induced Raw264.7 Macrophages*

The results (Figure 11) demonstrate that after the intervention of 25, 50, and 100 μmol/L derived from the Gorgon shell, the levels of gene expression of TNF-α, iNOS, and COX-2 were significantly reduced in a dose-dependent manner. Additionally, intervention by 100 μmol/L of corilagin significantly reduced the level of gene expression of IL-6. The nitric oxide synthase (NOS) can convert left-transverse arginine to left-transverse citrulline, thereby causing NO production, which can be classified into constitutive and inducible. Many studies have suggested that continuous and excessive NO production is primarily associated with the high expression level of inducible nitric oxide synthase (iNOS) [21]. The results indicate that the corilagin could reduce the secretion of NO by lowering the expression of iNOS, thereby exerting its anti-inflammatory effect. In addition, the level of COX-2 is increased in the inflammatory process, causing excessive production of PGE2, resulting in an excessive inflammatory response. The results also display that the corilagin could downregulate the expression of COX-2 to alleviate inflammation. TNF-α and IL-6 are the most common inflammatory cytokines, which play a vital role in inflammation. The results demonstrate that corilagin could reduce the gene expression of TNF-α and IL-6, thereby reducing the secretion of TNF-α and IL-6 and achieving an anti-inflammatory effect.

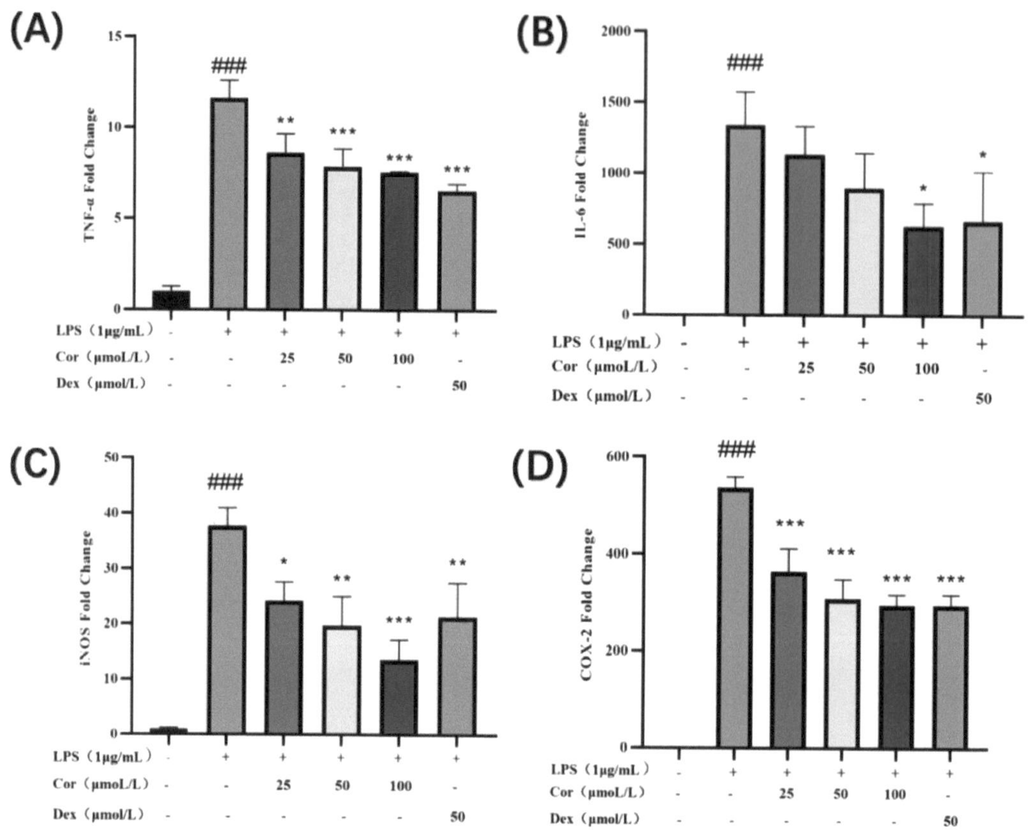

Figure 11. Effect of corilagin on TNF-α, IL-6, COX-2, and iNOS expression of Raw264.7 cells: expressions of (**A**) TNF-α, (**B**) IL-6, (**C**) iNOS, and (**D**) expression of COX-2. (Compared with the control group, ### $p < 0.001$; Compared with LPS group, * $p < 0.05$, ** $p < 0.01$ and *** $p < 0.001$).

3.10. The Effect of Corilagin on the Expression of Key Proteins in the NF-κB Signaling Pathway in Raw264.7 Cells

The effect of corilagin from the Gorgon shell on the expression of NF-κB pathway-related proteins in Raw264.7 cells is presented in Figure 12. The protein assay of IκB-α revealed that LPS stimulation and intervention by Gorgon shell-derived corilagin produced no significant effect on phosphorylation of IκB-α. Additionally, the phosphorylation of P65 was significantly downregulated following LPS stimulation, whereas the treatment with 50 μmol/L of corilagin resulted in upregulation of the phosphorylation of P65.

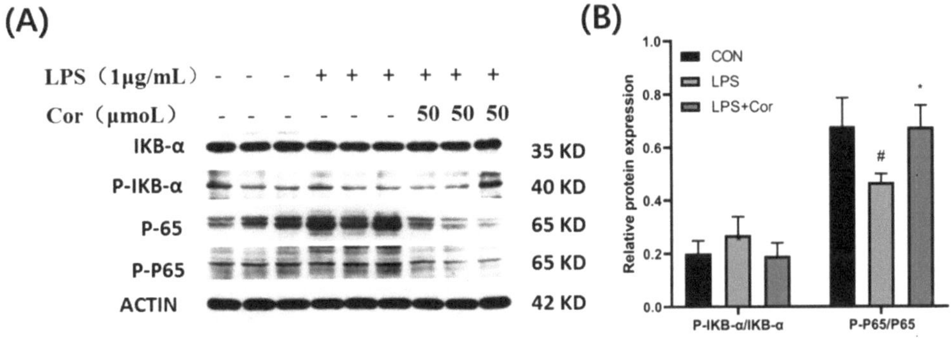

Figure 12. Effect of corilagin on the expression of NF-κB pathway-related proteins in Raw264.7 cells. (**A**) Western Blot results, (**B**) Relative protein expression (Compared with the control group, # $p < 0.05$; Compared with LPS group, * $p < 0.05$).

3.11. The Effect of Corilagin on the Expression of Key Proteins in the MAPK Signaling Pathway in Raw264.7 Cells

The effect of corilagin from the Gorgon shell on the expression of MAPK pathway-related proteins in Raw264.7 cells is presented in Figure 13. The ERK protein assay revealed that LPS stimulation produced no significant effect on the phosphorylation of ERK, whereas the phosphorylation of ERK was significantly upregulated after the intervention of corilagin at 50 μmol/L. It was also found that the phosphorylation of JNK was significantly downregulated following LPS stimulation, while the intervention by 50 μmol/L of corilagin resulted in upregulation of phosphorylation of JNK. Additionally, both LPS stimulation and treatment with 50 μmol/L of corilagin produced no significant effect on the phosphorylation of P38.

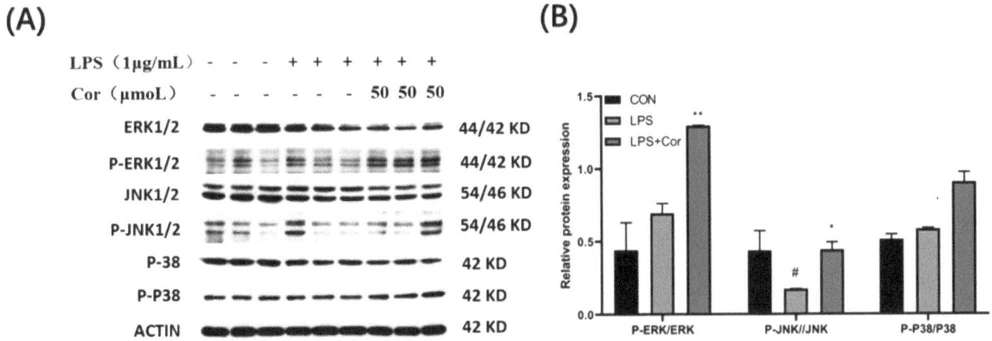

Figure 13. Effect of corilagin on the expression of MAPK pathway-related proteins in Raw264.7 cells. (**A**) Western Blot results, (**B**) Relative protein expression (Compared with the control group, # $p < 0.05$; Compared with LPS group, * $p < 0.05$ and ** $p < 0.01$).

3.12. The Effect of Corilagin on the Expression of Key Proteins in the PI3K-AKT Signaling Pathway in Raw264.7 Cells

The effect of corilagin from the Gorgon shell on the expression of PI3K-AKT pathway-related proteins in Raw264.7 cells is depicted in Figure 14. The PI3K protein assay revealed that both stimulations by LPS and intervention with 50 μmol/L of corilagin produced no significant effect on the phosphorylation of PI3K. Compared with the blank group, the phosphorylation of AKT protein was significantly downregulated following LPS stimulation, whereas the intervention with 50 μmol/L of corilagin produced no significant effect on the phosphorylation of AKT protein.

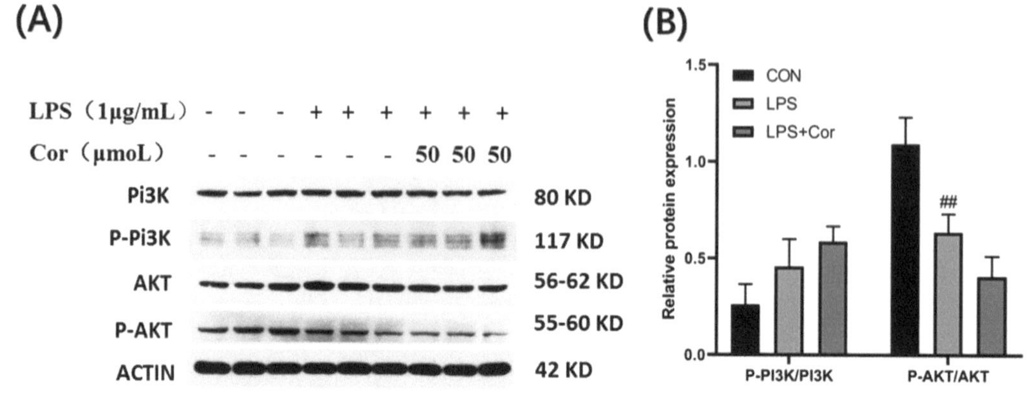

Figure 14. Effect of corilagin on the expression of PI3K-AKT pathway-related proteins in Raw264.7 cells. (**A**) Western Blot results, (**B**) Relative protein expression (Compared with the control group, ## $p < 0.01$).

The investigation of the effect of the corilagin intervention revealed that no significant effect was produced on the expression of key proteins in the NF-κB signaling pathway in Raw264.7 cells. Similarly, stimulation by LPS produced no significant effect on the phosphorylation of IκB-α but significantly downregulated the phosphorylation of P65. In the MAPK signaling pathway, LPS stimulation had no significant effect on the phosphorylation of ERK and P38 but significantly inhibited the phosphorylation of JNK. In addition, in the PI3K-AKT signaling pathway, LPS stimulation had no significant effect on the phosphorylation of PI3K but significantly downregulated the phosphorylation of AKT. Previous studies have demonstrated that when macrophages develop lipopolysaccharide tolerance, LPS stimulation results in reduced P65 phosphorylation in the NF-κB signaling pathway, as well as that of ERK, JNK, and P38 in the MAPK signaling pathway [21]. The results from Western blot analysis demonstrate that the macrophages were lipopolysaccharide-tolerant and that LPS at 50 μmol/L was tolerated in this study. Additionally, the intervention of corilagin originated from the Gorgon husk significantly upregulated the phosphorylation of P65 and JNK proteins and relieved the lipopolysaccharide tolerance state of Raw264.7 cells, indicating its potential immunoregulatory role [21].

4. Discussion

Corilagin is a polyphenolic tannin compound [16], which is widely found in geranium, bead, white clover, longan, *Phyllanthi fructus*, and other plants [22]. Structurally, ellagic acid is a dilactone of hexahydroxy biphenyl acid (HHDP). Corilagin, a kind of ellagitannin which is a condensation of ellagic acid, has good antioxidant potential. After optimizing the extraction process of ellagitannin, Anindita Paul et al. obtained the result that ellagitannin could combine well with catalase through calculation and analysis [23]. Studies have demonstrated that corilagin exhibits antitumor, antiviral, and antibacterial activities in addition to anti-inflammatory and antioxidant effects, suggesting its potential use as an agent in the preventive treatments of cardiovascular diseases [13,22,24–26]. Corilagin sourced from Gorgon husk was extracted and identified during the early stages of a study on Gorilla husk by our research team, but its anti-inflammatory properties have not yet been revealed.

Cyber-pharmacology efficiently integrates research content, utilizing high-throughput computing methods and software. By setting different screening conditions, proteins interacting with small molecules can be accurately predicted, to predict protein-related metabolic pathways [27]. In this study, we identified 307 targets of corilagin using PharmMapper. The subsequent GO and KEGG enrichment analyses revealed that the anti-inflammatory effects of corilagin are primarily associated with MAPK and TOLL-like receptor signaling pathways. Additionally, LPS and other pro-inflammatory factors can induce the phosphorylation of MAPK signaling pathway-associated proteins and trigger the expression of iNOS genes in the nucleus.

Findings from this study demonstrate that corilagin, a Gorgon fruit source, could significantly inhibit the production of reactive oxygen species induced by LPS in macrophages, as well as the related oxidative stress responses in cells. These results further confirm the findings reported in previous studies regarding the positive correlation between the degree of oxidative stress and the degree of inflammation [28] and that inhibiting excessive production of reactive oxygen species can suppress the inflammatory response. Analysis of NO secretion revealed that the intervention of corilagin from Gorgon shell significantly reduced NO content in the supernatant of LPS-induced Raw264.7 cells.

Analysis of the gene expression of inflammatory factors demonstrated that the corilagin intervention significantly reduced the expression of IL-6 and TNF-α genes compared with LPS-induced macrophage inflammation. Corilagin also significantly downregulated the expression of iNOScox-2, and nitric oxide synthase (NOS) could convert L-arginine to L-citrulline, thereby causing the release of NO. Numerous studies have suggested that persistent and excessive NO production is mainly attributed to the overexpression of inducible nitric oxide synthase (iNOS) [29]. In addition, the expression of COX-2 is elevated during the inflammation process, causing the overproduction of PGE2, which results in an excessive response to inflammation [30]. This proves that corilagin, a source of a gorgonian shell, could reduce the secretion of NO by downregulating the expression of iNOS, thereby exerting its anti-inflammatory effect.

Many physiological and pathological responses in mammalian cells and tissues are mediated by MAPK signaling, including stress responses, inflammation, and apoptosis. Phosphorylation of ERK1/2 and p38 promotes the production of inflammatory factors, including TNFα, IL-6, and IL-8 [31]. In addition, TNF-α also stimulates the MAPK cascade and promotes IL-8 secretion [32]. The analysis conducted in this study revealed that phosphorylation levels of ERK and JNK proteins increased following treatment with corilagin.

Related studies have reported that p38 of MAPK and PI3K-Akt can regulate LPS-induced gene expression by controlling the hyperphosphorylation and nuclear translocation of p65 of NF-κB [33]. Corilagin could also significantly upregulate the phosphorylation of P65 protein. This study also examined the expression of related proteins in the PI3K-Akt signaling pathway, which revealed that LPS induction significantly reduced the phosphorylation level of AKT.

Previous studies have demonstrated that when macrophages develop lipopolysaccharide tolerance, LPS stimulation can reduce the phosphorylation of P65 protein in the NF-κB signaling pathway, as well as that of ERK, JNK, and P38 in the MAPK signaling pathway [34]. Western blot analysis demonstrated that the intervention of 50 μmol/L of gorgonian shell-derived corilagin could significantly upregulate the phosphorylation of P65 and JNK, relieving the lipopolysaccharide tolerance state of Raw264.7 cells, thereby exerting an immune-regulating effect. The results indicate that macrophages developed lipopolysaccharide tolerance and that corilagin interfered with the phosphorylation of P65 and JNK proteins and relieved the macrophages of their tolerance. Notably, lipopolysaccharide tolerance plays an immunomodulatory role [24].

Supplementary Materials: The following are available online at https://www.mdpi.com/article/10.3390/foods12050979/s1, Figure S1. Mass spectra of corilagin; Figure S2. Claritin extraction process.

Author Contributions: Methodology, Y.J. and J.W.; Software, J.W. and T.L.; Validation, Y.J. and L.W.; Writing—original draft, M.W.; Supervision, T.L. and Y.Y.; Project administration, L.W.; Funding acquisition, Y.Y. and H.W. All authors have read and agreed to the published version of the manuscript.

Funding: This work was supported by Hubei Province Natural Science Foundation of China (2022CFB429) and the Primary Research and Development Plan of Hubei Province (2022BBA0023).

Data Availability Statement: The datasets generated for this study are available on request to the corresponding author.

Conflicts of Interest: The authors declare no conflict of interest.

References

1. Raza, A.; Crothers, J.W.; McGill, M.M.; Mawe, G.M.; Teuscher, C.; Krementsov, D.N. Anti-inflammatory roles of p38alpha MAPK in macrophages are context dependent and require IL-10. *J. Leukoc. Biol.* **2017**, *102*, 1219–1227. [CrossRef]
2. Yang, H.; Liu, C.; Wu, Y.; Yuan, M.; Huang, J.; Xia, Y.; Ling, Q.; Hoffmann, P.R.; Huang, Z.; Chen, T. Atherosclerotic plaque-targeted nanotherapeutics ameliorates atherogenesis by blocking macrophage-driven inflammation. *Nano Today* **2022**, *42*, 101351. [CrossRef]
3. Inoue, M.; Sakamoto, K.; Suzuki, A.; Nakai, S.; Ando, A.; Shiraki, Y.; Nakahara, Y.; Omura, M.; Enomoto, A.; Nakase, I.; et al. Size and surface modification of silica nanoparticles affect the severity of lung toxicity by modulating endosomal ROS generation in macrophages. *Part. Fibre Toxicol.* **2021**, *18*, 21. [CrossRef] [PubMed]
4. Ley, K. Inflammation and Atherosclerosis. *Cells* **2021**, *10*, 1197. [CrossRef] [PubMed]
5. Liberale, L.; Badimon, L.; Montecucco, F.; Lüscher, T.F.; Libby, P.; Camici, G.G. Inflammation, Aging, and Cardiovascular Disease: JACC Review Topic of the Week. *J. Am. Coll. Cardiol.* **2022**, *79*, 837–847. [CrossRef]
6. Yuan, H.; Gong, Z.; Meng, S.; He, G. Hypoglycemic and hypolipidemic effects of a triterpenoid-rich extract from Euryale shell on streptozotocin-induced diabetic mice. *Die Pharmazie* **2013**, *68*, 227–231. [PubMed]
7. Zhang, W.; Su, R.-N.; Gong, L.-L.; Yang, W.-W.; Chen, J.; Yang, R.; Wang, Y.; Pan, W.-J.; Lu, Y.-M.; Chen, Y. Structural characterization and in vitro hypoglycemic activity of a glucan from *Euryale ferox* Salisb. seeds. *Carbohydr. Polym.* **2019**, *209*, 363–371. [CrossRef]
8. Wu, P.; Liu, A.; Li, L. Metabolomics and transcriptome analysis of the biosynthesis mechanism of flavonoids in the seeds of *Euryale ferox* Salisb at different developmental stages. *Mol. Genet. Genom.* **2021**, *296*, 953–970. [CrossRef]
9. Liu, X.; He, Z.; Yin, Y.; Xu, X.; Wu, W.; Li, L. Transcriptome sequencing and analysis during seed growth and development in *Euryale ferox* Salisb. *BMC Genom.* **2018**, *19*, 343. [CrossRef]
10. Yang, L.; Yin, P.; Fan, H.; Xue, Q.; Li, K.; Li, X.; Sun, L.; Liu, Y. Response Surface Methodology Optimization of Ultrasonic-Assisted Extraction of Acer Truncatum Leaves for Maximal Phenolic Yield and Antioxidant Activity. *Molecules* **2017**, *22*, 232. [CrossRef] [PubMed]
11. Wu, C.Y.; Chen, R.; Wang, X.S.; Shen, B.; Yue, W.; Wu, Q. Antioxidant and Anti-Fatigue Activities of Phenolic Extract from the Seed Coat of *Euryale ferox* Salisb. and Identification of Three Phenolic Compounds by LC-ESI-MS/MS. *Molecules* **2013**, *18*, 11003–11021. [CrossRef]
12. Zhao, L.; Zhang, S.-L.; Tao, J.-Y.; Pang, R.; Jin, F.; Guo, Y.-J.; Dong, J.-H.; Ye, P.; Zhao, H.-Y.; Zheng, G.-H. Preliminary exploration on anti-inflammatory mechanism of Corilagin (beta-1-O-galloyl-3,6-(R)-hexahydroxydiphenoyl-D-glucose) in vitro. *Int. Immunopharmacol.* **2008**, *8*, 1059–1064. [CrossRef]
13. Yeo, S.G.; Song, J.H.; Hong, E.-H.; Lee, B.-R.; Kwon, Y.S.; Chang, S.-Y.; Kim, S.H.; Lee, S.W.; Park, J.-H.; Ko, H.-J. Antiviral effects of Phyllanthus urinaria containing corilagin against human enterovirus 71 and Coxsackievirus A16 in vitro. *Arch. Pharmacal Res.* **2015**, *38*, 193–202. [CrossRef]
14. Reddy, B.U.; Mullick, R.; Kumar, A.; Sharma, G.; Bag, P.; Roy, C.L.; Sudha, G.; Tandon, H.; Dave, P.; Shukla, A.; et al. A natural small molecule inhibitor corilagin blocks HCV replication and modulates oxidative stress to reduce liver damage. *Antivir. Res.* **2018**, *150*, 47–59. [CrossRef] [PubMed]
15. Jia, L.; Jin, H.; Zhou, J.; Chen, L.; Lu, Y.; Ming, Y.; Yu, Y. A potential anti-tumor herbal medicine, Corilagin, inhibits ovarian cancer cell growth through blocking the TGF-beta signaling pathways. *BMC Complement Altern. Med.* **2013**, *13*, 33. [CrossRef] [PubMed]
16. Li, H.R.; Liu, J.; Zhang, S.-L.; Luo, T.; Wu, F.; Dong, J.-H.; Guo, Y.-J.; Zhao, L. Corilagin ameliorates the extreme inflammatory status in sepsis through TLR4 signaling pathways. *BMC Complement Altern. Med.* **2017**, *17*, 18. [CrossRef] [PubMed]
17. Tong, F.; Zhang, J.; Liu, L.; Gao, X.; Cai, Q.; Wei, C.; Dong, J.; Hu, Y.; Wu, G.; Dong, X. Corilagin Attenuates Radiation-Induced Brain Injury in Mice. *Mol. Neurobiol.* **2016**, *53*, 6982–6996. [CrossRef] [PubMed]
18. Asiimwe, N.; Yeo, S.G.; Kim, M.-S.; Jung, J.; Jeong, N.Y. Nitric Oxide: Exploring the Contextual Link with Alzheimer's Disease. *Oxidative Med. Cell. Longev.* **2016**, *2016*, 7205747. [CrossRef]
19. Kong, L.B.; Smith, W.; Hao, D. Overview of Raw264.7 for osteoclastogensis study: Phenotype and stimuli. *J. Cell. Mol. Med.* **2019**, *23*, 3077–3087. [CrossRef] [PubMed]
20. Predonzani, A. Spotlights on immunological effects of reactive nitrogen species: When inflammation says nitric oxide. *World J. Exp. Med.* **2015**, *5*, 64–76. [CrossRef] [PubMed]
21. Murakami, A.; Ohigashi, H. Targeting NOX, INOS and COX-2 in inflammatory cells: Chemoprevention using food phytochemicals. *Int. J. Cancer* **2007**, *121*, 2357–2363. [CrossRef] [PubMed]
22. Li, X.; Deng, Y.; Zheng, Z.; Huang, W.; Chen, L.; Tong, Q.; Ming, Y. Corilagin, a promising medicinal herbal agent. *Biomed. Pharmacother.* **2018**, *99*, 43–50. [CrossRef]
23. Paul, A.; Dutta, A.; Kundu, A.; Singh, S.B.; Banerjee, K.; Saha, S. Response surface methodology driven ultrasonic-assisted extraction of ellagitannins from pomegranate rind: Optimization of parameters and in silico molecular interaction with catalase. *Biomass Convers. Biorefinery* **2022**. [CrossRef]
24. Yang, L.J.; Chen, R.H.; Hamdoun, S.; Coghi, P.; Ng, J.P.; Zhang, D.W.; Guo, X.; Xia, C.; Law, B.Y.; Wong, V.K. Corilagin prevents SARS-CoV-2 infection by targeting RBD-ACE2 binding. *Phytomedicine* **2021**, *87*, 153591. [CrossRef] [PubMed]

25. Puljula, E.; Walton, G.; Woodward, M.J.; Karonen, M. Antimicrobial Activities of Ellagitannins against Clostridiales perfringens, Escherichia coli, Lactobacillus plantarum and Staphylococcus aureus. *Molecules* **2020**, *25*, 3714. [CrossRef]
26. Tao, Y.; Zhang, L.; Yang, R.; Yang, Y.; Jin, H.; Zhang, X.; Hu, Q.; He, B.; Shen, Z.; Chen, P. Corilagin ameliorates atherosclerosis by regulating MMP-1, -2, and -9 expression in vitro and in vivo. *Eur. J. Pharmacol.* **2021**, *906*, 174200. [CrossRef]
27. de Anda-Jauregui, G.; Guo, K.; McGregor, B.A.; Hur, J. Exploration of the Anti-Inflammatory Drug Space Through Network Pharmacology: Applications for Drug Repurposing. *Front. Physiol.* **2018**, *9*, 151. [CrossRef] [PubMed]
28. Peritore, A.F.; Impellizzeri, D.; Gugliandolo, E.; Siracusa, R.; Fusco, R.; D'Amico, R.; Cordaro, M.; Crupi, R.; Cuzzucrea, S.; Di Paola, R. Comicronized PEA and Rutin reduces inflammation and oxidative stress in a mouse model of vascular injury. *FASEB J.* **2020**, *34*, 1. [CrossRef]
29. Vasicek, O.; Lojek, A.; Ciz, M. Serotonin and its metabolites reduce oxidative stress in murine Raw264.7 macrophages and prevent inflammation. *J. Physiol. Biochem.* **2020**, *76*, 49–60. [CrossRef] [PubMed]
30. Liu, J.; Zong, Z.; Zhang, W.; Chen, Y.; Wang, X.; Shen, J.; Yang, C.; Liu, X.; Deng, H. Nicotinamide Mononucleotide Alleviates LPS-Induced Inflammation and Oxidative Stress via Decreasing COX-2 Expression in Macrophages. *Front. Mol. Biosci.* **2021**, *8*, 702107. [CrossRef]
31. Guan, C.; Xiao, Y.; Li, K.; Wang, T.; Liang, Y.; Liao, G. MMP-12 regulates proliferation of mouse macrophages via the ERK/P38 MAPK pathways during inflammation. *Exp. Cell Res.* **2019**, *378*, 182–190. [CrossRef] [PubMed]
32. Zhong, J.; Wang, F.; Wang, Z.; Shen, C.; Zheng, Y.; Ma, F.; Zhu, T.; Chen, L.; Tang, Q.; Zhu, J. Aloin attenuates cognitive impairment and inflammation induced by d-galactose via down-regulating ERK, p38 and NF-κB signaling pathway. *Int. Immunopharmacol.* **2019**, *72*, 48–54. [CrossRef] [PubMed]
33. Fang, L.; Wang, K.-K.; Huang, Q.; Cheng, F.; Huang, F.; Liu, W.-W. Nucleolin Mediates LPS-induced Expression of Inflammatory Mediators and Activation of Signaling Pathways. *Curr. Med. Sci.* **2020**, *40*, 646–653. [CrossRef] [PubMed]
34. Yuan, Z.; Matias, F.B.; Wu, J.; Liang, Z.; Sun, Z. Koumine Attenuates Lipopolysaccaride-Stimulated Inflammation in Raw264.7 Macrophages, Coincidentally Associated with Inhibition of NF-kappaB, ERK and p38 Pathways. *Int. J. Mol. Sci.* **2016**, *17*, 430. [CrossRef]

foods

MDPI

Article

Physicochemical and Functional Changes in Lotus Root Polysaccharide Associated with Noncovalent Binding of Polyphenols

Qiulan Liu [1], Xiaoqin Zou [1], Yang Yi [1,*], Ying Sun [2,*], Hongxun Wang [3], Xueyu Jiang [1] and Kaidi Peng [1]

[1] College of Food Science and Engineering, Wuhan Polytechnic University, Wuhan 430023, China
[2] Hubei Key Laboratory for Processing and Transformation of Agricultural Products, Wuhan Polytechnic University, Wuhan 430023, China
[3] College of Life Science and Technology, Wuhan Polytechnic University, Wuhan 430023, China
* Correspondence: yiy86@whpu.edu.cn (Y.Y.); sunying7535@163.com (Y.S.); Tel.: +86-138-8615-2207 (Y.Y.); +86-151-7150-7535 (Y.S.)

Abstract: To promote the functional applications of lotus root polysaccharides (LRPs), the effects of noncovalent polyphenol binding on their physicochemical properties, as well as antioxidant and immunomodulatory activities, were investigated. Ferulic acid (FA) and chlorogenic acid (CHA) were spontaneously bound to the LRP to prepare the complexes LRP-FA_1, LRP-FA_2, LRP-FA_3, LRP-CHA_1, LRP-CHA_2 and LRP-CHA_3, and their mass ratios of polyphenol to LRP were, respectively, 121.57, 61.18, 34.79, 2359.58, 1276.71 and 545.08 mg/g. Using the physical mixture of the LRP and polyphenols as a control, the noncovalent interaction between them in the complexes was confirmed by ultraviolet and Fourier-transform infrared spectroscopy. The interaction increased their average molecular weights by 1.11~2.27 times compared to the LRP. The polyphenols enhanced the antioxidant capacity and macrophage-stimulating activity of the LRP depending on their binding amount. Particularly, the DPPH radical scavenging activity and FRAP antioxidant ability were positively related to the FA binding amount but negatively related to the CHA binding amount. The NO production of the macrophages stimulated by the LRP was inhibited by the co-incubation with free polyphenols; however, the inhibition was eliminated by the noncovalent binding. The complexes could stimulate the NO production and tumor necrosis factor-α secretion more effectively than the LRP. The noncovalent binding of polyphenols may be an innovative strategy for the structural and functional modification of natural polysaccharides.

Keywords: lotus root; polysaccharide; polyphenol; noncovalent interaction; bioactivity

Citation: Liu, Q.; Zou, X.; Yi, Y.; Sun, Y.; Wang, H.; Jiang, X.; Peng, K. Physicochemical and Functional Changes in Lotus Root Polysaccharide Associated with Noncovalent Binding of Polyphenols. *Foods* **2023**, *12*, 1049. https://doi.org/10.3390/foods12051049

Academic Editor: Barry J. Parsons

Received: 18 January 2023
Revised: 19 February 2023
Accepted: 27 February 2023
Published: 1 March 2023

1. Introduction

A high intake of plant-based natural foods is associated with the prevention of various degenerative and chronic diseases, such as obesity, diabetes, coronary heart disease, etc., and polysaccharide and polyphenol are two important components contributing to the health benefits [1]. As biological macromolecules, polysaccharides play multiple roles in the life process. Those from natural products have attracted more and more attention due to their wide variety of pharmacological activities, non-toxicity and high stability. In particular, as an immunomodulator, polysaccharides can be recognized by various immune cell receptors to trigger the different signaling pathways of the immunological response [2]. Both researchers and consumers have shown increasing interest in polysaccharides and their products for disease alleviation and health benefits [3]. Meanwhile, polyphenols have a variety of physiological functions as natural antioxidants, such as ferulic acid (FA) and chlorogenic acid (CHA). They are widely applied in food, medicine and cosmetic fields due to their excellent performances in antioxidant, hypoglycemic, anti-inflammatory and other activities [4,5].

Polysaccharides and polyphenols generally coexist in vegetables, fruits, legumes, grains, tea and other plant foods. The tissue destruction of plant foods by processing may result in the mixture of polysaccharides and polyphenols, followed by the spontaneous noncovalent interaction between them. The formation of polysaccharide–polyphenol complexes can sequentially affect the physicochemical and nutritional properties of processed foods [6]. For example, the noncovalent binding of phenolic acids to a polysaccharide reduced their bioavailability in the small intestine, and thus they could be transported to the large intestine and suffer from fermentation and metabolism by intestinal bacteria [7]. This binding was also reported to significantly affect the processing, storage, texture and edible quality of foods [8–10]. Therefore, understanding the occurrence and influence of the noncovalent interaction between a polysaccharide and polyphenol will help to improve various qualities of processed plant foods and is of great significance for guiding food processing.

As an important aquatic vegetable in China, lotus root possesses the cultivation area of 4.0×10^5 hm^2 and the annual yield of 1.2×10^7 t [11]. It contains a variety of bioactive compounds such as polysaccharides and polyphenols, and their antioxidant, anti-inflammatory, immunomodulatory activities have attracted considerable research interest [12–14]. A previous study confirmed the noncovalent binding of the lotus root polysaccharide (LRP) to various polyphenols, which generate complexes with different contents of a polyphenol, as well as significant differences in the physicochemical and functional properties [15]. The noncovalent interaction principally induced by hydrogen bonding and the hydrophobic effect is closely related to the conditions of the temperature, pH, ionic strength and substrate concentration [6,16]. It allows for the accurate preparation of complexes with a desired mass ratio of polysaccharide to polyphenol, to investigate the effect of different binding degrees on the structural and functional properties of the complexes.

In this work, CHA and FA were used to prepare the LRP–polyphenols complexes by using the equilibrium dialysis method, which had certain mass ratios of the LRP to the polyphenol. Their structural features were analyzed by spectroscopic and chromatographic methods. Meanwhile, their antioxidant ability and macrophage-stimulating activity were evaluated in vitro. The physicochemical and functional effects of polyphenol binding on the LRP were further explored.

2. Materials and Methods

2.1. Materials and Chemicals

Fresh lotus roots (cultivar Elian No. 5), which were purchased from Wuhan Jinshui Qiliang Agricultural and Sideline Products Co., Ltd. (Wuhan, China), were manually washed and sliced. The slices (thickness 5 mm) were dried by 65 °C hot air to the moisture content of about 12%, followed by grinding with a multi-functional grinder (Suzhou Jiangdong Precision Instrument CO., Ltd., Suzhou, China). The powder samples were then stored in a desiccator at room temperature.

FA (≥95%) and CHA (≥95%) were purchased from Aladdin Biochemical Technology Co., Ltd. (Shanghai, China). Ferric reducing antioxidant power (FRAP) assay kit was purchased from Biyuntian Biotechnology Co., Ltd. (Shanghai, China). Folin-Ciocalteu reagent and other analytical reagents were purchased from Sinopharm Chemical Reagent Co., Ltd. (Shanghai, China). Lipopolysaccharide (LPS) was gained from Beijing Boaotuo Technology Co., Ltd. (Beijing, China). Dulbecco's modified eagle medium (DMEM) and phosphate-buffered saline (PBS, pH 7.4) were obtained from HyClone (Logan, UT, USA). A cell counting kit (CCK-8) was acquired from Langeco Technology Co., Ltd. (Beijing, China). The mouse TNF-α ELISA kit was supplied from Dakewe Bio-engineering Co., Ltd. (Shenzhen, China).

2.2. Extraction of LRP

LRP was extracted from lotus root powder using a previous method [15], with slight modifications. Briefly, the powder (100 g) was dispersed in 1000 mL distilled water at room temperature, accompanied with 400 W ultrasound treatment for 10 min. The mixture was then transferred to a 90 °C water bath. After 2 h hot-water extraction, the mixture was water cooled and added in absolute ethanol (final volume concentration 30%) to precipitate starches at 4 °C for 3 h. The supernatant was isolated by centrifugation (4500 r/min, 10 min) and added in absolute ethanol (final volume concentration 75%) to sequentially precipitate polysaccharides at 4 °C overnight. Finally, the precipitate obtained by centrifugation (4500 r/min, 10 min) was redissolved in 30 mL distilled water, and then vacuum freeze-dried to collect powdery LRP.

2.3. Composition Analysis of LRP

The contents of polysaccharide and protein in LRP were determined by the phenol-sulfuric acid method [17] and Coomassie blue assay kit, respectively. In addition, the monosaccharide composition of LRP was analyzed by a reversed-phase high-performance liquid chromatography method established by Yi et al. [14].

2.4. Preparation of LRP–Polyphenol Complexes

LRP–polyphenol complexes were prepared by an equilibrium dialysis method [15], with slight modifications. The aqueous solutions of LRP (2 mg/mL) and polyphenol (1 mg/mL) were mixed at certain conditions according to Table 1. After stirring at 120 r/min for 30 min, the mixture was transferred to a dialysis bag (MWCO 500 Da) and dialyzed against distilled water for 72 h to remove free polyphenols. Macromolecules remaining in the bag were vacuum freeze-dried and named as complex LRP-FA or LRP-CHA. In addition, LRP and polyphenols were directly mixed to obtain their physical mixtures, which were named as LRP&FA (61.18 mg FA/g LRP) and LRP&CHA (1276.71 mg CHA/g LRP).

Table 1. Preparation conditions of LRP–polyphenol complexes and their polyphenol binding amount.

Complexes	Preparation Conditions			Polyphenol Binding Amount (mg/g LRP)
	Temperature (°C)	pH	Mass Ratio of LRP to Polyphenol	
LRP-FA_1	0	6	2:1	121.57
LRP-FA_2	15	5	2:1	61.18
LRP-FA_3	30	5	2:1	34.79
LRP-CHA_1	15	5	1:4	2359.58
LRP-CHA_2	15	5	1:2	1276.71
LRP-CHA_3	15	6	1:1	545.08

2.5. Spectral Analyses

Powdery sample was dissolved in distilled water and diluted to the concentration of 50 μg/mL. The sample solution was then scanned in the range of 190~400 nm using an ultraviolet (UV) and visible spectrophotometer (TU-1810, Puxi, Shanghai, China), with the baseline established by distilled water.

Sample (1 mg) mixed with dried KBr (200 mg) was ground and pressed into tables for Fourier-transform infrared spectroscopy (FTIR) analysis. The analysis was performed with an FTIR spectrophotometer (Nexus 5DXC FT-IR, Thermo Nicolet, Madison, WI, USA), using the frequency range of 4000~400 cm^{-1}, the scanning number of 32 and the resolution of 4 cm^{-1} [18].

2.6. Measurement of Molecular Weight Distribution

The high-performance size-exclusion chromatography coupled with multi-angle laser light scattering and refractive index detection (HPSEC-MALLS-RI) was used to analyze the molecular weight (Mw) distribution of samples according to the method of Yi et al. [19].

2.7. Evaluation of Antioxidant Capacity

The 1,1-diphenyl-2-picrylhydrazyl (DPPH) radical scavenging activity of samples was assessed using the method of Han et al. [20], with slight modifications. The aqueous solution of sample (1 mL, 0.1~0.8 mg/mL) was mixed with the alcohol solution of DPPH (4.0 mL, 0.2 mmol/L), followed by reaction in the dark for 30 min at room temperature. The mixture of 1 mL distilled water and 4.0 mL DPPH solution was used as the control. The absorbance of reaction mixture was finally measured at 517 nm. The DPPH radical scavenging rate (S, %) of sample solution was calculated as its decrease percentage in absorbance relative to the control. Meanwhile, the FRAP capacity of samples was evaluated using a FRAP assay kit according to its instruction, expressed as millimoles of Fe^{2+} equivalents per 1 g of sample.

2.8. Evaluation of Macrophage-Stimulating Activity

The effect of samples on the cell viability of RAW264.7 macrophages was evaluated using a CCK-8 kit according to its instruction. In brief, cells were harvested in the logarithmic phase of growth and dispersed in complete culture medium (DMEM containing 10% fetal bovine serum) to a cell density of 2×10^5 cells/mL. Then, 200 μL/well of cell suspension was plated in 96-well culture plates for 24 h incubation at 37 °C in a humidified 5% CO_2 incubator (MCO-17 AIC, SANYO, Tokyo, Japan). The adherent cells were washed twice with PBS after removal of culture supernatant, followed by 24 h incubation in 200 μL medium containing samples (50, 100, 200 or 400 μg/mL, four replicates for each concentration). In addition, the cells incubated without sample stimulation were used as control, and the wells containing 200 μL culture medium only were set as blank. After 2 h reaction with CCK-8 solution (20 μL/well), the optical density (OD) of each well was measured at 450 nm on a microplate reader (Westchemy Technology Co., Ltd., Beijing, China). The cell viability (%) was calculated as follows: $(OD_s - OD_b)/(OD_c - OD_b) \times 100\%$, where OD_s, OD_c, OD_b are the OD values of sample, control and blank, respectively.

Moreover, the effects of samples on the nitrogen oxide (NO) production and tumor necrosis factor-alpha (TNF-α) secretion of macrophages were evaluated using the method of Yi et al. [14]. Their test concentrations were 50, 100 and 200 μg/mL, while 500 ng/mL LPS was used as a positive control. After incubation with the stimulants for 24 or 48 h, the NO production in culture supernatant was analyzed by the Griess method and expressed as sodium nitrite equivalents (μmol/L) [14]. Meanwhile, the TNF-α concentration (pg/mL) was determined using a mouse TNF-α ELISA kit according to its instruction.

2.9. Statistical Analysis

Experimental data were expressed as "mean ± standard deviation". Duncan method was adopted to analyze the significant differences between groups at the 0.05 level by SPSS19.0 software (Chicago, IL, USA).

3. Results and Discussion

3.1. Basic Composition of LRP

It has been reported that polysaccharides from various cultivars and parts of lotus root differed significantly in content, composition and structure [21]. Of the 13 cultivars, Elian No. 5, which showed a relatively high polysaccharide content, was selected to prepare the LRP. The polysaccharide content and protein content of the LRP were 83.27% and 1.46%, respectively. The polysaccharide content was obviously higher than the previous result of 71.31% [21], which might be related to the preparation difference for the polysaccharides. It was suggested that the method of two-step alcohol precipitation in the present work was more effective for isolating the LRP with a higher purity, as compared to the traditional combination of amylase-based enzymolysis, the Sevage method and 75% alcohol precipitation. The LRP was composed of rhamnose, glucose, galactose and arabinose at the molar ratio of 0.12:7.63:1.30:0.95. This composition was approximately in accord with the previous results [17,21].

3.2. Factors Influencing the Formation of LRP–Polyphenol Complexes

LRP–polyphenol complexes were designed by controlling the conditions of the temperature, pH and polysaccharide-polyphenol mass ratio for a noncovalent interaction. The polyphenol binding amount (mg/g LRP) in the serial complexes of both the LRP-FA and LRP-CHA showed an approximate ratio of 4:2:1 (Table 1). Compared to the FA, the CHA showed a stronger affinity to the LRP, which might be closely related to its galloyl group [22]. Tang et al. [22] indicated that the galloyl group was the major functional group in the cellulose–tannin interactions. Its three hydroxyl groups potentially contributed to the hydrogen bonding, and the aromatic moiety of the tannin was important for the hydrophobic interactions. At the same time, because the formation of a hydrogen bond is an exothermic process, the decrease in the binding amount caused by a temperature rise might be due to the weakening of the hydrogen bond [23,24]. In addition, when the temperature increased, the hydrophobic interaction between the benzene rings was conducive to the formation of FA polymers [25], which may hinder the noncovalent binding between the FA and LRP. The mixing ratio of polysaccharides and polyphenols is also an important factor affecting their interaction [26]. With the increase in the mass ratio of CHA to LRP in the mixed system, the availability of the active sites of LRP molecules was conducive to increasing the binding amount [27]. Moreover, the change in the pH would lead to the change in the potential on the molecular surface of polysaccharides and polyphenols, affecting the electrostatic interaction between them, and thus also affecting the noncovalent binding of them [28].

3.3. Spectral Features of LRP–Polyphenol Complexes

The characteristic UV peaks of polyphenols mainly exist in the wavelength range of 240~380 nm, while most polysaccharides have no absorption in this range [29]. Therefore, it is possible to confirm the successful binding by comparing the UV spectra of polyphenols, polysaccharides and their corresponding complexes. FA was successfully adsorbed by arabinan-rich pectic polysaccharide in Zhang's study, which was determined via UV spectra [25]. The formation of noncovalent complexes between them can result in the weakness or disappearance of the UV characteristics belonging to the polyphenols [30]. As shown in Figure 1, the LRP had a strong absorption only in the range of 190~220 nm, which was related to its unsaturated carbonyl and carboxyl groups [25]. The characteristic peaks of the FA and CHA at 312, 288 and 218 nm could be assigned to the π-π^* transition of the double bond, phenol group and phenyl ring, respectively [19,25]. They were found in the spectra of the LRP&FA and LRP&CHA mixtures but were obviously weakened (even disappeared) in the spectra of the LRP-FA and LRP-CHA complexes. Thus, the results implied that the interaction of the LRP and polyphenols exists, which obviously altered the ultraviolet properties of the FA and CHA. The results confirmed the intermolecular interaction between the LRP and polyphenols in the complexes.

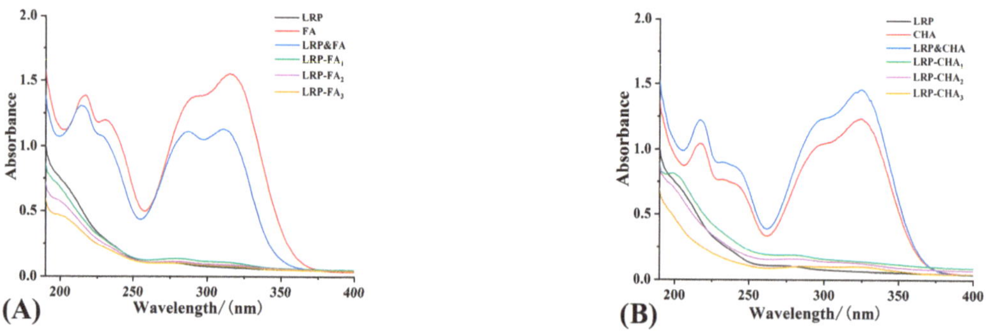

Figure 1. UV spectra of LRP-FA complexes (**A**) and LRP-CHA complexes (**B**).

FTIR spectroscopy can provide structural information related to polysaccharide–phenol interactions [16]. As seen in Figure 2, the typical polysaccharide peaks of the LRP were confirmed [31], which included the stretching vibration (SV) of O–H at 3382 cm^{-1}, the SV of C–H at 2931 cm^{-1}, the SV of C=O at 1618 cm^{-1} and the SV of C–O at 1410 cm^{-1}. Among them, the broad peak at 2931 cm^{-1} was attributed to the C–H stretching vibration of the LRP, which was also observed in the FTIR spectrum of the LRP–polyphenol complexes and mixtures. In addition, the SV of O–H in the LRP–polyphenol complex spectrum (at 3417 cm^{-1} and 3416 cm^{-1} for LRP-FA$_2$ and LRP-CHA$_2$, respectively) were broader than that in the LRP spectrum. It has been reported that the O–H stretching band was a sensitive indicator of the strength of the hydrogen bond, the O–H stretching band also shows a drastic change in its wave number and intensity depending on the formation of the hydrogen bonds and the broadening of the O–H stretching band was often observed in strong hydrogen bonds [32]. Here, it was suggested that the hydrogen bond was formed between the LRP and FA/CHA.

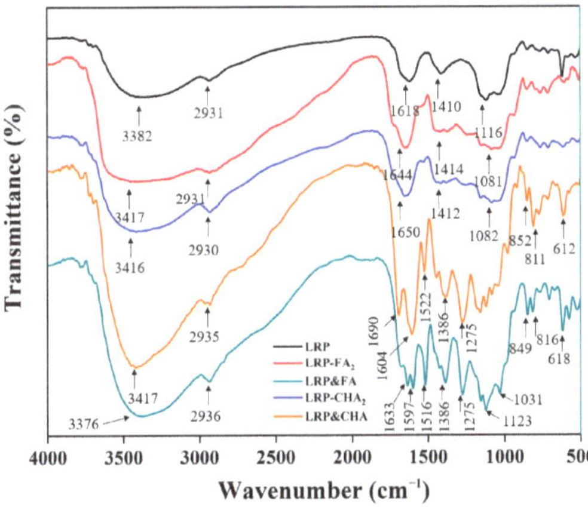

Figure 2. FTIR spectra of LRP and LRP–polyphenol complexes.

In addition, the LRP&FA and LRP&CHA showed the characteristic bands of the aromatic nucleus in the wavenumber range of 1630~1516 cm^{-1}, which were assigned to the SV of the aromatic ring C=C [33]. The LRP&FA and LRP&CHA mixtures spectrum had characteristic peaks at 1604 cm^{-1} and 1522 cm^{-1} and at 1633 cm^{-1}, 1597 cm^{-1} and 1516 cm^{-1}, respectively, which were weakened or even disappeared in the corresponding complex. Moreover, the bands at 1275 cm^{-1} and 1031 cm^{-1} were both related to the C−O−C vibration, and the bands near 1123 cm^{-1} and 618 (or 612) cm^{-1}, respectively, belonged to the bending stretching (BS) of the phenolic hydroxyls [34] and the C–H bonds of the out-of-plane flexion in the aromatic compounds [19]. By comparing with the mixtures, the LRP–polyphenol complexes showed significantly attenuated bands at 1123 cm^{-1} and 617 cm^{-1}. Moreover, the absorption intensity of the bands at 852 cm^{-1}, 811 cm^{-1}, 849 cm^{-1} and 816 cm^{-1} was decreased. The attenuated signals of the phenolic hydroxyls and rings indicated the noncovalent interaction between the LRP and polyphenols, probably via the hydrogen bonding and hydrophobic interaction.

3.4. Molecular Weight Distribution of LRP–Polyphenol Complexes

For the LRP, the increase in the Mw is one of the typical features associated with the noncovalent binding of polyphenols [15]. The effect of the polyphenol-binding ratio on the Mw distribution of the LRP–polyphenol complexes was determined by HPSEC-MALLS-RI. The chromatograms are shown in Figure S1, and the detailed information is collected in Table 2. The LRP consisted of seven fractions, with an average Mw of 146.3 kDa, of which the faction with the highest Mw of 1144 kDa accounted for 8.4% of the total peak area; by comparison, the LRP-FA and LRP-CHA complexes showed obviously higher values of the average Mw. Of all the fractions, the one with the highest Mw (1627 kDa, 14.1%) was found in LRP-FA_3. The LRP contained 52.5% fractions within the low Mw range of $3.0 \leq$ lg Mw < 4.0 (Figure 3). The low Mw fractions in the LRP-CHA_1, LRP-CHA_2 and LRP-CHA_3, respectively, accounted for 4.7%, 2.8% and 5.6%, while those in the LRP-FA complexes were not detected. Moreover, the relative content of the fractions within the range of $4.0 \leq$ lg Mw < 4.5 and $5.0 \leq$ lg Mw < 5.5 significantly increased in all the complexes. No fraction below 1 kDa was found in the complexes, indicating that they did not contain free FA and CHA.

Table 2. Molecular weight distribution of LRP and LRP–polyphenol complexes.

Samples	Retention Time (min)	Molecular Weight (Da)	Peak Area Percentage (%)	Average Molecular Weight (Da)
LRP	7.297–9.107	1.144×10^6 (±0.791%)	8.4	1.463×10^5
	9.077–11.490	1.359×10^5 (±1.050%)	33.1	
	11.490–12.184	2.628×10^4 (±4.650%)	6.3	
	12.184–13.934	9.036×10^3 (±3.661%)	20.6	
	13.934–14.718	3.695×10^3 (±5.083%)	12.7	
	14.748–18.398	6.798×10^3 (±4.744%)	18.3	
	21.868–24.402	2.494×10^3 (±51.551%)	0.9	
LRP-FA_1	7.026–8.926	1.465×10^6 (±0.723%)	13.5	2.978×10^5
	8.926–11.400	1.730×10^5 (±0.858%)	50.3	
	11.400–12.637	2.898×10^4 (±4.228%)	19.5	
	12.637–13.994	4.509×10^4 (±4.033%)	7.5	
	13.994–15.291	3.066×10^4 (±3.972%)	7.2	
	15.291–15.985	7.438×10^4 (±4.292%)	1.2	
	17.373–19.002	1.283×10^5 (±3.532%)	0.7	
LRP-FA_2	7.237–8.896	1.603×10^6 (±0.721%)	13.1	3.102×10^5
	8.896–11.340	1.820×10^5 (±0.824%)	48.7	
	11.340–12.516	2.681×10^4 (±3.786%)	19.3	
	12.516–13.873	3.354×10^4 (±4.439%)	8.8	
	13.873–15.261	2.407×10^4 (±3.966%)	8.1	
	15.261–16.076	5.990×10^4 (±3.650%)	1.4	
	17.312–18.821	1.179×10^5 (±4.124%)	0.6	
LRP-FA_3	7.297–8.926	1.627×10^6 (±0.688%)	13.1	3.321×10^5
	8.926–11.309	1.894×10^5 (±0.735%)	48.7	
	11.309–12.516	2.956×10^4 (±2.849%)	19.3	
	12.516–13.904	3.674×10^4(±2.926%)	8.8	
	13.904–15.261	2.723×10^4 (±2.523%)	8.1	
	15.261–16.226	7.541×10^4 (±2.588%)	1.4	
	17.343–18.549	2.141×10^5 (±2.951%)	0.6	
LRP-CHA_1	7.207–8.956	1.157×10^6 (±0.660%)	7.8	1.621×10^5
	8.956–11.370	1.365×10^5 (±0.960%)	47.2	
	11.400–12.576	1.840×10^4 (±5.402%)	19.8	
	12.576–13.934	2.211×10^4 (±5.505%)	8.2	
	13.934–15.020	1.253×10^4 (±5.063%)	7.5	
	15.020–16.317	1.676×10^4 (±4.756%)	4.8	
	17.011–18.700	4.540×10^3 (±23.036%)	4.7	
LRP-CHA_2	7.177–8.931	1.184×10^6 (±0.671%)	9.2	1.913×10^5
	8.931–11.392	1.424×10^5 (±0.955%)	52.4	
	11.392–12.561	2.015×10^4 (±5.290%)	20.8	
	12.561–13.953	2.674×10^4 (±4.907%)	7.8	
	13.953–15.485	1.875×10^4 (±3.172%)	7.0	
	16.837–18.773	7.451×10^3 (±25.715%)	2.8	
	7.177–8.931	1.184×10^6 (±0.671%)	9.2	
LRP-CHA_3	7.237–8.866	1.545×10^6 (±0.691%)	9.4	2.433×10^5
	8.866–11.340	1.724×10^5 (±0.791%)	51.2	
	11.340–12.546	2.422×10^4 (±4.184%)	20.4	
	12.546–13.964	3.613×10^4 (±4.401%)	7.5	
	13.964–15.291	2.728×10^4 (±3.526%)	6.0	
	16.528–19.122	8.365×10^3 (±22.813%)	5.6	

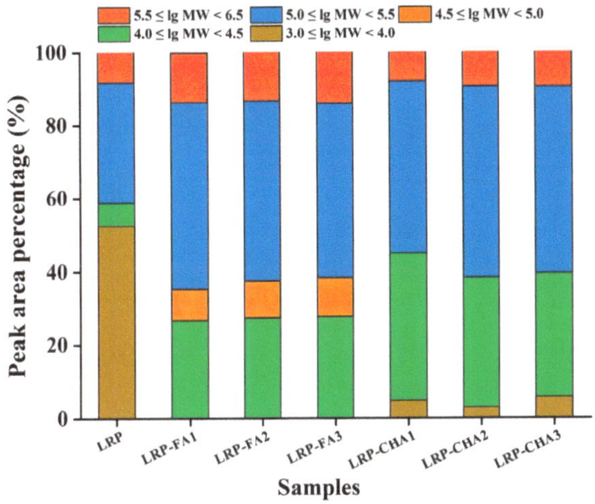

Figure 3. Comparison of the molecular weight distribution of LRP and LRP–polyphenol complexes.

Compared to the CHA, the FA with a lower affinity for the LRP could form complexes with higher Mws. Moreover, the increase in the polyphenol binding reduced the Mw of both the LRP-CHA and LRP-FA complexes. It was suggested that more polyphenols combined to the single molecule of the LRP might form the cluster structure with a relatively strong effect of the steric hindrance, inhibiting the unutilized active groups of polyphenols noncovalently bound to other polysaccharide molecules.

3.5. Antioxidant Activity of LRP–Polyphenol Complexes

Previous studies indicated that the antioxidant activity of the LRP was relatively weak. Its IC_{50} value of DPPH radical scavenging was 1.65 mg/mL [17]. As natural antioxidants, phenolic compounds still retained their antioxidant activity when combined with soluble dietary fiber [35], and the antioxidant activity of their complex might be positively correlated with the phenolic binding amount [15]. As shown in Figure 4A,B, the noncovalent binding of polyphenols could effectively enhance the DPPH radical scavenging activity of the LRP. The scavenging capacity of the LRP-FA was significantly higher than that of the LRP ($p < 0.05$) and was positive to the FA binding amount. It might be attributed to the increased number of active hydroxyl groups, providing more protons to block the free-radical reaction [36]. Alejandra et al. also obtained similar results that a phenolic antioxidant (3,4-Dihydroxyphenylglycol, DHPG) linked to soluble dietary fiber maintains some antioxidant activity, and it was explained that the additional -OH group of the DHPG provided the availability of the catechol group and contributed to the antioxidant activity of the soluble fiber–DHPG complex. By comparison, the LRP-CHA complexes with higher polyphenol proportions did not show a prominent advantage in the DPPH radical scavenging. Moreover, their scavenging ability decreased with the increase in the CHA binding amount. The results implied that the active hydroxyl groups of the CHA in the complexes were significantly consumed due to the formation of intermolecular hydrogen bonds or inhibited by a steric hindrance. In addition, the LRP–polyphenol complexes showed a lower scavenging capacity than their physical mixtures, due to the reduction in active hydroxyl groups by the hydrogen bond formation [37]. The DPPH radical scavenging IC_{50} values of the LRP, LRP-FA_1, LRP-CHA_3, LRP&FA and LRP&CHA were 1.04, 0.47, 0.58, 0.17 and 0.10 mg/mL, respectively.

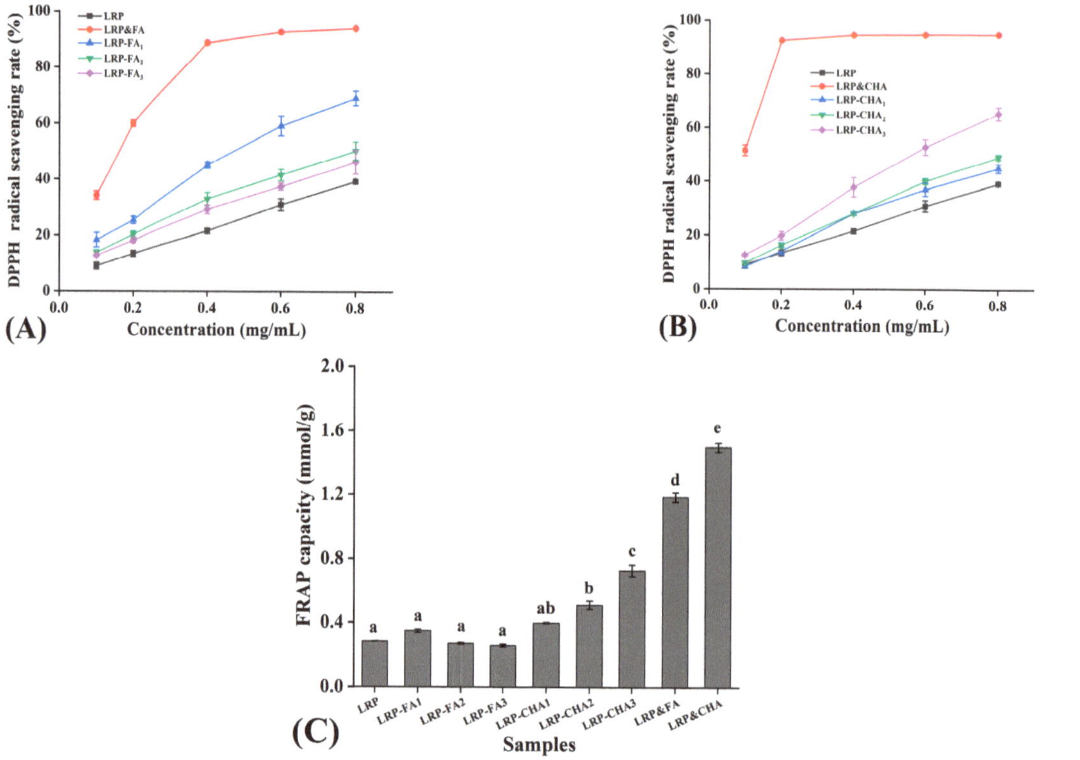

Figure 4. Antioxidant activity of LRP and LRP–polyphenol complexes. (**A**,**B**) are the DPPH radical scavenging activity, and (**C**) is the FRAP total antioxidant capacity. (a–e indicates that there are significant differences between different letters, $p < 0.05$).

The FRAP capacities of the LRP and LRP–polyphenol complexes were also evaluated, as seen in Figure 4C. The FRAP value of the LRP was 0.29 mmol/g, which showed no significant difference with the LRP-FA complexes ($p > 0.05$). In comparison to them, the LRP-CHA_2 and LRP-CHA_3 possessed stronger FRAP capacities ($p < 0.05$), and the capacity showed a negative correlation with the CHA binding amount which was similar to the DPPH radical scavenging activity. In addition, the physical mixtures of the LRP and polyphenols exhibited higher FRAP values than their corresponding complexes ($p < 0.05$), indicating that the noncovalent interaction significantly decreased their FRAP capacity.

3.6. Macrophage-Stimulating Activity of LRP–Polyphenol Complexes

Macrophages have a unique role in the immune system because they not only elicit an innate immune response but are also effector cells in inflammation and infection [38]. After incubation with the LRP or LRP–polyphenol complexes (50~400 μg/mL) for 24 h, the proliferation rates of the macrophages indicated that they had no cytotoxicity. Their immunostimulatory effects on the cells were evaluated in the concentration ranges of 50~200 μg/mL.

Macrophages kill pathogens not only directly through phagocytosis but also indirectly through the secretion of pro-inflammatory cytokines. When macrophages are activated, they release NO to kill microorganisms, parasites and tumor cells while inducing an inflammatory response to protect the body [39]. As seen in Figure 5A,B, the LRP and LRP–polyphenol complexes all stimulated NO production as compared to the control ($p < 0.05$), and the 48 h incubation had almost double the NO production of the 24 incubation. The FA and CHA had relatively weak effects on the NO production, and they both showed inhibitory effects on the immunostimulation of the LRP. However, the inhibition disappeared after the noncovalent binding of polyphenols to the LRP.

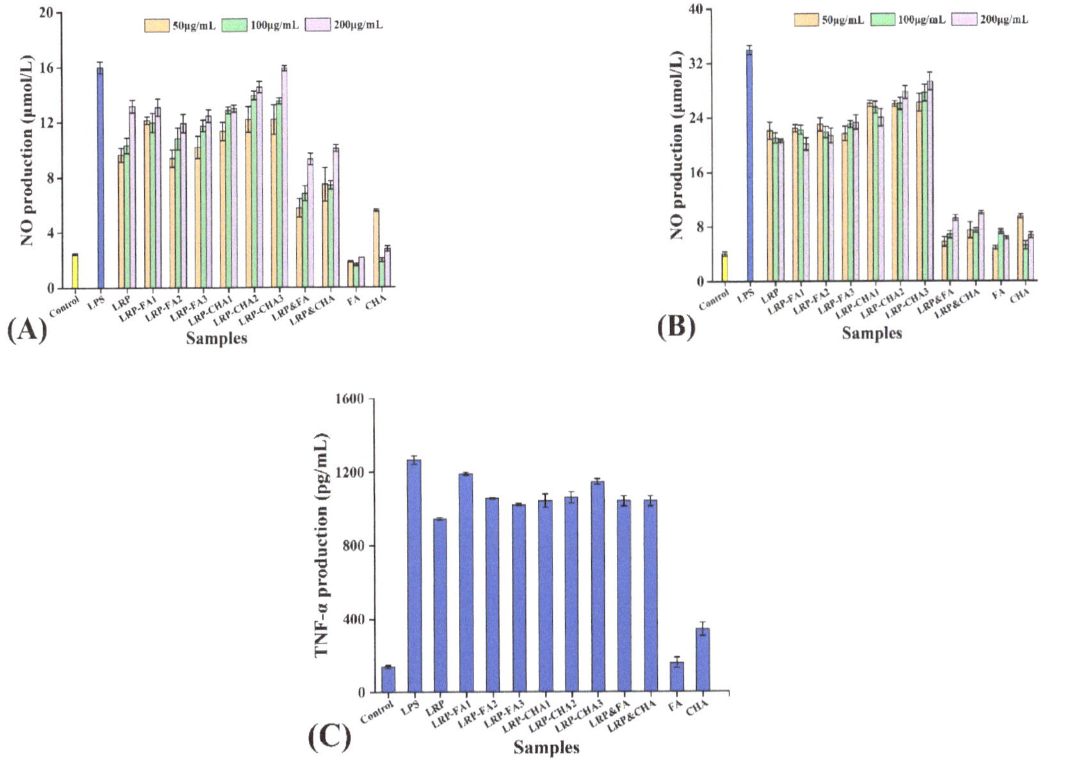

Figure 5. Effect of LRP and LRP–polyphenol complexes on macrophage activation. (**A**,**B**) are the NO production after 24 h and 48 incubations, respectively. (**C**) is the TNF-α secretion after 24 h incubation.

TNF-α is an immunoactive factor responsible for regulating the response of inflammation and the host defense, and it plays a role in the removal of pathogens and the promotion of wound healing [40]. As seen in Figure 5C, the LRP could also stimulate TNF-α secretion as compared to the control ($p < 0.05$), but the effect was significantly weaker than those of the LRP–polyphenol complexes ($p < 0.05$). Overall, the LRP-FA_1 and LRP-CHA_3 exhibited relatively strong immunostimulation on the macrophages. The difference may be attributable to discrepancies in the active compounds (polyphenol) [41].

4. Conclusions

LRPs, as the main bioactive macromolecules in lotus root, can spontaneously combine with FA/CHA through a noncovalent interaction. The resulting complexes showed significant differences in the structural and functional features as compared to the LRP. The average Mw of the LRP–polyphenol complexes was 1.11~2.27 times higher than that of the LRP. The LRP-FA/CHA complexes and their mixtures had a significantly stronger DPPH radical scavenging activity and FRAP capacity than the LRP, which was due to the increased number of active hydroxyl groups of the FA/CHA, and the effects were closely related to the structure and binding amount of the polyphenols. However, the LRP–polyphenol complexes showed lower antioxidant activity than their mixtures, due to the reduction in active hydroxyl groups by the hydrogen bond formation. Moreover, the binding had positive effects on the macrophage activation, by stimulating the production of NO and TNF-α. Free polyphenols could inhibit the LRP-induced NO production, but this inhibitory effect disappeared after the formation of the noncovalent binding. This work provides a promising approach to improve the functional activity of polysaccharides and the bioavailability of polyphenols by modulating the noncovalent interactions between them. However, endogenous and exogenous factors affecting the bioactivity and bioavailability of polysaccharide–polyphenol complexes, as well as the structure–activity relationship, need to be additionally explored.

Supplementary Materials: The following supporting information can be downloaded at: https://www.mdpi.com/article/10.3390/foods12051049/s1, Figure S1: HPSEC-MALLS-RI chromatogram of LRP, and polyphenol complex.

Author Contributions: Conceptualization, Q.L., Y.Y. and H.W.; Methodology, Q.L., X.Z., Y.S. and X.J.; Data curation, X.Z. and Y.S.; Writing—original draft, Q.L.; Writing—review & editing, Q.L. and Y.Y.; Supervision, X.Z., Y.Y., Y.S., H.W., X.J. and K.P.; Project administration, K.P.; Funding acquisition, Y.Y. All authors have read and agreed to the published version of the manuscript.

Funding: This work was supported by the Science and Technology Department of Hubei Province [grant number 2022BBA0023], the Young Top-notch Talent Cultivation Program of Hubei Province, and the Wuhan Science and Technology Bureau [grant number 2022020801010389].

Data Availability Statement: Data is contained within the article.

Conflicts of Interest: The authors declare no conflict of interest.

References

1. Zhu, F. Interactions between cell wall polysaccharides and polyphenols. *Crit. Rev. Food Sci. Nutr.* **2018**, *58*, 1808–1831. [CrossRef]
2. Kojima, H.; Akaki, J.; Nakajima, S.; Kamei, K.; Tamesada, M. Structural analysis of glycogen-like polysaccharides having macrophage-activating activity in extracts of Lentinula edodes mycelia. *J. Nat. Med.* **2010**, *64*, 16–23. [CrossRef]
3. Yan, Y.Y.; Yuan, S.; Zhao, S.; Xu, C.Y.; Zhang, X.F. Preparation and application of phosphorylated Lotus root polysaccharide. *Food Sci. Technol.* **2021**, *42*. [CrossRef]
4. Li, D.; Rui, Y.X.; Guo, S.D.; Luan, F.; Liu, R.; Zeng, N. Ferulic acid: A review of its pharmacology, pharmacokinetics and derivatives. *Life Sci.* **2021**, *284*, 119921. [CrossRef] [PubMed]
5. Mattila, P.; Hellström, J. Phenolic acids in potatoes, vegetables, and some of their products. *J. Food Compos. Anal.* **2007**, *20*, 152–160. [CrossRef]
6. Renard, C.M.; Watrelot, A.A.; Le Bourvellec, C. Interactions between polyphenols and polysaccharides: Mechanisms and consequences in food processing and digestion. *Trends Food Sci. Technol.* **2017**, *60*, 43–51. [CrossRef]
7. Padayachee, A.; Netzel, G.; Netzel, M.; Day, L.; Zabaras, D.; Mikkelsen, D.; Gidley, M.J. Binding of polyphenols to plant cell wall analogues—Part 2: Phenolic acids. *Food Chem.* **2012**, *135*, 2287–2292. [CrossRef]
8. Le Bourvellec, C.; Le Quere, J.M.; Renard, C.M. Impact of noncovalent interactions between apple condensed tannins and cell walls on their transfer from fruit to juice: Studies in model suspensions and application. *J. Agric. Food Chem.* **2007**, *55*, 7896–7904. [CrossRef]
9. Troszyńska, A.; Narolewska, O.; Robredo, S.; Estrella, I.; Hernández, T.; Lamparski, G.; Amarowicz, R. The effect of polysaccharides on the astringency induced by phenolic compounds. *Food Qual. Prefer.* **2010**, *21*, 463–469. [CrossRef]
10. Liu, J.; Chen, F.; Tian, W.; Ma, Y.; Li, J.; Zhao, G. Optimization and characterization of curcumin loaded in octenylsuccinate oat beta-glucan micelles with an emphasis on degree of substitution and molecular weight. *J. Agric. Food Chem.* **2014**, *62*, 7532–7540. [CrossRef]
11. Wu, M.; Zong, Y.; Zhao, B.; Zhu, H. Development status, problems and development ideas of aquatic vegetables industry in China. *J. Changjiang Veg.* **2019**, *2*, 35–41.
12. Yi, Y.; Sun, J.; Xie, J.; Min, T.; Wang, L.M.; Wang, H.X. Phenolic Profiles and Antioxidant Activity of Lotus Root Varieties. *Molecules* **2016**, *21*, 863. [CrossRef] [PubMed]
13. Hu, W.; Jiang, Y.; Xue, Q.; Sun, F.; Zhang, J.; Zhou, J.; Niu, Z.; Li, Q.; Li, F.; Shen, T. Structural characterisation and immunomodulatory activity of a polysaccharide isolated from lotus (*Nelumbo nucifera* Gaertn.) root residues. *J. Funct. Foods* **2019**, *60*, 103457. [CrossRef]
14. Yi, Y.; Lamikanra, O.; Sun, J.; Wang, L.M.; Min, T.; Wang, H.X. Activity diversity structure-activity relationship of polysaccharides from lotus root varieties. *Carbohydr. Polym.* **2018**, *190*, 67–76. [CrossRef] [PubMed]
15. Yi, Y.; Tang, H.S.; Sun, Y.; Xu, W.; Min, T.; Wang, H.X. Comprehensive characterization of lotus root polysaccharide-phenol complexes. *Food Chem.* **2022**, *366*, 130693. [CrossRef]
16. Liu, X.; Bourvellec, C.L.; Renard, C. Interactions between cell wall polysaccharides and polyphenols: Effect of molecular internal structure. *Compr. Rev. Food Sci. Food Saf.* **2020**, *19*, 3574–3617. [CrossRef]

17. DuBois, M.; Gilles, K.A.; Hamilton, J.K.; Rebers, P.T.; Smith, F. Colorimetric Method for Determination of Sugars and Related Substances. *Anal. Chem.* **1956**, *28*, 350–356. [CrossRef]
18. Bermúdez-Oria, A.; Rodríguez-Gutiérrez, G.; Fernández-Prior, Á.; Knicker, H.; Fernández-Bolaños, J. Confirmation by solid-state NMR spectroscopy of a strong complex phenol-dietary fiber with retention of antioxidant activity in vitro. *Food Hydrocoll.* **2020**, *102*, 105584. [CrossRef]
19. Yi, Y.; Huang, X.Y.; Zhong, Z.T.; Huang, F.; Li, S.Y.; Wang, L.M.; Min, T.; Wang, H.X. Structural and biological properties of polysaccharides from lotus root. *Int. J. Biol. Macromol.* **2019**, *130*, 454–461. [CrossRef]
20. Han, M.M.; Yi, Y.; Wang, H.X.; Huang, F. Investigation of the Maillard Reaction between Polysaccharides and Proteins from Longan Pulp and the Improvement in Activities. *Molecules* **2017**, *22*, 938. [CrossRef]
21. Wang, H.X.; Yi, Y.; Sun, J.; Lamikanra, O.; Min, T. Fingerprint profiling of polysaccharides from different parts of lotus root varieties. *RSC Adv.* **2018**, *8*, 16574–16584. [CrossRef]
22. Tang, H.R.; Covington, A.D.; Hancock, R.A. Structure-activity relationships in the hydrophobic interactions of polyphenols with cellulose and collagen. *Biopolymers* **2003**, *70*, 403–413. [CrossRef]
23. Wu, Z.; Li, H.; Ming, J.; Zhao, G. Optimization of adsorption of tea polyphenols into oat β-glucan using response surface methodology. *J. Agric. Food Chem.* **2011**, *59*, 378–385. [CrossRef]
24. Jakobek, L.; Matic, P. Non-covalent dietary fiber—Polyphenol interactions and their influence on polyphenol bioaccessibility. *Trends Food Sci. Technol.* **2019**, *83*, 235–247. [CrossRef]
25. Zhang, D.; Zhu, J.; Ye, F.; Zhao, G. Non-covalent interaction between ferulic acid and arabinan-rich pectic polysaccharide from rapeseed meal. *Int. J. Biol. Macromol.* **2017**, *103*, 307–315. [CrossRef] [PubMed]
26. Wang, L.; Li, F.; Yang, Y. Interaction mechanism between polyphenols and polysaccharides and effect on polyphenolic properties: A review. *Food Sci.* **2017**, *38*, 276–282.
27. Zou, X.; Liu, Q.; Huang, F.; Xiao, J.; Yi, Y. Preparation and Characterization of Lotus Root Polysaccharide-Chlorogenic Acid Complex. *Mod. Food Sci. Technol.* **2023**, *39*, 1–11.
28. Le Bourvellec, C.; Renard, C.M. Non-covalent interaction between procyanidins and apple cell wall material. Part II: Quantification and impact of cell wall drying. *Biochim. Biophys. Acta* **2005**, *1725*, 1–9. [CrossRef] [PubMed]
29. Liu, J.; Wang, X.; Yong, H.; Kan, J.; Jin, C. Recent advances in flavonoid-grafted polysaccharides: Synthesis, structural characterization, bioactivities and potential applications. *Int. J. Biol. Macromol.* **2018**, *116*, 1011–1025. [CrossRef]
30. Li, S.; Li, J.; Zhu, Z.; Cheng, S.; He, J.; Lamikanra, O. Soluble dietary fiber and polyphenol complex in lotus root: Preparation, interaction and identification. *Food Chem.* **2020**, *314*, 126219. [CrossRef]
31. Yi, Y.; Zhang, M.W.; Liao, S.T.; Zhang, R.F.; Deng, Y.Y.; Wei, Z.C.; Tang, X.J.; Zhang, Y. Structural features and immunomodulatory activities of polysaccharides of longan pulp. *Carbohydr. Polym.* **2012**, *87*, 636–643. [CrossRef]
32. Wu, Y.; Lin, Q.; Chen, Z.; Xiao, H. The interaction between tea polyphenols and rice starch during gelatinization. *Food Sci Technol. Int.* **2011**, *17*, 569–577. [CrossRef]
33. Singh, A.; Dutta, P.K.; Kumar, H.; Kureel, A.K.; Rai, A.K. Improved antibacterial and antioxidant activities of gallic acid grafted chitin-glucan complex. *J. Polym. Res.* **2019**, *26*, 1–11. [CrossRef]
34. Zhou, P.; Feng, R.; Luo, Z.; Li, X.; Wang, L.; Gao, L. Synthesis, identification and bioavailability of *Juglans regia* L. polyphenols-Hohenbuehelia serotina polysaccharides nanoparticles. *Food Chem.* **2020**, *329*, 127158. [CrossRef] [PubMed]
35. Corrêa, R.C.; Haminiuk, C.W.; Barros, L.; Dias, M.I.; Calhelha, R.C.; Kato, C.G.; Correa, V.G.; Peralta, R.M.; Ferreira, I.C. Stability and biological activity of Merlot (*Vitis vinifera*) grape pomace phytochemicals after simulated in vitro gastrointestinal digestion and colonic fermentation. *J. Funct. Foods* **2017**, *36*, 410–417. [CrossRef]
36. Aytekin, A.O.; Morimura, S.; Kida, K. Synthesis of chitosan–caffeic acid derivatives and evaluation of their antioxidant activities. *J. Biosci. Bioeng.* **2011**, *111*, 212–216. [CrossRef] [PubMed]
37. Guo, Z.; Xing, R.; Liu, S.; Yu, H.; Wang, P.; Li, C.; Li, P. The synthesis and antioxidant activity of the Schiff bases of chitosan and carboxymethyl chitosan. *Bioorg. Med. Chem. Lett.* **2005**, *15*, 4600–4603. [CrossRef]
38. Wang, X.; Hunter, D.; Xu, J.; Ding, C. Metabolic triggered inflammation in osteoarthritis. *Osteoarthr. Cartil.* **2011**, *41*, 90–91. [CrossRef]
39. Gong, W.; Han, R.; Li, H.; Song, J.; Yan, H.; Li, G.; Liu, A.; Cao, X.; Guo, J.; Zhai, S.; et al. Agronomic Traits and Molecular Marker Identification of Wheat-Aegilops caudata Addition Lines. *Front. Plant Sci.* **2017**, *8*, 1743. [CrossRef]
40. Meng, Q.; Wang, Y.; Chen, F.; Xiao, T.; Zhang, L. Polysaccharides from Diaphragma juglandis fructus: Extraction optimization, antitumor, and immune-enhancement effects. *Int. J. Biol. Macromol.* **2018**, *115*, 835–845. [CrossRef]
41. Wang, K.; Zhang, J.; Ping, S.; Ma, Q.; Chen, X.; Xuan, H.; Shi, J.; Zhang, C.; Hu, F. Anti-inflammatory effects of ethanol extracts of Chinese propolis and buds from poplar (*Populus × canadensis*). *J. Ethnopharmacol.* **2014**, *155*, 300–311. [CrossRef] [PubMed]

foods

MDPI

Article

Three Phenolic Extracts Regulate the Physicochemical Properties and Microbial Community of Refrigerated Channel Catfish Fillets during Storage

Jian Huang [1,2], Lan Wang [2], Zhenzhou Zhu [1,3], Yun Zhang [4], Guangquan Xiong [2,*] and Shuyi Li [1,3,*]

[1] National R&D Center for Se-Rich Agricultural Products Processing, College of Food Science and Engineering, Wuhan Polytechnic University, Wuhan 430023, China
[2] Key Laboratory of Cold Chain Logistics Technology for Agro-Product, Institute of Agricultural Products Processing and Nuclear-Agricultural Technology, Hubei Academy of Agricultural Sciences, Ministry of Agriculture and Rural Affairs, Wuhan 430064, China
[3] Hubei Engineering Research Center for Deep Processing of Green Se-Rich Agricultural Products, School of Modern Industry for Selenium Science and Engineering, Wuhan Polytechnic University, Wuhan 430023, China
[4] College of Tourism and Hotel Management, Hubei University of Economics, Wuhan 430205, China
* Correspondence: xiongguangquan@163.com (G.X.); lishuyisz@sina.com (S.L.)

Abstract: It has been demonstrated that polyphenols have the potential to extend the shelf life of fish products. Thus, the effects of phenolic extracts from grape seed (GSE), lotus seedpod (LSPC), and lotus root (LRPE) were investigated in this study, focusing on the physicochemical changes and bacterial community of refrigerated channel catfish fillets during storage at 4 °C, using ascorbic acid (AA) as reference. As a result, GSE, LSPC, LRPE and AA inhibit the reproduction of microbials in catfish fillets during storage. According to the microbial community analysis, the addition of polyphenols significantly reduced the relative abundance of *Proteobacterial* in the early stage and changed the distribution of the microbial community in the later stage of storage. After 11 days of storage, the increase in total volatile base nitrogen (TVB-N) in fish was significantly reduced by 25.85%, 25.70%, 22.41%, and 39.31% in the GSE, LSPC, LRPE, and AA groups, respectively, compared to the control group (CK). Moreover, the lipid oxidation of samples was suppressed, in which thiobarbituric acid-reactive substances (TBARS) decreased by 28.77% in the GSE group, compared with the CK. The centrifugal loss, LF-NMR, and MRI results proved that GSE significantly delayed the loss of water and the increase in immobilized water flowability in catfish fillets. The polyphenol-treated samples also showed less decrease in shear force and muscle fiber damage in histology, compared to the CK. Therefore, the dietary polyphenols including GSE, LSPC, and LRPE could be developed as natural antioxidants to protect the quality and to extend the shelf life of freshwater fish.

Keywords: polyphenol; catfish fillets; physicochemical changes; microbial community; storage

Citation: Huang, J.; Wang, L.; Zhu, Z.; Zhang, Y.; Xiong, G.; Li, S. Three Phenolic Extracts Regulate the Physicochemical Properties and Microbial Community of Refrigerated Channel Catfish Fillets during Storage. *Foods* **2023**, *12*, 765. https://doi.org/10.3390/foods12040765

Academic Editor: Filomena Nazzaro

Received: 29 December 2022
Revised: 1 February 2023
Accepted: 6 February 2023
Published: 9 February 2023

1. Introduction

Native to Canada and Mexico, the channel catfish (*Ictalurus puctactus*) is a major aquaculture species in the United States. Since the introduction of channel catfish in China in 1984, the aquaculture scale has been expanding, and the market potential has been gradually explored [1]. From 2016, China has become the main channel catfish producer around the world with an annual yield of channel catfish reaching 280,000 tons [2]. Even though channel catfish lack intramuscular spines and scales, they are abundant in vitamins, unsaturated fatty acids, and high-quality protein [3]. As the main processing products, the refrigerated channel catfish fillets are easy to cook and store, which makes them popular with consumers [4]. However, the storage and transportation of refrigerated catfish fillets are limited due to their susceptibility to bacterial contamination and decay during storage, which greatly reduces their economic benefits [3]. Consequently, it is crucial to develop an effective preservation stratagem that can delay the quality decline of catfish fillets and prolong the shelf life of catfish fillets.

In recent decades, plant polyphenols have been widely used in the preservation of aquatic products because of their excellent antioxidative and bacteriostatic efficacy [5]. Grape seed extract (GSE), as a typical natural plant extract, contains abundant polyphenols, such as catechin, epicatechin, and proanthocyanidins. It not only has excellent antioxidant activities by scavenging free radicals, but it also has antibacterial properties due to the core structures containing 3,4,5-trihydroxyphenyl groups [6]. Shi, et al. [7], for example, discovered that grape seed extracts can prevent lipid and protein oxidation and extend the shelf life of silver carp fillets in chilled storage. Lotus seedpod proanthocyanidins (LSPC) are complex polyphenols composed of flavan-3-ol units. It has been reported that LSPC not only has high safety and antioxidant activity (including scavenging superoxide anion inhibiting the formation of malondialdehyde and advanced glycation end products), but it also has strong antibacterial activity [8–11]. Li, Wang, Gao, Xie and Sun [8] investigated the effects of lotus seedpod procyanidins on the quality of chilled beef during refrigerated storage and observed that LSPC could not only inhibit lipid oxidation but could also delay the denaturation of proteins. The main components of lotus root polyphenol extract (LRPE) include chlorogenic acid, B-type procyanidin dimer·H_2O, (+)-Catechin, (−)-Epicatechin, propyl gallate·H_2O, caffeic acid, (−)-Epicatechin-3-gallate, and rutin [12], which has also exhibited high antioxidative capacity [13].

On an account of the abundant available nutrients, high water content and neutral pH, catfish fillets became the suitable substrate for microorganism growth and reproduction. According to statistics, about 30% of economic losses were caused by microorganisms [3,14]. Even though the mechanism of how polyphenols kill or inhibit bacteria is still unclear, studies have shown that the interactions between polyphenols and nonspecific force, including hydrogen-bonding, hydrophobic interactions, and the formation of covalent bonds, were related to microbial membranes, enzymes, and cell envelope transport proteins [15,16]. The polyphenols' capacity to interact with bacterial cell wall components and the bacterial cell membrane allows them to inhibit and control biofilm formation, as well as inhibit microbial enzymes, interfere with protein regulation, and deprive bacterial cell enzymes of substrates [17]. Meanwhile, polyphenols have the capacity to chelate metal ions, vital for survival of almost all bacteria, which may be an important reason for their antibacterial properties [17,18]. Therefore, polyphenols are effective antimicrobials.

Previous research has demonstrated the composition and antioxidant activity of grape seed extract (GSE), lotus seedpod procyanidins (LSPC) and lotus root polyphenol extract (LRPE), all of which have high potential as preservatives. However, there is no study to confirm and compare their fresh-keeping capacity. As a consequence, the overall objective of the present study was to identify how the addition of GSE, LSPC, and LRPE, as well as ascorbic acid (AA) can retard both lipid and protein oxidation, inhibit the growth of meat spoilage bacteria and extend the shelf life of catfish fillets. Moreover, this study provided greater insight into the potential of lotus root, grape seed, and lotus seedpod natural extract as natural and effective sources of antioxidants and antibacterials for fish processing.

2. Materials and Methods

2.1. Preparation of Samples

The lotus root polyphenol extract (LRPE) was extracted from rhizome knots according to Zhu, Li, He, Thirumdas, Montesano and Barba [12]. Following that, the crude extract solution was loaded onto the AB-8 macroporous resin (0.3–1.25 mm particle size, Macklin, Shanghai, China) for purification as described by Wu, et al. [19]. The rhizome knot was purchased from a local market (Wuhan, China). The concentration of total polyphenol in LRPE determined by the method of Zhu, Li, He, Thirumdas, Montesano and Barba [12] was 68.73 mg/100 mg. The ascorbic acid (AA) (99.7%) was purchased from Sinopharm Chemical Reagent Co., Ltd. (Shanghai, China). The lotus seedpod procyanidins (LSPC) (50%) were provided by College of Food Science and Technology, Huazhong Agricultural

University. The grape seed extract (GSE) (95%) was obtained from Tianjin Jianfeng Natural Products Research and Development Co., Ltd. (Tianjin, China).

Fresh and alive channel catfish, each 3 ± 0.2 kg, were purchased from a local market (Wuhan, China). All channel catfish were transported using water tanks to the laboratory with 30 min. The fish head, bone, viscera and skin were immediately removed. The back muscles of the fish were cut into approximately 2 cm thick and approximately 20 g fillets and then randomly separated into five groups: a CK group treated with sterile distilled water and four other treatment groups including AA, GSE, LSPC and LRPE at a concentration of 2 g/L. The concentration of each treatment group was determined according to the results of previous studies. Each solution was prepared just before use and precooled to 4 °C. Fillets were immersed in each treatment group for 10 min at room temperature before being drained thoroughly. The fillets were then placed in food packaging boxes at a cold storage temperature (4 °C) and were collected at random for analysis at 0, 1, 3, 5, 7, 9, and 11 days.

2.2. *Total Viable Counts (TVC)*

The TVC of the fish meat sample was determined by using the method of plate counting described by Zhang, et al. [20] with some modifications. Briefly, samples (1 mL) of serial dilutions (1: 9, sterile 0.85% saline) of homogenates were mixed with the plate count agar (PCA). The plate was cultured upside down at 30 °C for 72 h after the agar solidified. The microbial counts were expressed as log CFU/g.

2.3. *High-Throughput Sequencing on the Illumina Platform*

Extraction of bacterial DNA followed the method described by Li, et al. [21] with some modifications. Homogenates (10 mL, obtained in Section 2.2) were centrifuged at 13,000× *g* for 10 min. After that, the supernatants were removed and sediments were washed by 1 mL sterile water. Bacterial DNA was extracted from the sediments with the assay kit of bacterial DNA extraction (Biomed Biological Technology Co., Ltd., Beijing, China). The V3 and V4 variable regions of bacterial 16S rDNA gene were amplified with the 338F (5′-ACTCCTACGGGAGGCAGCA-3′) and 806R (5′-GGACTACHVGGGTWTCTAA-3′) primers with different barcodes. The TruSeq Nano DNA LT library preparation kit (Illumina Inc., San Diego, CA, USA) was used to produce sequencing library. Operational taxonomic units (OTU) were picked only if they had similarity values of 97% or higher. R software (version 2.15.3, The University of Auckland, Auckland, New Zealand) was used to create a heat map based on the relative abundance of microbial genera.

2.4. *Total Volatile Base Nitrogen (TVB-N)*

The TVB-N of the fish meat sample was determined by using an Automatic Kjeldahl Nitrogen Determinator (K9860, Hai Neng Instrument Co., Ltd., Shanghai, China), according to the method described by Sun, et al. [22] with some modifications. Briefly, 5 g of minced fish meat was mixed with 75 mL distilled water and soaked for 30 min after full vibration at 4 °C. The mixture was then filtered. Then, 5 mL of filtrate was mixed with 5 mL of 10 g/L MgO solution and was distilled for 6 min with kjeldahl. The receiving solution was titrated with 0.01 mol/L hydrochloric acid standard titration solution, with blue-purple as the end point of titration. All chemicals were provided by sigma (Steinheim, Germany).

2.5. *Thiobarbituric Acid Reactive Substance (TBARS)*

The TBARS value was determined according to the method described by Sun, et al. [23]. Briefly, 2 g of minced fish meat was mixed with 200 μL butyl hydroxyanisole (BHA, 7.2%, *w*/*v* 98% ethanol) and 7 mL trichloroacetic acid (TCA, 5%). Filter after homogenization. The filtrate was then diluted ten times, and 5 mL of diluent was mixed with 5 mL 0.2 M TBA, followed by heated at 80 °C for 60 min. The absorbance at 532 nm was recorded after cooling.

2.6. Shear Force

The shear force of samples was determined according to the method described by Pinheiro, et al. [24]. Briefly, five cores in cubes shape (2.0 × 3.0 × 3.0 cm) vertical to the fiber direction were obtained. The cores were then sheared to a cross-section of the muscle fiber by a WarnerBratzer shear blade, using a TA-XT Plus Texure Analyzer (Stable Micro System Ltd., UK). The parameters in shear force (g) measurement were set as follows: testing speed = 1.0 mm/s, distance = 25 mm, and trigger force = 5 g.

2.7. Histology

The histology of catfish samples was preformed according to the method described by Shao, et al. [25]. The samples were fixed, dehydrated, and embedded in paraffin. After hematoxylin–eosin staining, the microstructure of the tissue sections was observed using an optical microscope (Eclipse Ci, Nikon, Tokyo, Japan).

2.8. Centrifugal Loss

The centrifugal loss was measured according to the method described by Li, et al. [21] with some modifications. Then, 2 g of minced sample wrapped by absorbent paper was put into a 50 mL centrifuge tube and centrifuged at 4000 rpm using a refrigerated centrifuge (CR21N, Hitachi, Japan) at 4 °C for 10 min. Subsequently, the sample was put out and weighed.

2.9. Low-Field Nuclear Magnetic Resonance (LF-NMR) and Magnetic Resonance Imaging (MRI)

Low-field nuclear magnetic resonance (LF-NMR) relaxation measurement and magnetic resonance imaging (MRI) analysis were preformed according to the method described by Xia, et al. [26] using an NMR analyzer (NIUMAG, Shanghai, China). Briefly, the catfish samples (2.0 × 2.0 × 2.0 cm) were put into the NMR tubes for determination. The parameters in moisture measurement were set as follows: SW = 100, SF = 20, NS = 4, TW = 1000 ms, TE = 1.00 ms, NECH = 1000, PRG = 1.

2.10. ATP-Related Compounds

Adenosine triphosphate (ATP)-related compounds were extracted according to Shao, et al. [25] and were analyzed by high-performance liquid chromatograph (HPLC) (Waters e2695 Separations Module and 2998 PDA Detector, USA) equipped with an SPD-10A (V) detector and a VP-ADS C18 column (4.6 mm i.d. × 250 mm, 5 μm particle size). The mobile phase was a phosphate buffer (pH 6.0) consisting of 0.02 mol/L K2HPO4 solution and 0.02 mol/L KH2PO4 solution (1: 1, v/v), and the flow rate was 1.0 mL/min. The samples were filtered through a 0.22 μm membrane, analysis was carried out by injecting 20 μL of sample, and detection was monitored at 254 nm. The temperature of the column oven was set at 35 °C and maintained. Each sample was run for 45 min. All reagents were of chromatographic grade.

2.11. Statistical Analysis

All data were expressed as mean ± standard deviation and analysis of variance. The measurements for each parameter were repeated three times. The statistical significance was identified at the 95% confidence level ($p < 0.05$) and was calculated by SPSS 26 (Chicago, IL, USA) software. Origin 2018 (Origin-Lab, Northampton, MA, USA) software was used for data processing and chart plotting.

3. Results and Discussion

3.1. Microbial Analysis

The changes in TVC values of catfish fillets during storage at 4 °C are shown in Figure 1A. The initial microbial counts of fresh catfish fillets were 4.88 ± 0.08 log CFU/g. A lag phase of 3 days was observed in all groups. From day 3 to 5, the TVC values of GSE, LRPE, AA and the CK group increased significantly ($p < 0.05$), with a nonsignificant increase ($p > 0.05$) observed in the LSPC group. After being stored at 4 °C for 5 days, an exponential augment occurred in the TVC value of each group. On day 7, the highest TVC value was observed in CK (8.58 ± 0.07 log CFU/g), followed by the LRPE, AA and GSE groups (7.62 ± 0.14, 7.48 ± 0.12 and 7.03 ± 0.06 log CFU/g, respectively), while the LSPC group had the lowest TVC value (6.71 ± 0.17 log CFU/g), indicating that GSE, LSPC, LRPE and AA could effectively inhibit the growth of microorganisms in catfish fillets during storage. The foremost among them was LSPC, which showed excellent antibacterial activity. Wang, Xie and Sun [10] believed that the antibacterial activity of LSPC was associated with the interaction of LSPC and the cell membrane of spoilage organisms. LSPC contains rich B-type procyanidins; among them, the hydrophobic domain in catechins could band with the surface choline group of the lipid membrane [27]. As a result, the cell membrane is damaged, resulting in the escape of cell contents. These stresses destroy the intracellular homeostasis environment, while other studies have reported that GSE revealed broadly antimicrobial activities against both Gram-positive and Gram-negative bacteria [28,29]. According to the International Commission on Microbiological Specifications for Foods, the upper acceptability limit of microorganisms in freshwater fish is 7.0 log CFU/g [30], implying that after 7 days of storage at 4 °C, catfish fillets of all groups except for LSPC were rotten and unfit for consumption. At the end of storage, the bacterial counts in all samples exceeded 9 log CFU/g, which was far beyond the upper tolerable limit for fresh and refrigerated catfish fillets.

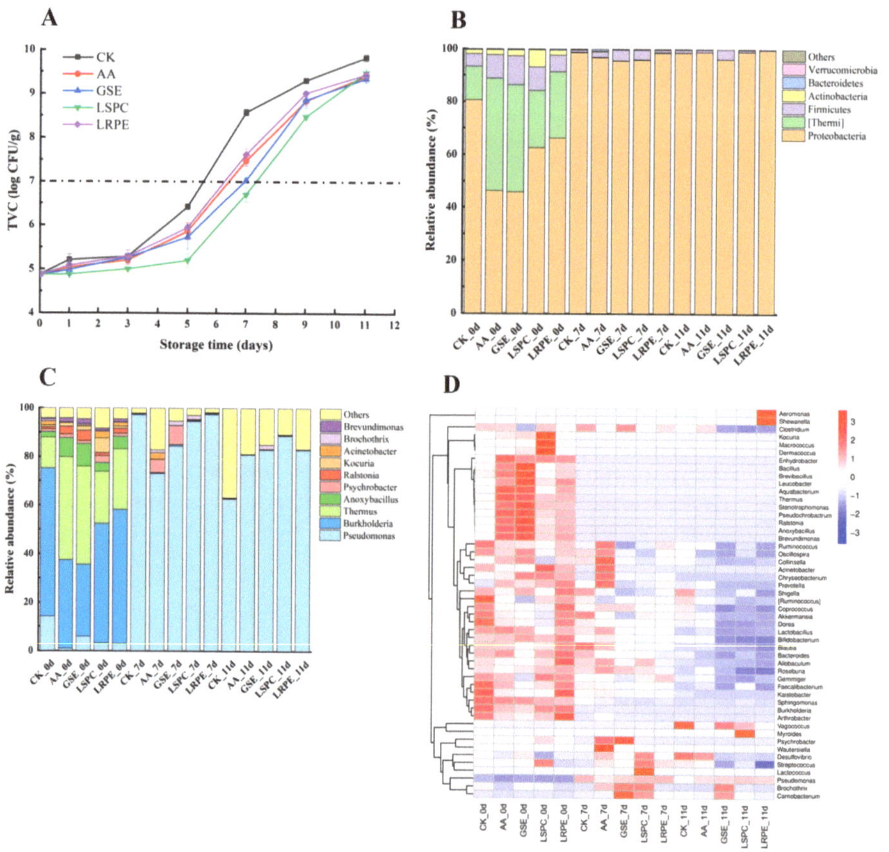

Figure 1. Changes in TVC (**A**), relative abundance of bacterial at phylum (**B**) and genus (**C**) level and heatmap (**D**) at genus level based on 16S rDNA sequencing. The category "others" represents a collection of genus with relative abundance less than 1%. (0 d: samples on day 0; 7 d: samples on day 7; 11 d: samples on day 11.).

Furthermore, the microbiota composition of the initial (day 0), intermediate stage (day 7), and spoiled (day 11) catfish fillets were analyzed through high-throughput sequencing based on the Illumina MiSeq platform. After filtering the low-quality sequences, sequencing on the Illumina MiSeq platform resulted in a total of 1,730,008 effective sequences (Table 1). The Good's coverage was ≥0.999, suggesting that almost all microbial phylotypes in catfish fillets were identified.

Table 1. Alpha diversity estimation of bacterial communities in CK-, GSE-, LSPC-, LRPE- and AA-treated catfish fillets during storage at 4 °C for the indicated times.

Samples	Sequences	OTUs	Chao	Good's Coverage	Shannon	Simpson
CK-0 d	96,878	287	374	0.9993	2.81	0.73
AA-0 d	105,787	252	350	0.9990	2.92	0.68
GSE-0 d	110,338	275	371	0.9992	3.25	0.77
LSPC-0 d	117,002	256	391	0.9991	3.14	0.74
LRPE-0 d	96,209	267	377	0.9991	2.60	0.63
CK-7 d	112,132	185	289	0.9991	3.35	0.83
AA-7 d	104,106	192	294	0.9992	3.49	0.83
GSE-7 d	126,224	169	299	0.9993	3.46	0.79
LSPC-7 d	129,438	176	312	0.9991	2.83	0.64
LRPE-7 d	116,472	187	234	0.9994	1.53	0.33
CK-11 d	127,863	165	250	0.9995	2.73	0.69
AA-11 d	128,558	151	237	0.9994	2.30	0.52
GSE-11 d	124,207	98	152	0.9997	1.76	0.41
LSPC-11 d	107,973	104	132	0.9998	1.81	0.42
LRPE-11 d	126,801	83	135	0.9998	1.93	0.45

The composition and relative abundance of bacterial communities at the phylum level are shown in Figure 1B. The dominant bacteria in fresh samples of the five groups are *Proteobacteria*, which accounted for 80.64%, 46.21, 45.83%, 62.75%, and 66.33% in the CK, AA, GSE, LSPC, and LRPE groups, respectively. The addition of bacteriostatic substances significantly ($p < 0.05$) reduced the relative abundance of *Proteobacteria* on day 0. Among them, the antibacterial activity of GSE could be attributed to its phenolic acids, catechins and proanthocyanidins, which were confirmed to be the most potent chemicals against bacteria [31,32]. Other bacterial phyla, such as *Thermales*, *Firmicutes*, *Actinobacteria*, and *Bacteroidetes*, were identified in CK with a proportion of 12.62%, 4.90%, 1.57%, and 0.17%, respectively. The diversity of the microbiota community in refrigerated catfish fillets dropped drastically with time, and only a tiny portion of bacteria phylum engaged in the spoiling process. On days 7 and 11, the proportion of *Proteobacteria* in all groups reached more than 95%, followed by the *Firmicutes* of 0.33~4.20%. The remaining phyla were very low in abundance, including *Verrucomicrobia*, *Fusobacteria*, *Chloroflexi*, *Cyanobacteria*, *Acidobacteria*, *Gemmatimonadetes*, *Chlorobi*, *Deferribacteres* and *Tenericutes* (less than 0.3% of total sequences).

The composition and relative abundance of different genera are shown in Figure 1C. At the genus level, the dominant strains of fresh catfish fillets are *Pseudomonas*, *Burkholderia*, *Thermus*, and *Anoxybacillus*, accounting for 14.21%, 61.13%, 12.67%, and 2.19%, respectively, in the CK group. The addition of bacteriostatic substances significantly reduced the relative abundance of *Pseudomonas* at the beginning of storage ($p < 0.05$). Li, et al. [8] also observed that *Pseudomonas* showed an increasing trend during the 17-day storage period of refrigerated beef, and the addition of LSPC could retard the reproduction of *Pseudomonas*. However, in the later stage of storage (day 11), the relative abundance of *Pseudomonas* in the CK, AA, GSE, LSPC, and LRPE groups grew significantly to 97.25%, 80.87%, 83.04%, 88.77%, and 82.90%, respectively.

A heatmap was used to exhibit the microbiota differences between the five groups. According to Figure 1D, the color reflects the relative abundances of the microbiological genera, with the redder and the bluer colors illustrating the higher and the lower relative abundances, respectively. The cluster trees on the left and top were added based on the similarity of genera abundances. The most diversified bacterial composition was observed in the fresh sample. As previous research described, the *Aeromonas*, *Acinetobacter*, *Moraxella*, *Pseudomonas*, *Shewanella* and other gramnegative bacteria were usually contained in freshwater fish from temperate waters [33,34]. In the present study, *Burkholderia* and *Thermus* were predominant in the samples stored on day 0. Despite that the genus richness of each group decreased sharply with increasing storage time, the genus richness of the spoiled CK (CK_11d) was still substantially higher than the other groups. The relative abundance of *Pseudomonas* in CK sharply increased from 14.21% (day 0) to 62.88% (day 11).

3.2. TVB-N and TBARS

The changes in TVB-N values of catfish fillets during storage at 4 °C are displayed in Figure 2A. After 3 days, the TVB-N values of each group changed slightly, which was similar to the trends of TVC. From day 3 to 7, the TVB-N value of CK increased and was higher than in other groups ($p < 0.05$). Subsequently, the TVB-N value of each group increased dramatically, and the TVB-N value of CK exceeded the limit level on day 9 (27.55 ± 0.96 mg/100 g), considered as serious spoilage. The rapid augment of the TVB-N value occurred in the late storage period, which can be attributed to the essential long degradation process for the generation of TVB-N from convent nitrogen-containing macromolecules to volatile small molecular compounds under the action of microbes [35]. Liu, et al. [36] and Yu, et al. [37] believe that the sharp increase in TVB-N in the later stages of storage was also associated with the increase in pH. At the end of storage (day 11), the TVB-N value of the GSE, LSPC, LRPE and AA groups was 27.99 ± 1.01, 28.05 ± 1.02, 29.29 ± 1.06 and 22.91 ± 0.84 mg/100 g, respectively, all of which were significantly lower ($p < 0.05$) than CK (37.75 ± 1.34 mg/100 g), indicating that the GSE, LSPC, LRPE groups had a similar effect on reducing TVB-N production. Furthermore, in later storage, the effect of AA on lowering TVB-N formation was better than that of GSE, LSPC and LRPE, most likely because AA, as a small molecule substance, was more easily absorbed by the catfish fish matrix and played an antioxidant role for a long time. In summary, GSE, LSPC and LRPE are conducive in reducing TVB-N production, thereby extending the storage time of catfish samples.

In the present study, TBARS values of different samples during storage are shown in Figure 2B. The results suggest that TBA values of the five groups increased continuously from the initial 0.12 ± 0.03 mg MDA/kg to 0.73 ± 0.02, 0.61 ± 0.02, 0.52 ± 0.00, 0.58 ± 0.02, and 0.59 ± 0.09 mg MDA/kg after 11 days of storage for the CK, AA, GSE, LSPC and LRPE group, respectively, indicating that lipid oxidation occurred during the whole storage process. According to Fan, et al. [38], the increase in TBARS values during refrigerated storage could be attributed to the partial dehydration of fish as well as to the enhanced oxidation of unsaturated fatty acids. Different from TVB-N, a one day of lag phase was observed in TBARS values of all groups except for CK in the early stage of storage. As for CK, the TBARS value was significantly higher than that of other groups and increased rapidly throughout storage ($p < 0.05$), further confirming the anti-lipid oxidation activity of GSE, LSPC, LRPE and AA. Polyphenols, according to Li, et al. [26], may play a role in protecting the endogenous enzyme antioxidant system, thus blocking the chain reaction of lipid oxidation. Among the four groups, the GSE-treated group had the best effect. It was reported that the GSE demonstrated high anti-lipid oxidant activity in rainbow trout meat [39]. It was probably because that GSE contained 98% total flavanols, including 89% proanthocyanidins, which had high antioxidant activity [40]. Because of the antioxidant properties, polyphenol-rich plant extracts such as GSE could be widely used in food processing and preservation [28].

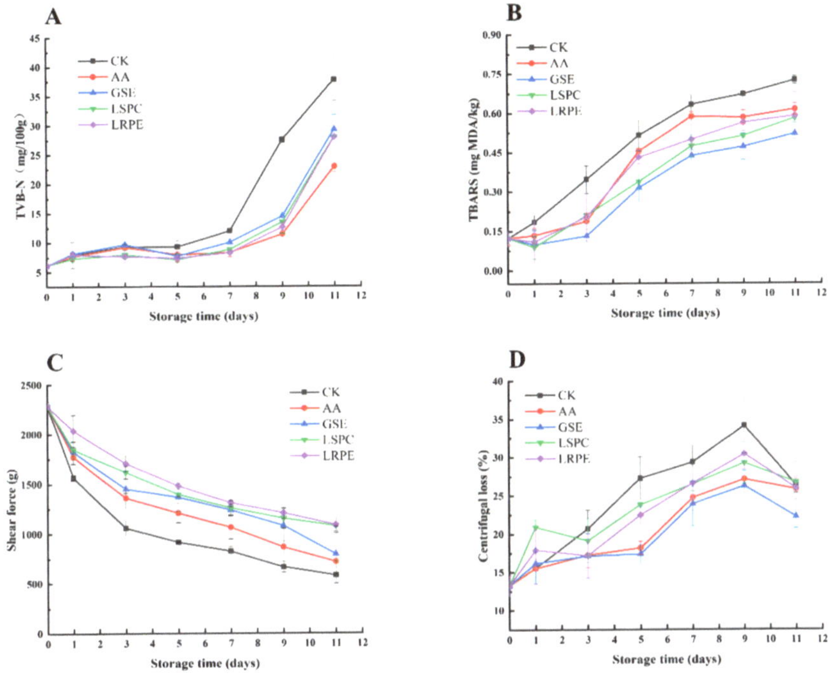

Figure 2. Changes in TVB-N (**A**), TBARS (**B**), centrifugal loss (**C**), and shear force (**D**) of CK-, GSE-, LSPC-, LRPE- and AA-treated catfish fillets during storage at 4 °C for the indicated times.

3.3. Shear Force and Histology

The initial shear force of fresh catfish fillets was 2287 ± 49 g, as shown in Figure 2C, and a steady lower trend was seen in all groups during storage at 4 °C. It was obvious that the changes in shear force of each group, except for LRPE and LSPC, could be roughly divided into two stages, namely, the sharp decline in the first three days at the early storage period and the slow decline from the third day to the end of storage. As for the LRPE and LSPC groups, the texture deterioration was slower than in the other groups. Meanwhile, the shear force of the LRPE group remained the highest from days 3–11, indicating that LRPE effectively inhibited texture deterioration of the catfish fillets.

The changes of the microstructure in catfish fillets during storage are depicted in Figure 3. It was obvious that, as storage time progressed, cracks appeared in the muscle fibers of each group, and the fiber gaps widened. At the initial stage of storage, the sarcomeres of the catfish fillets treated with GSE, LSPC, and LRPE were closely arranged, and no obvious damage and breakage were found, while that of the CK and AA group were somewhat fractured, with some space between the tissues. On day 7, the integrity of the cell morphologies was better in the AA group than in the GSE, LSPC and LRPE groups, demonstrating that the addition of AA may alleviate tissue structure deterioration in catfish fillets better than in GSE, LSPC and LRPE during the middle storage period. After 11 days of storage at 4 °C, the samples in each group revealed varying degrees of fiber damage, among which CK had the most extensive damage, with cells dispersed and vacuolated, whereas the GSE, LSPC, LRPE, and AA groups displayed relatively intact histomorphology. The microstructure alterations of meat and meat products are intimately connected to their quality features, particularly water-holding capacity [41]. The retention of water in catfish fillets by polyphenols may be one of the reasons for delayed muscle fiber damage, which corresponds to the results of centrifugal loss (Figure 2D).

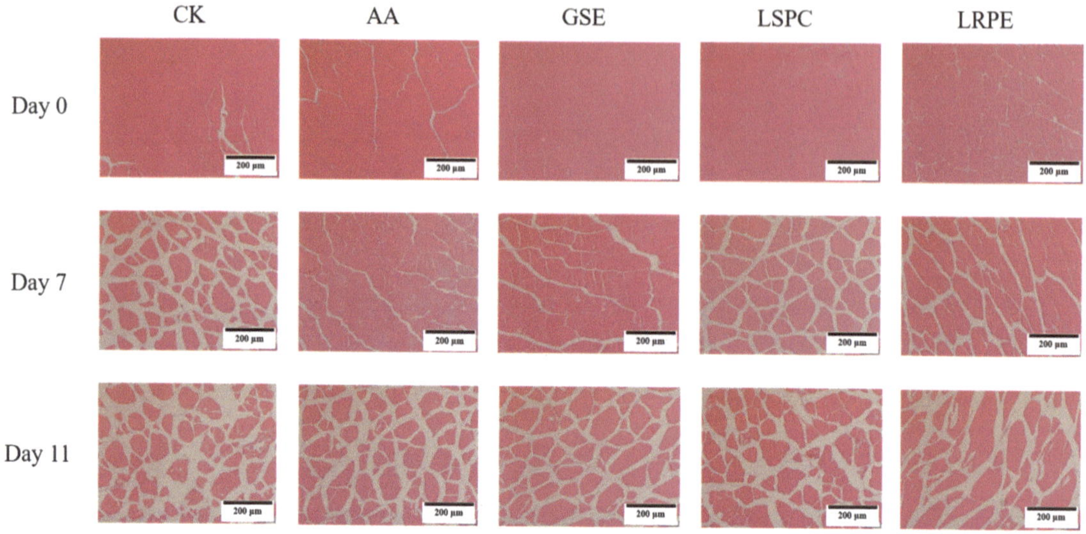

Figure 3. Changes in microstructure profiles of CK-, GSE-, LSPC-, LRPE- and AA-treated catfish fillets during storage at 4 °C for the indicated times.

3.4. Centrifugal Loss, Water Molecules Distribution, and MRI Analysis

As shown in Figure 2D, the reduction of WHC could be ascribed to the denaturation and aggregation of myofibrillar protein in fish during cold storage, which led to the increase in centrifugation loss during the first 9 days storage. The centrifugal loss decreased in the last 2 days of the storage period, which may be attributed to the large amount of loss of water in the first 9 days, resulting in less free water in the catfish fillets. During the storage period, the centrifugal loss of the CK was significantly higher than that of the other groups ($p < 0.05$), indicating that GSE, LSPC, LRPE and AA could effectively inhibit water loss in the catfish fillets. GSE produced the greatest effects of all, which was probably because the proanthocyanidins in GSE have a significant effect in delaying myofibrillar protein denaturation and aggregation.

Three water components were identified according to different water activities, referred to as T_{2b} (0–10 ms), T_{21} (10–100 ms), and T_{22} (>100 ms), which represented bound water, immobilized water and free water, respectively (Table 2). In general, T_{2b} does not alter much during storage because the bound water does not relate to changes in mechanical stress and micro- or macrostructure in the meat matrix and is usually tightly attached to muscle protein. However, T_{2b} increased with storage time in the present research, which could be attributed to discrepancies in the individual catfish and experimental settings. With the extension of storage time, a transformation from long relaxation time to short relaxation time was observed in T_{21}, and the most pronounced trend was seen in the CK, indicating that the water was more and more active and the binding capacity of muscle tissue to water molecules was becoming weaker and weaker, which could be conjectured as the hydrolysis of the catfish fillet muscle proteins resulting from bacteria or enzymes [42]. With the addition of polyphenols, the process of protein oxidative denaturation was delayed, creating a binding between protein and water molecules that was more difficult to destroy. As a result, the T_{21} of the GSE, LSPC, and LRPE groups moved from a short to long relaxation time more slowly than the CK. Moreover, the structures of these proteins can be changed by lactic acid fermentation of meat [43]. The increase in T_{22} could be attributed to the dissociation of immobilized water [44], which could be confirmed from the reduction of P_{21} in Table 3. As shown in Table 3, at any given period, the combination of P_{21} and P_{22} accounted for more than 95% of the total area, suggesting that the majority of water was present in catfish fillets as free water and immobilized water [43].

Table 2. Changes in relaxation time of CK-, GSE-, LSPC-, LRPE- and AA-treated catfish fillets during storage at 4 °C for the indicated times. Different capital letters at the same day of storage are significantly different ($p < 0.05$) between different groups. Different lowercase letters at the same group are significantly different ($p < 0.05$) between different storage time.

Relaxation Time	Samples	Storage Time (Days)						
		0	1	3	5	7	9	11
T_{21} (ms)	CK	0.89 ± 0.08 ^{Ad}	1.02 ± 0.11 ^{Ad}	1.26 ± 0.09 ^{Ac}	1.31 ± 0.06 ^{Ac}	1.38 ± 0.08 ^{Abc}	1.51 ± 0.11 ^{Ab}	2.25 ± 0.14 ^{Aa}
	AA	0.89 ± 0.08 ^{Ae}	1.09 ± 0.08 ^{Acd}	0.98 ± 0.02 ^{Bde}	1.17 ± 0.12 ^{Abc}	1.24 ± 0.08 ^{Aabc}	1.28 ± 0.11 ^{Cab}	1.34 ± 0.10 ^{Ba}
	GSE	0.89 ± 0.08 ^{Ac}	1.13 ± 0.13 ^{Ab}	1.20 ± 0.04 ^{Aab}	1.18 ± 0.12 ^{Aab}	1.29 ± 0.07 ^{Aab}	1.27 ± 0.06 ^{Cab}	1.34 ± 0.05 ^{Ba}
	LSPC	0.89 ± 0.08 ^{Ac}	1.04 ± 0.13 ^{Ac}	1.08 ± 0.04 ^{Bbc}	1.11 ± 0.19 ^{Abc}	1.30 ± 0.19 ^{Aab}	1.74 ± 0.04 ^{ABa}	1.73 ± 0.03 ^{Ba}
	LRPE	0.89 ± 0.08 ^{Ac}	1.05 ± 0.10 ^{Ab}	1.31 ± 0.07 ^{Aa}	1.32 ± 0.12 ^{Aa}	1.36 ± 0.08 ^{Aa}	1.32 ± 0.04 ^{BCa}	1.42 ± 0.10 ^{Ba}
T_{22} (ms)	CK	53.54 ± 0.05 ^{Ad}	61.10 ± 0.33 ^{Ac}	63.29 ± 5.30 ^{Abc}	63.29 ± 4.52 ^{Abc}	64.87 ± 0.41 ^{Abc}	66.46 ± 2.04 ^{Ab}	78.21 ± 0.08 ^{Aa}
	AA	53.54 ± 0.05 ^{Ab}	61.20 ± 0.33 ^{Aa}	61.57 ± 2.22 ^{Aa}	57.17 ± 2.60 ^{Bab}	63.16 ± 6.34 ^{Aa}	63.13 ± 2.59 ^{Ba}	63.17 ± 4.11 ^{Ba}
	GSE	53.54 ± 0.05 ^{Ac}	60.61 ± 0.41 ^{Ab}	60.31 ± 0.37 ^{Ab}	63.50 ± 2.68 ^{Aab}	63.95 ± 3.28 ^{Aa}	61.46 ± 1.00 ^{Bab}	60.51 ± 0.73 ^{Bb}
	LSPC	53.54 ± 0.05 ^{Ad}	56.25 ± 1.13 ^{Bcd}	63.16 ± 0.90 ^{Aab}	59.27 ± 1.41 ^{ABbc}	60.42 ± 3.99 ^{Abc}	63.34 ± 0.47 ^{Bab}	66.48 ± 4.93 ^{Ba}
	LRPE	53.54 ± 0.05 ^{Ad}	55.86 ± 2.07 ^{Bcd}	60.11 ± 0.89 ^{Aab}	57.24 ± 2.26 ^{Bbc}	63.13 ± 1.87 ^{Aa}	62.84 ± 1.10 ^{Ba}	62.84 ± 3.22 ^{Ba}
T_{23} (ms)	CK	550.73 ± 12.14 ^{Ae}	691.73 ± 2.01 ^{Ad}	695.99 ± 19.12 ^{Ad}	709.62 ± 12.95 ^{Ad}	815.47 ± 25.18 ^{Ac}	926.52 ± 21.49 ^{Ab}	1259.95 ± 21.71 ^{Aa}
	AA	550.73 ± 12.14 ^{Ae}	635.71 ± 17.08 ^{Bd}	654.40 ± 8.66 ^{Bd}	699.58 ± 6.25 ^{Ac}	718.55 ± 16.19 ^{Bc}	827.21 ± 14.68 ^{Bb}	860.71 ± 10.86 ^{Ba}
	GSE	550.73 ± 12.14 ^{Ad}	592.51 ± 6.66 ^{Cc}	610.87 ± 20.60 ^{Cbc}	618.77 ± 8.97 ^{Bb}	629.26 ± 13.57 ^{Db}	724.20 ± 16.21 ^{Ca}	740.49 ± 5.78 ^{CDa}
	LSPC	550.73 ± 12.14 ^{Ae}	551.27 ± 17.50 ^{De}	582.79 ± 18.47 ^{Cd}	628.44 ± 9.16 ^{Bc}	654.74 ± 2.88 ^{CDb}	705.71 ± 12.19 ^{CDa}	726.23 ± 15.47 ^{Da}
	LRPE	550.73 ± 12.14 ^{Af}	595.52 ± 5.68 ^{Ce}	607.93 ± 6.30 ^{Cde}	617.33 ± 3.70 ^{Bd}	662.58 ± 15.62 ^{Cc}	683.92 ± 3.34 ^{Db}	757.46 ± 17.12 ^{Ca}

Table 3. Changes in water distribution of CK-, GSE-, LSPC-, LRPE- and AA-treated catfish fillets during storage at 4 °C for the indicated times. Different capital letters at the same day of storage are significantly different ($p < 0.05$) between different groups. Different lowercase letters at the same group are significantly different ($p < 0.05$) between different storage time.

Water Distribution	Samples	Storage Time (Days)						
		0	1	3	5	7	9	11
P_{21} (%)	CK	1.01 ± 0.44 ^{Ac}	1.65 ± 0.26 ^{BCc}	2.74 ± 0.48 ^{Ab}	2.76 ± 0.29 ^{Ab}	2.52 ± 0.37 ^{Ab}	3.04 ± 0.25 ^{Aab}	3.60 ± 0.72 ^{Aa}
	AA	1.01 ± 0.44 ^{Ac}	1.00 ± 0.25 ^{Cc}	1.93 ± 0.16 ^{Ab}	2.15 ± 0.86 ^{Aab}	2.64 ± 0.42 ^{Aab}	2.86 ± 0.20 ^{Aa}	2.78 ± 0.25 ^{ABa}
	GSE	1.01 ± 0.44 ^{Ac}	1.84 ± 0.79 ^{Bbc}	2.30 ± 0.77 ^{Aab}	2.37 ± 0.34 ^{Aab}	2.52 ± 0.21 ^{Aab}	2.87 ± 0.26 ^{Aa}	2.47 ± 0.36 ^{Bab}
	LSPC	1.01 ± 0.44 ^{Ac}	1.61 ± 0.09 ^{BCbc}	1.99 ± 0.66 ^{Aabc}	3.19 ± 1.64 ^{Aa}	3.02 ± 0.89 ^{Aa}	2.83 ± 0.42 ^{Aab}	2.56 ± 0.32 ^{Bab}
	LRPE	1.01 ± 0.44 ^{Ab}	2.67 ± 0.27 ^{Aa}	2.67 ± 0.23 ^{Aa}	2.97 ± 0.64 ^{Aa}	2.60 ± 0.34 ^{Aa}	2.50 ± 0.49 ^{Aa}	1.63 ± 0.48 ^{Cb}
P_{22} (%)	CK	98.70 ± 0.47 ^{Aa}	97.94 ± 0.06 ^{ABa}	96.82 ± 0.68 ^{Ab}	96.66 ± 0.56 ^{Ab}	95.95 ± 0.29 ^{Cbc}	96.03 ± 0.20 ^{Abc}	95.29 ± 0.72 ^{Bc}
	AA	98.70 ± 0.47 ^{Aa}	98.74 ± 0.33 ^{Aa}	97.95 ± 0.21 ^{Aa}	97.13 ± 0.73 ^{Ab}	96.96 ± 0.61 ^{ABb}	96.72 ± 0.16 ^{Abc}	96.02 ± 0.35 ^{ABc}
	GSE	98.70 ± 0.47 ^{Aa}	97.49 ± 1.15 ^{Bb}	97.00 ± 0.61 ^{Abc}	96.58 ± 0.39 ^{Abcd}	96.53 ± 0.45 ABCbcd	96.22 ± 0.49 ^{Acd}	95.54 ± 0.50 ^{ABd}
	LSPC	98.70 ± 0.47 ^{Aa}	97.92 ± 0.21 ^{ABab}	97.07 ± 1.08 ^{Abc}	96.21 ± 1.41 ^{Ac}	96.03 ± 0.70 ^{BCc}	96.40 ± 0.43 ^{Ac}	96.49 ± 0.43 ^{Ac}
	LRPE	98.70 ± 0.47 ^{Aa}	96.94 ± 0.19 ^{Bb}	97.09 ± 0.23 ^{Ab}	96.13 ± 0.66 ^{Ab}	97.07 ± 0.34 ^{Ab}	96.47 ± 0.84 ^{Ab}	96.37 ± 0.38 ^{Ab}
P_{23} (%)	CK	0.28 ± 0.08 ^{Ad}	0.40 ± 0.29 ^{Acd}	0.44 ± 0.20 ^{ABcd}	0.58 ± 0.28 ^{Abcd}	1.53 ± 0.63 ^{Aa}	0.96 ± 0.01 ^{Abc}	1.11 ± 0.04 ^{Bab}
	AA	0.28 ± 0.08 ^{Acd}	0.27 ± 0.10 ^{Acd}	0.12 ± 0.12 ^{Bd}	0.72 ± 0.15 ^{Ab}	0.40 ± 0.24 ^{Bc}	0.42 ± 0.06 ^{Bc}	1.20 ± 0.10 ^{Ba}
	GSE	0.28 ± 0.08 ^{Ac}	0.33 ± 0.11 ^{Ac}	0.91 ± 0.52 ^{Ab}	1.06 ± 0.38 ^{Ab}	0.95 ± 0.28 ^{ABb}	0.90 ± 0.27 ^{ABb}	1.99 ± 0.18 ^{Aa}
	LSPC	0.28 ± 0.08 ^{Ac}	0.50 ± 0.09 ^{Aabc}	0.94 ± 0.46 ^{Aa}	0.46 ± 0.03 ^{Abc}	0.95 ± 0.22 ^{ABa}	0.77 ± 0.01 ^{ABab}	0.96 ± 0.28 ^{Ba}
	LRPE	0.28 ± 0.08 ^{Ac}	0.39 ± 0.08 ^{Ac}	0.24 ± 0.01 ^{Bc}	0.90 ± 0.49 ^{Ab}	0.31 ± 0.03 ^{Bc}	1.03 ± 0.53 ^{Ab}	2.00 ± 0.16 ^{Aa}

MRI was used to visualize the internal structure of the catfish fillets during refrigerated storage in a lossless way (Figure 4). In the MRI, red represents high proton density regions and green represents low proton density regions. The stronger the water proton signal, that is, the redder the image color, the higher the water content. As shown in Figure 4, the MRI of all samples at day 0 revealed yellow red, suggesting that the water content of catfish fillets was high at the beginning of storage. As storage time progressed, the MRI of catfish fillets altered from red to green, suggesting that the moisture content steadily reduced. The hue of the CK was the greenest at the end of storage, indicating that its water loss was the most significant and that the water loss could be inhibited by polyphenols.

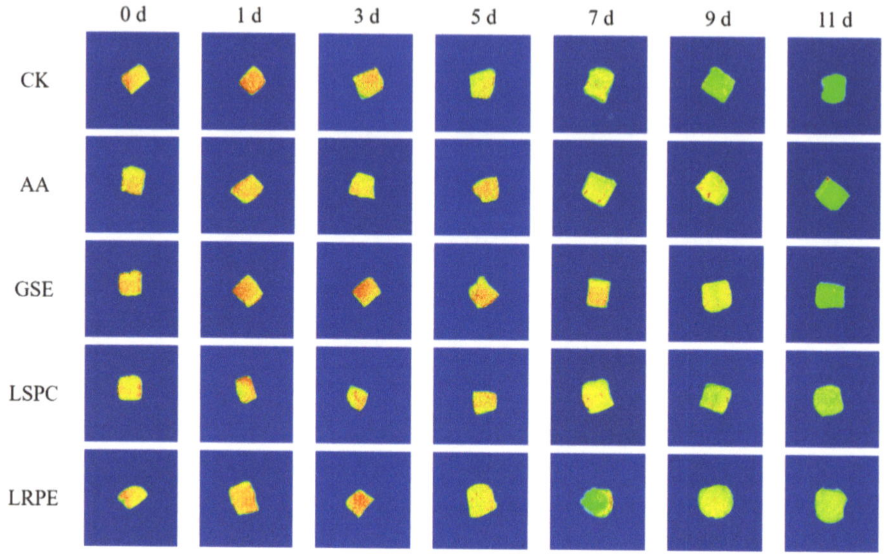

Figure 4. Changes in false-color image of the water proton density of CK-, GSE-, LSPC-, LRPE- and AA-treated catfish fillets during storage at 4 °C for the indicated times.

3.5. ATP-Related Compounds

In general, ATP is degraded by autolytic breakdown, endogenous enzymes, as well as by bacterial activity, including *Pseudomonas* spp., *S. putrefaciens*, and *P. phosphoreum* and follows the following process: adenosine triphosphate (ATP) → adenosine diphosphate (ADP) → adenosine monophosphate (AMP) → inosine monophosphate (IMP) → and hypoxanthine riboside (HxR) → hypoxanthine (Hx) → xanthine → uric acid [45,46]. The changes in ATP-related compounds and K values are depicted in Figure 5. Regarding the changes in ATP-related compounds, the levels of IMP, HxR, and Hx varied substantially. The contents of ATP, ADP, and AMP showed a significant downward trend in the early storage period and then slowly decreased. The rapid decrease in the three ATP-related compounds during the early storage period may correspond to the expeditious degradation of ATP. The levels of ATP, ADP, and AMP in the CK were reduced by 89.95%, 48.22%, and 97.49%, respectively, at the end of storage, further confirming the degradation process of ATP. The most significant declining trend was noticed in the CK, indicating that GSE, LSPC, LRPE, and AA delayed the degradation of ATP to a certain extent.

As shown in Figure 5C, the initial level of IMP was 6.77 ± 0.25 μmol/g, which was significantly ($p < 0.05$) higher than the concentration of other ATP-related compounds at the early stage of storage. The IMP content in all groups demonstrated a progressive decreasing tendency with the passage of time. The IMP value in the LSPC group was significantly ($p < 0.05$) higher than that of the other groups during the middle storage period (day 1 to 7), especially on the third day of storage, whenever the IMP value of the LSPC group was nearly twice (1.89 times) as much as that of the CK, indicating that LSPC could delay ATP degradation and IMP accumulation to a considerable extent. The IMP value of each group remained steady after storage for 9 days, with no significant ($p > 0.05$) change. Additionally, GSE, Zhao, et al. [47] discovered a similar result when storing tilapia (*Oreochromis niloticus*) fillets, namely that the IMP content of GSE group was greater than that of the CK, of which it was possible that IMP degradation was restrained by GSE addition. Li, Zhuang, Liu, Zhang, Liu, Cheng, Liu, Shu and Luo [31] further explained that it was probably because of the inhibitory effect of GSE on the related enzymes activities.

As displayed in Figure 5D, the level of HxR in all groups increased at the early stage and then declined. The HxR content of the CK increased from 0.05 ± 0.01 to 2.71 ± 0.05 μmol/g during the first 9 days of storage and decreased to 2.23 ± 0.01 μmol/g on day 11. Compared with CK, the three polyphenol extracts and AA treatment prevented the accumulation of HxR by 0.93~1.20 μmol/g at the end of storage. Along with the accumulation of HxR, the content of Hx in the fillets also increased, and the discrepancy between the groups became increasingly noticeable. On day 11, the Hx levels in the GSE, LSPC, LRPE, and AA groups were 1.98 ± 0.28, 2.41 ± 0.07, 2.79 ± 0.47 and 3.31 ± 0.41 μmol/g, respectively, considerably ($p < 0.05$) lower than that of the CK (3.77 ± 0.07 μmol/g). Given that the breakdown process from IMP to Hx was significantly influenced by spoilage bacteria, one possible explanation for the decreased levels of Hx in the GSE, LSPC and LRPE groups might be due to the polyphenols' antibacterial activity [47]. It could be deduced that the reduced accumulation of HxR and Hx in the GSE-, LSPC-, LRPE- and AA-treated groups will result in a beneficial effect on fillet edibility via increasing flavor quality after refrigeration.

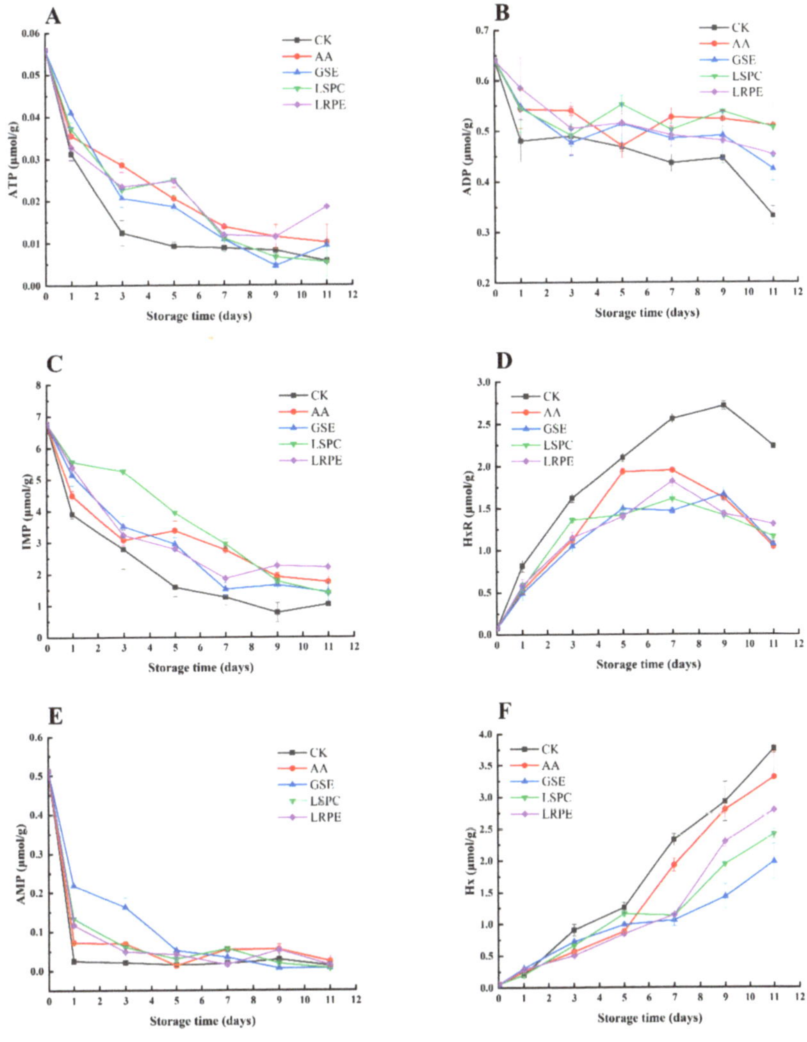

Figure 5. Changes in ATP concentration (**A**), ADP concentration (**B**), IMP concentration (**C**), HxR concentration (**D**), AMP concentration (**E**), and Hx concentration (**F**) of CK-, GSE-, LSPC-, LRPE- and AA-treated catfish fillets during storage at 4 °C for the indicated times.

As depicted in Figure 6, the initial K value was far below 10%, suggesting that the catfish fillets were quite fresh. For CK, the K value increased steadily, reaching 81.70% on day 9 before decreasing slightly ($p > 0.05$). At the end of storage (day 11), the K value of the CK was approximately 19.2~25.6% higher than that of the other groups, demonstrating the capacities of GSE, LSPC, LRPE, and AA to inhibit the degradation of ATP-related compounds. The LSPC group had the strongest impact in suppressing the increase in K value, especially on day 7, which was 40.79% less than the CK. This might be attributable to the excellent bacteriostatic action of LSPC, which has been confirmed by Li, et al. [8].

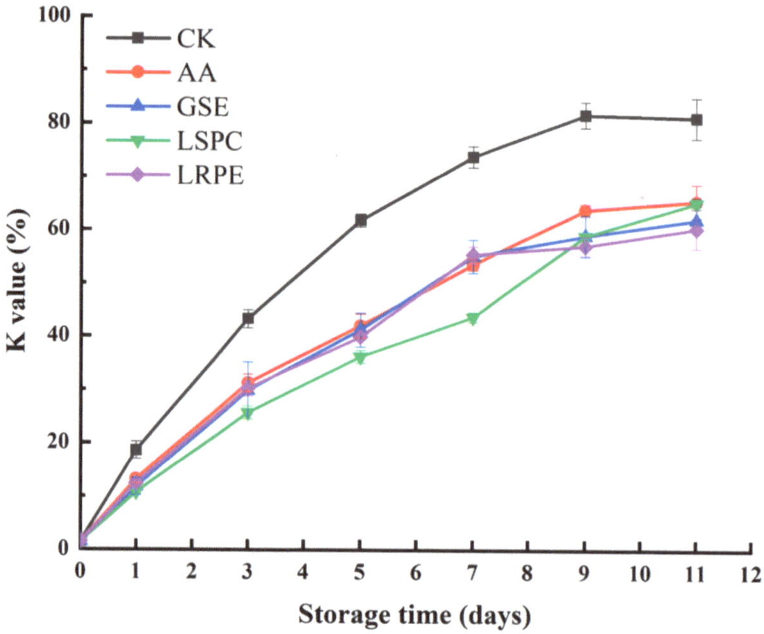

Figure 6. Changes in K value of CK-, GSE-, LSPC-, LRPE- and AA-treated catfish fillets during storage at 4 °C for the indicated times.

4. Conclusions

GSE, LSPC, LRPE, and AA in this study showed strong effects on delaying the quality deterioration of catfish fillets during storage at 4 °C. Among them, GSE showed a good effect in inhibiting fat oxidation and water loss, while LSPC represented great antibacterial activity. Additionally, the effect of AA on lowering TVB-N formation was better than that of GSE, LSPC and LRPE, while LRPE had a strong ability to delay the decline of catfish shear force. These results provide a theoretical basis for the application of GSE, LSPC, LRPE, and AA as natural preservative in meat preservation.

Author Contributions: J.H.: writing—original draft preparation, visualization, investigation. L.W.: conceptualization, methodology. Z.Z.: funding acquisition, supervision. Y.Z.: visualization. S.L.: supervision, writing—reviewing and editing. G.X.: funding acquisition, supervision. We confirm that the manuscript or other forms of its contents have not been previously published or submitted by any author. All authors have read and agreed to the published version of the manuscript.

Funding: The research was funded by the China Agriculture Research System (CARS-46), Major project of Scientific and Technological R&D of Hubei Agricultural Scientific and Technological Innovation Center (2020–620-000–002-03), the National Youth Talent Program In the Food Industry of China and Outstanding Young and Middle-Aged Science and Technology Innovation Team in Hubei Province (T2020012), and the Science and Technology Research Project of the Hubei Provincial Department Of Education (B2022138).

Institutional Review Board Statement: Not applicable.

Informed Consent Statement: Not applicable.

Data Availability Statement: The authors confirm that all data included in this study are available upon request by contacting the corresponding author.

Conflicts of Interest: The authors declare that there are no conflicts of interest regarding the publication of this paper.

References

1. Zhong, L.; Song, C.; Chen, X.; Deng, W.; Xiao, Y.; Wang, M.; Qin, Q.; Luan, S.; Kong, J.; Bian, W. Channel catfish in China: Historical aspects, current status, and problems. *Aquaculture* **2016**, *465*, 367–373.
2. Yearbook, C.F.S. *China Fishery Statistic Yearbook*; China Agriculture Press: Beijing, China, 2018.
3. Huang, H.; Sun, W.; Xiong, G.; Shi, L.; Jiao, C.; Wu, W.; Li, X.; Qiao, Y.; Liao, L.; Ding, A.; et al. Effects of HVEF treatment on microbial communities and physicochemical properties of catfish fillets during chilled storage. *LWT* **2020**, *131*, 109667. [CrossRef]
4. Shi, L.; Yin, T.; Xiong, G.; Ding, A.; Li, X.; Wu, W.; Qiao, Y.; Liao, L.; Wang, J.; Wang, L. Microstructure and physicochemical properties: Effect of pre-chilling and storage time on the quality of Channel catfish during frozen storage. *LWT* **2020**, *130*, 109606.
5. Guo, A.; Jiang, J.; True, A.D.; Xiong, Y.L. Myofibrillar Protein Cross-Linking and Gelling Behavior Modified by Structurally Relevant Phenolic Compounds. *J. Agric. Food Chem.* **2021**, *69*, 1308–1317. [CrossRef]
6. Shen, X.; Su, Y.-C. Application of grape seed extract in depuration for decontaminating *Vibrio parahaemolyticus* in Pacific oysters (*Crassostrea gigas*). *Food Control.* **2017**, *73*, 601–605. [CrossRef]
7. Shi, C.; Cui, J.; Yin, X.; Luo, Y.; Zhou, Z. Grape seed and clove bud extracts as natural antioxidants in silver carp (*Hypophthalmichthys molitrix*) fillets during chilled storage: Effect on lipid and protein oxidation. *Food Control.* **2014**, *40*, 134–139. [CrossRef]
8. Li, X.; Wang, J.; Gao, X.; Xie, B.; Sun, Z. Inhibitory effects of lotus seedpod procyanidins against lipid and protein oxidation and spoilage organisms in chilled-storage beef. *LWT* **2022**, *160*, 113247. [CrossRef]
9. Tang, C.; Xie, B.; Sun, Z. Antibacterial activity and mechanism of B-type oligomeric procyanidins from lotus seedpod on enterotoxigenic *Escherichia coli*. *J. Funct. Foods* **2017**, *38*, 454–463. [CrossRef]
10. Wang, J.; Xie, B.; Sun, Z. The improvement of carboxymethyl β-glucan on the antibacterial activity and intestinal flora regulation ability of lotus seedpod procyanidins. *LWT* **2021**, *137*, 110441. [CrossRef]
11. Wu, Q.; Zhao, K.; Chen, Y.; Ouyang, Y.; Feng, Y.; Li, S.; Zhang, L.; Feng, N. Effect of lotus seedpod oligomeric procyanidins on AGEs formation in simulated gastrointestinal tract and cytotoxicity in Caco-2 cells. *Food Funct.* **2021**, *12*, 3527–3538. [CrossRef]
12. Zhu, Z.; Li, S.; He, J.; Thirumdas, R.; Montesano, D.; Barba, F.J. Enzyme-assisted extraction of polyphenol from edible lotus (*Nelumbo nucifera*) rhizome knot: Ultra-filtration performance and HPLC-MS2 profile. *Food Res. Int.* **2018**, *111*, 291–298. [CrossRef]
13. Hu, M.; Skibsted, L.H. Antioxidative capacity of rhizome extract and rhizome knot extract of edible lotus (*Nelumbo nuficera*). *Food Chem.* **2002**, *76*, 327–333. [CrossRef]
14. Ceylan, Z.; Sengor, G.F.U.; Yilmaz, M.T. Nanoencapsulation of liquid smoke/thymol combination in chitosan nanofibers to delay microbiological spoilage of sea bass (*Dicentrarchus labrax*) fillets. *J. Food Eng.* **2018**, *229*, 43–49. [CrossRef]
15. Cowan, M.M. Plant Products as Antimicrobial Agents. *Clin. Microbiol. Rev.* **1999**, *12*, 564–582. [CrossRef]
16. Kumar, S.; Pandey, A.K. Chemistry and Biological Activities of Flavonoids: An Overview. *Sci. World J.* **2013**, *2013*, 162750. [CrossRef]
17. Papuc, C.; Goran, G.V.; Predescu, C.N.; Nicorescu, V.; Stefan, G. Plant Polyphenols as Antioxidant and Antibacterial Agents for Shelf-Life Extension of Meat and Meat Products: Classification, Structures, Sources, and Action Mechanisms. *Compr. Rev. Food Sci. Food Saf.* **2017**, *16*, 1243–1268. [CrossRef]
18. Field, J.; Lettinga, G. Toxicity of tannic compounds to microorganisms. In *Plant Polyphenols*; Hemingway, R., Laks, P., Eds.; Springer: Boston, MA, USA, 1992; pp. 673–692.
19. Wu, Q.; Chen, H.; Lv, Z.; Li, S.; Hu, B.; Guan, Y.; Xie, B.; Sun, Z. Oligomeric procyanidins of lotus seedpod inhibits the formation of advanced glycation end-products by scavenging reactive carbonyls. *Food Chem.* **2013**, *138*, 1493–1502. [CrossRef]
20. Zhang, J.; Li, Y.; Liu, X.; Lei, Y.; Regenstein, J.M.; Luo, Y. Characterization of the microbial composition and quality of lightly salted grass carp (*Ctenopharyngodon idellus*) fillets with vacuum or modified atmosphere packaging. *Int. J. Food Microbiol.* **2019**, *293*, 87–93. [CrossRef]
21. Li, N.; Zhang, Y.; Wu, Q.; Gu, Q.; Chen, M.; Zhang, Y.; Sun, X.; Zhang, J. High-throughput sequencing analysis of bacterial community composition and quality characteristics in refrigerated pork during storage. *Food Microbiol.* **2019**, *83*, 86–94. [CrossRef]
22. Sun, Q.; Sun, F.; Xia, X.; Xu, H.; Kong, B. The comparison of ultrasound-assisted immersion freezing, air freezing and immersion freezing on the muscle quality and physicochemical properties of common carp (Cyprinus carpio) during freezing storage. *Ultrason. Sonochem.* **2019**, *51*, 281–291. [CrossRef]
23. Sun, Y.; Luo, H.; Cao, J.; Pan, D. Structural characteristics of Sheldrake meat and secondary structure of myofibrillar protein: Effects of oxidation. *Int. J. Food Prop.* **2017**, *20*, 1553–1566. [CrossRef]
24. Pinheiro, R.S.; Francisco, C.L.; Lino, D.M.; Borba, H. Meat quality of Santa Inês lamb chilled-then-frozen storage up to 12 months. *Meat Sci.* **2018**, *148*, 72–78. [CrossRef] [PubMed]

25. Shao, Y.; Xiong, G.; Ling, J.; Hu, Y.; Shi, L.; Qiao, Y.; Yu, J.; Cui, Y.; Liao, L.; Wu, W.; et al. Effect of ultra-high pressure treatment on shucking and meat properties of red swamp crayfish (*Procambarus clarkia*). *LWT* **2018**, *87*, 234–240. [CrossRef]
26. Xia, M.; Chen, Y.; Guo, J.; Huang, H.; Wang, L.; Wu, W.; Xiong, G.; Sun, W. Water distribution and textual properties of heat-induced pork myofibrillar protein gel as affected by sarcoplasmic protein. *LWT* **2019**, *103*, 308–315. [CrossRef]
27. Nakayama, M.; Shimatani, K.; Ozawa, T.; Shigemune, N.; Tsugukuni, T.; Tomiyama, D.; Kurahachi, M.; Nonaka, A.; Miyamoto, T. A study of the antibacterial mechanism of catechins: Isolation and identification of Escherichia coli cell surface proteins that interact with epigallocatechin gallate. *Food Control.* **2013**, *33*, 433–439. [CrossRef]
28. Guan, W.; Ren, X.; Li, Y.; Mao, L. The beneficial effects of grape seed, sage and oregano extracts on the quality and volatile flavor component of hairtail fish balls during cold storage at 4 °C. *LWT* **2019**, *101*, 25–31. [CrossRef]
29. Trošt, K.; Klančnik, A.; Vodopivec, B.M.; Lemut, M.S.; Novšak, K.J.; Raspor, P.; Možina, S.S. Polyphenol, antioxidant and antimicrobial potential of six different white and red wine grape processing leftovers. *J. Sci. Food Agric.* **2016**, *96*, 4809–4820. [CrossRef]
30. ICMSF. Sampling for microbiological analysis: Principles and specific applications. In *Microorganisms in Foods 2*; International Commission on Microbiological Specifications for Foods, University of Toronto Press: Toronto, ON, Canada, 1986; pp. 181–196.
31. Li, Y.; Zhuang, S.; Liu, Y.; Zhang, L.; Liu, X.; Cheng, H.; Liu, J.; Shu, R.; Luo, Y. Effect of grape seed extract on quality and microbiota community of container-cultured snakehead (*Channa argus*) fillets during chilled storage. *Food Microbiol.* **2020**, *91*, 103492. [CrossRef]
32. Silván, J.M.; Mingo, E.; Hidalgo, M.; de Pascual-Teresa, S.; Carrascosa, A.V.; Martinez-Rodriguez, A.J. Antibacterial activity of a grape seed extract and its fractions against *Campylobacter* spp. *Food Control.* **2013**, *29*, 25–31. [CrossRef]
33. Li, Q.; Zhang, L.; Luo, Y. Changes in microbial communities and quality attributes of white muscle and dark muscle from common carp (Cyprinus carpio) during chilled and freeze-chilled storage. *Food Microbiol.* **2018**, *73*, 237–244. [CrossRef]
34. Zang, J.; Xu, Y.; Xia, W.; Yu, D.; Gao, P.; Jiang, Q.; Yang, F. Dynamics and diversity of microbial community succession during fermentation of Suan yu, a Chinese traditional fermented fish, determined by high throughput sequencing. *Food Res. Int.* **2018**, *111*, 565–573. [CrossRef]
35. Zhuang, S.; Li, Y.; Jia, S.; Hong, H.; Liu, Y.; Luo, Y. Effects of pomegranate peel extract on quality and microbiota composition of bighead carp (*Aristichthys nobilis*) fillets during chilled storage. *Food Microbiol.* **2019**, *82*, 445–454. [CrossRef]
36. Liu, D.; Liang, L.; Xia, W.; Regenstein, J.M.; Zhou, P. Biochemical and physical changes of grass carp (*Ctenopharyngodon idella*) fillets stored at −3 and 0 degrees C. *Food Chem.* **2013**, *140*, 105–114. [CrossRef]
37. Yu, D.; Regenstein, J.M.; Zang, J.; Jiang, Q.; Xia, W.; Xu, Y. Inhibition of microbial spoilage of grass carp (*Ctenopharyngodon idellus*) fillets with a chitosan-based coating during refrigerated storage. *Int. J. Food Microbiol.* **2018**, *285*, 61–68. [CrossRef]
38. Fan, W.; Chi, Y.; Zhang, S. The use of a tea polyphenol dip to extend the shelf life of silver carp (*Hypophthalmicthys molitrix*) during storage in ice. *Food Chem.* **2008**, *108*, 148–153. [CrossRef]
39. Raeisi, M.; Tajik, H.; Aliakbarlu, J.; Mirhosseini, S.H.; Hosseini, S.M.H. Effect of carboxymethyl cellulose-based coatings incorporated with *Zataria multiflora* Boiss. essential oil and grape seed extract on the shelf life of rainbow trout fillets. *LWT* **2015**, *64*, 898–904. [CrossRef]
40. Shirahigue, L.D.; Plataoviedo, M.; de Alencar, S.M.; D'Arce, M.A.B.R.; Ferreira de Souza Vieira, T.M.; Oldoni, T.; Contreras-Castillo, C.J. Wine industry residue as antioxidant in cooked chicken meat. *Int. J. Food Sci. Technol.* **2010**, *45*, 863–870. [CrossRef]
41. Jiang, Q.; Jia, R.; Nakazawa, N.; Hu, Y.; Osako, K.; Okazaki, E. Changes in protein properties and tissue histology of tuna meat as affected by salting and subsequent freezing. *Food Chem.* **2018**, *271*, 550–560. [CrossRef]
42. Wang, X.; Geng, L.; Xie, J.; Qian, Y.-F. Relationship Between Water Migration and Quality Changes of Yellowfin Tuna (*Thunnus albacares*) During Storage at 0 °C and 4 °C by LF-NMR. *J. Aquat. Food Prod. Technol.* **2017**, *27*, 35–47. [CrossRef]
43. Zhang, Q.Q.; Li, W.; Li, H.K.; Chen, X.H.; Jiang, M.; Dong, M.S. Low-field nuclear magnetic resonance for online determination of water content during sausage fermentation. *J. Food Eng.* **2017**, *212*, 291–297. [CrossRef]
44. Andersen, C.M.; Rinnan, Å. Distribution of water in fresh cod. *LWT* **2002**, *35*, 687–696. [CrossRef]
45. Hong, H.; Regenstein, J.M.; Luo, Y. The Importance of ATP-related Compounds for the Freshness and Flavor of Post-mortem Fish and Shellfish Muscle: A Review. *Crit. Rev. Food Sci. Nutr.* **2017**, *57*, 1787–1798. [CrossRef] [PubMed]
46. Özoğul, Y.; Özoğul, F. Degradation products of adenine nucleotide in rainbow trout (*Oncorhynchus mykiss*) stored in ice and in modified atmosphere packaging. *Turk. J. Zool.* **2002**, *26*, 127–130.
47. Zhao, X.; Wu, J.; Chen, L.; Yang, H. Effect of vacuum impregnated fish gelatin and grape seed extract on metabolite profiles of tilapia (*Oreochromis niloticus*) fillets during storage. *Food Chem.* **2019**, *293*, 418–428. [CrossRef]

MDPI AG
Grosspeteranlage 5
4052 Basel
Switzerland
Tel.: +41 61 683 77 34
www.mdpi.com

Foods Editorial Office
E-mail: foods@mdpi.com
www.mdpi.com/journal/foods

www.ingramcontent.com/pod-product-compliance
Lightning Source LLC
Chambersburg PA
CBHW041349210926
51850CB00013B/3027
* 9 7 8 3 7 2 5 8 2 3 0 3 1 *